Black Holes to Big Bang:
Cosmo Waves

LARRY J CHAMBER

Abstract

The growing catalog of gravitational-wave signals from compact object mergers has allowed us to study the properties of black holes and neutron stars more precisely than ever before and has opened a new window through which to probe the earliest moments in our universe's history. Population-level measurements of the masses and spins of compact objects can reveal how these systems form and evolve. Multimessenger observations of compact object mergers can shed light on the properties of the electromagnetic counterparts of these systems, such as short gamma-ray bursts and kilonovae. Finally, observations of the stochastic gravitational-wave background can constrain early-universe physics inaccessible with other means.

In this book, I demonstrate how we can leverage such observations of gravitational waves and their electromagnetic counterparts to learn about astrophysics and cosmology. The first p art f ocuses o n m ethods f or f acilitating t he d etection o f electromagnetic counterparts and the simultaneous analysis of gravitational-wave and electromagnetic data for mergers including a neutron star. I then transition to a detailed study of black hole spin, including characterizing the measurability of spin in individual systems with current gravitational-wave detectors and presenting novel population-level analyses. This work is complemented by the development of new methods for increasingly detailed gravitational-wave data analysis. Such analyses will be critical to the astrophysical interpretation of the growing catalog of compact-object binaries and will enable the future detection of the cosmological stochastic gravitational-wave background.

Acknowledgments

Thinking back to when I started my academic journey ten years ago, it seems nearly inconceivable that I would now be preparing my PhD book in astrophysics at MIT. In fact, until I started my undergraduate studies, I was convinced that I would become a professional violinist. Making the transition from seeing myself as a musician to seeing myself as a physicist would not have been possible without the guidance and support of mentors, family, and friends along the way.

Thank you to Prof. Richard Robinett, my undergraduate advisor at Penn State and the first person who believed that I could be a physicist before I believed it myself. With his guidance, I reached out to Prof. Miguel Mostafá my freshman year to ask about research opportunities. Despite the fact that I knew nothing about particle astrophysics or coding, Miguel saw something in me. I am so grateful not only for the research and networking skills that Miguel gave me over the course of our four years of working together, but the example he set as a teacher, researcher, group leader, and mentor.

My first experience with gravitational-wave research came in the Summer of 2015 through an international program at Monash University in Australia where I worked with Prof. Eric Thrane. Within a month of the program ending, gravitational waves were directly detected for the first time. I am grateful to Eric for the opportunity to work in this field at the time when it was most exciting and just starting off and have felt privileged to continue our collaboration for the last eight years. Thank you to Eric for always pushing me to think independently as a scientist, to try new research avenues, and to pursue the projects that I find most interesting.

When I first met my advisor, Prof. Salvatore Vitale, what struck me most was his enthusiasm for his work and the field of gravitational-wave astronomy. Over the last five years of working together, I am so grateful that he has believed in me and valued my scientific ideas, allowing our research interests to grow and evolve together. Thank you to Salvo for teaching me how to drive my own research program and how to be a mentor and teacher myself.

I have really appreciated learning from and interacting with so many amazing collaborators and scientists in the MIT LIGO lab and Kavli Institute over the course of my PhD. Thank you to Tom Callister for helping me find my footing in this field and to Carl-Johan Haster for always being willing to answer my random questions. Thank you to my friends and office mates, Geoffrey, Tri, and Jack, who saved me from sitting alone in a small windowless office. Thank you to all the other MKI grad students for your friendship, cat pictures, and homework collaboration over the years.

I was lucky enough to be able to continue my musical pursuits at MIT, taking lessons and playing in chamber music. Thank you to my teacher, Prof. Rictor Noren, for taking a random MIT graduate student into his studio and giving me the freedom and flexibility to drive the course of my learning. Thank you also to Marcus Thompson and the other Chamber Music Society coaches that I had during my time at MIT for giving me the opportunity to grow as a chamber musician. I am especially grateful to Calvin for our many musical collaborations over the years, and I hope there will be more to come!

I would not have gotten to this point without my family. My parents immigrated to the US from Romania a few years before I was born in search of increased economic and political freedoms, leaving behind their families and their careers. I know I would not have ended up pursuing a PhD in Physics at MIT without their sacrifice, for which I am immensely grateful. Sunt foarte recunoscătoare şi pentru sacrificiul făcut de familia mea in România, care ne-a simţit lipsa toţi ani aceştia. Thank you to my mom, the original Dr. Biscoveanu, for giving me the best example of a professional woman that can balance both her career and her family. Thank you to my dad for sparking my interest in math and science and for helping me with my homework up to college linear algebra! Thank you to my brother Eric, for being my first student and teaching me how to be a patient teacher. Finally, thank you to Colm for teaching me everything I know about Bayesian inference, for always being willing to answer my incessant coding and computing questions, for keeping me grounded, and for supporting my dreams.

Contents

II Binary black holes

8 Measuring the spins of heavy binary black holes

9 A new spin on LIGO-Virgo binary black holes

List of Figures

List of Tables

Chapter 1

Introduction

Nearly one hundred years after Einstein predicted their existence theoretically, gravitational waves were directly detected by the LIGO instruments for the first time in 2015. This first gravitational-wave event, dubbed GW150914, originated from the merger of two black holes at a distance of ~ 400 Mpc, each approximately 30 times the mass of the sun. In the eight years since, the LIGO detectors—along with the Virgo detector in Italy and the KAGRA detector in Japan—have observed the mergers of nearly one hundred such compact-object binaries including both black holes and neutron stars. The first binary neutron star (BNS) event, GW170817, was accompanied by electromagnetic emission observed across the band from radio to gamma-ray wavelengths. These detections have provided a novel laboratory for probes of fundamental physics—like precision tests of General Relativity in the strong-field regime and independent measurements of the Hubble constant—and of the astrophysical processes governing the formation and evolution of compact-object binaries and their electromagnetic counterparts.

1.1 Gravitational-wave sources

1.1.1 From stars to compact-object binaries

Compact objects

Compact objects like black holes and neutron stars are the remnants left behind when massive stars end their lives [e.g., 452]. Stars with initial masses up to $\sim 9~M_\odot$ will end their lives as white dwarfs, supported by electron degeneracy pressure once nuclear fusion stops in their cores. With radii on the order of $\mathcal{O}(1000)$ km and masses up to 1.44 $M_\odot$, white dwarfs in binaries are not compact enough to be detectable with ground-based gravitational-wave experiments like LIGO, Virgo, and KAGRA. However, such systems within the Milky Way will flood the sensitive band of the planned space-based detector LISA in the mHz regime [654, 652, 653, 774, 957, 561, 79].

Stars with initial masses between $\sim 9-25~M_\odot$ will undergo a supernova explosion once iron is fused in their core and collapse to a neutron star with mass $\sim 1.4~M_\odot$ and radius ~ 12 km. The mass-radius relation and maximum mass up to which a neutron star is supported against gravitational collapse are both unknown due to their dependence on the unknown nuclear equation of state (EoS) [540]. The mass distribution of galactic neutron stars detected as pulsars—highly magnetized rotating neutron stars that emit beams of electromagnetic radiation along their magnetic poles—is narrowly peaked at 1.33 $M_\odot$ [77, 88, 859, 337, 802, 801], although the most massive pulsar has a mass of 2.35 $M_\odot$ [760]. In addition to the radius, neutron stars are alternatively characterized by their compactness or tidal deformability:

$$C_{\mathrm{NS}} = \frac{Gm_{\mathrm{NS}}}{R_{\mathrm{NS}}c^2}, \quad \Lambda_{\mathrm{NS}} = \frac{2k_2 c^{10}}{3G^5}\left(\frac{R_{\mathrm{NS}}}{M_{\mathrm{NS}}}\right)^5, \tag{1.1}$$

where $R_{\mathrm{NS}}, M_{\mathrm{NS}}$ are the neutron star radius and mass, respectively, and k_2 is the dimensionless tidal love number that determines the rigidity of the neutron star and the response of its shape to changes in the tidal potential.

Black holes form from the collapse of stars initially $\sim 25 - 150$ times the mass of the sun. Unlike neutron stars and white dwarfs, black holes are completely characterized by their mass and spin, and we assume $\Lambda_{BH} = 0$. More massive stars are thought to undergo a pair instability supernova, where their cores reach such high temperatures that they become unstable to electron-positron pair production which triggers a runaway thermonuclear reaction that releases enough energy to completely disrupt the star, leaving no black hole remnant [452]. This process implies that there should be a maximum black hole mass that can form from direct stellar collapse, which is predicted to be $43 - 96\ M_\odot$ [331], with the large uncertainty due to its dependence on the poorly constrained rate of the $^{12}\mathrm{C}(\alpha, \gamma)^{16}\mathrm{O}$ reaction that occurs during stellar helium burning. Stars with initial masses between $\sim 100 - 150\ M_\odot$ will go through the pair instability process episodically, leading to pulses of mass loss but not complete disruption. These pulsational pair instability supernovae should produce an excess of black holes with masses just below the maximum mass. Stars initially more massive than $\sim 250\ M_\odot$ may again directly collapse to black holes as the photodisintegration instability absorbs the energy released by pair production, preventing the runaway thermonuclear reaction that disrupts the star [384, 941]. However, black holes with masses on the order of $\mathcal{O}(100)\ M_\odot$ predicted to form from the direct collapse of the most massive stars have remained elusive observationally [e.g., 436].

In addition to the upper mass gap for black hole masses between $\sim 45 - 125\ M_\odot$ predicted by pair-instability supernovae, there may also be a lower mass gap between the most massive neutron stars and the least massive black holes. The presence or absence of this gap depends on the time between initial stellar collapse and the eventual supernova explosion. For supernovae driven on a rapid ($\lesssim 100 - 200$ ms) timescale, the explosion is strong, resulting in a break in the compact object mass distribution between about $3 - 5\ M_\odot$ [385]. Delayed supernova explosions, on the other hand, can be weak. This leads to increased fallback onto the proto-neutron star that forms during the initial collapse, resulting a continuous spectrum of compact object masses with no gap [136]. The mass distribution of the neutron stars and black holes observed electromagnetically before the direct detection of gravitational waves did feature a gap [676, 335] and hence provided support for the rapid supernova mechanism.

Compact binary evolution

There are two primary theories for the formation channels of compact-object binaries. Broadly, compact objects in binaries can evolve together their whole lives from a binary star system, or they can dynamically encounter each other later in life in a dense stellar environment (see, e.g., Refs. [709, 583] for reviews). If stars in binaries evolved like they do in isolation, the components would have to start out so close together that their radii would overlap, leading to an immediate stellar merger [583]. Thus, the main obstacle that both formation channels must overcome is bringing the component compact objects of an initially wide binary close enough together so they can merge within a Hubble time due to angular momentum dissipation via gravitational radiation.

A schematic of the canonical picture of isolated binary evolution in field environments is shown in Fig. 1-1 [817, 889, 557, 146, 653, 135, 925, 700, 289, 493, 672, 299, 609, 134, 320]. Two stars are born together in a binary (step a). When the initially more massive star, henceforth called the primary star, begins to evolve off the main sequence, mass transfer may occur (step b). Usually the primary star becomes the more massive compact object, but if enough mass is transferred onto the secondary star, the more massive star can become the less massive compact object in a process called mass ratio reversal [404, 837, 962]. The primary will eventually undergo a supernova explosion (step c) and collapse to either a neutron star or black hole (step d). If the supernova is asymmetric, a natal kick will be imparted to the newly-formed compact object in order to conserve linear momentum [503, 491, 492]. Depending on the velocity and direction of this kick, the binary may be disrupted. Such disruption is also possible during step g) following the second supernova, described below.

At this point, the system may be observable as an X-ray binary due to the electromagnetic emission produced by the accretion disk that forms around the compact object as it pulls matter away from the secondary star (step e). This mass transfer episode may result in a common envelope phase [484, 562], where the outer layers of the secondary expand so much that they engulf the compact object in addition to its own core (step f). Dynamical friction during this phase decreases the orbital separation by two or more orders of magnitude [583]. Whether or not the envelope is successfully blown off (step g) depends on its binding energy and the efficiency with which orbital energy is deposited into the envelope [481]. If the envelope is successfully ejected, the core of the secondary star may also undergo a supernova explosion, leaving behind a compact-object binary that could merge within a Hubble time depending on its resulting orbital separation. Alternatively, the envelope ejection process may lead to a merger between the first-formed compact object and the core of the secondary star [845, 383], which could potentially leave behind a Thorne-Zytgow object with a neutron star core surrounded by a diffuse stellar envelope [868, 869].

Besides the common envelope channel described above, isolated binary evolution may also proceed via stable mass transfer and chemically homogeneous evolution. Depending on the mass ratio at the onset of the second mass transfer event in step e), the mass transfer may remain dynamically stable, avoiding the common envelope phase [899, 694, 902]. If the stellar donor is more massive than the compact object accretor, conservation of angular momentum will drive the binary closer together. Eventually the envelope of the secondary is stripped away and it can undergo a supernova explosion to produce the second-born compact object in the binary, which may merge within a Hubble time depending on the orbital separation.

Alternatively, massive stars in low-metallicity binaries may avoid significant radial expansion due to efficient chemical mixing and could hence be born in tight enough orbits so that no additional mechanism is needed to bring the binary closer together [582, 596, 288]. Because of the initial tightness of the orbit, the stars become tidally locked, rotating at the same period as the orbit. Such rapid rotation enables efficient chemical mixing so that eventually all the star's hydrogen is fused to helium, leaving a Wolf-Rayet star at the end of its main sequence evolution [313, 841, 321, 573, 843]. This helium star contracts rather than expanding and eventually collapses to a black hole. This process, called *chemically homogenous evo-*

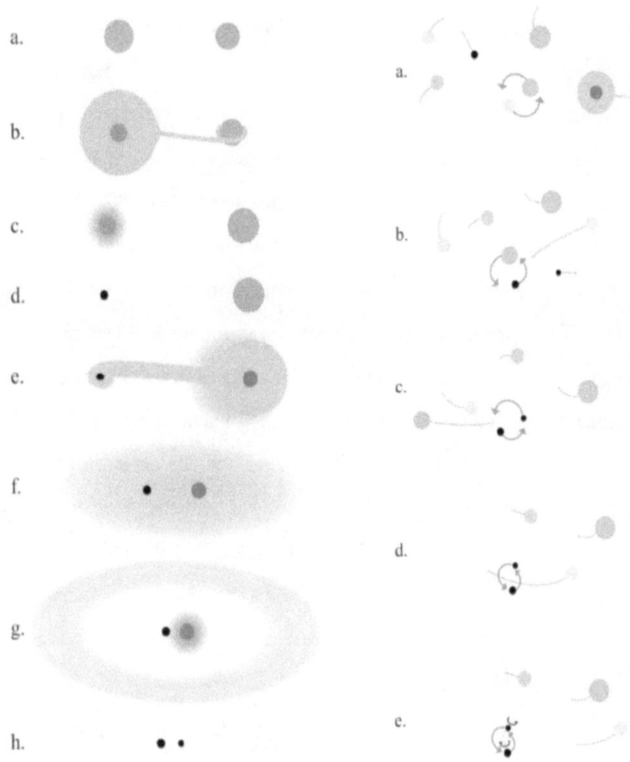

Figure 1-1: *Left:* Schematic of binary black hole formation via the isolated binary evolution channel including a common envelope phase. *Right:* Schematic of binary black hole formation via the dynamical channel. Both diagrams from Ref. [583].

lution is not a viable channel for the production of binaries containing neutron stars as the initial stars are too massive. However, it may be responsible for the formation of the most massive binary black hole (BBH) systems in low-metallicity environments, which are required to ensure the inefficiency of stellar winds that would widen the binary and break tidal locking [287, 304, 750].

While the components of compact-object binaries that form via isolated binary evolution spend their whole lives together, those that form via dynamical assembly may form from individual stars that encountered each other later in life. In a dense stellar environment like a globular cluster or nuclear star cluster, the black holes will tend towards the center of the cluster due to mass segregation [831, 377]. There they can form binaries via repeated three-body interactions. The lighter interlopers will leave the interaction with a higher speed and are more likely to be ejected from the cluster [460]. If the orbital speed of the remaining binary is larger than the typical stellar speed in the cluster, these repeated dynamical interactions will "harden" the binary, reducing the orbital separation with each encounter [454, 108, 754]. The binary may eventually get close enough to merge via the emission of gravitational radiation within a Hubble time if the environment is dense enough to sustain the interaction rate [812, 527, 706, 666, 300, 140, 636, 755, 98, 691, 371, 594]. This channel is not expected to contribute significantly to the rate or mergers including a neutron star, as these lighter objects are likely to be ejected from the cluster in three-body interactions [954].

In addition to the natal kicks received due to the asymmetry of the supernova explosion at the birth of the component compact objects, the merger remnant will also receive a recoil kick due to the conservation of linear momentum carried away by the gravitational waves emitted over the course of the binary's evolution [162, 201, 569, 830, 904]. If this recoil kick is smaller than the escape velocity of the cluster, the merger remnant can pair up with another compact object to merge again via a *hierarchical merger* [515, 403].

Besides stellar clusters, the accretion disks of active galactic nuclei have been proposed as a promising environment for dynamically-formed binary black holes [604,

117, 838, 603, 846]. Compact objects can either form directly in the disc or become captured due to gas drag [844]. Within the disc, their motion is governed by viscous interactions; they migrate radially until they are caught in a "migration trap", where the viscous torque changes sign, hence attracting compact objects at both larger and smaller disc radii [138]. These locations serve as ideal meeting points for compact objects to couple up into binaries via dynamical encounters. Because of the large escape velocity of the active galactic nucleus, this environment leads to larger retention fractions and hence larger rates of hierarchical mergers [952, 605, 403].

Predicted binary properties

The various formation channels above predict distinct features in the distributions of the masses, spins, and redshifts of merging compact-object binaries. Spin is a measure of the angular momentum of the spacetime for a black hole and a proxy for the object's rotation rate around its own axis for a neutron star. The dimensionless spin vector typical in the gravitational-wave literature is defined as $\chi = \mathbf{S}c/(Gm^2)$, where $\mathbf{S}$ is the object's angular momentum and m is its mass. For black holes in general relativity, $|\chi|$ can range from [0,1] [507], while for neutron stars the theoretical breakup spin at which the centrifugal force produced by rotation exceeds the gravitational force holding the star together is $|\chi| \approx 0.7$ for most EoSs [802, 639].

If the observed population of sources is dominated by systems formed via isolated binary evolution, the limits of the mass distribution should be set by supernova physics. The presence of a lower mass gap between the most massive neutron stars and least massive black holes would imply the activity of the rapid supernova engine rather than the delayed engine. Additionally, there should be no black holes with masses in the upper mass gap, and there should be a preference for equal-mass binaries due to the pairing function of the initial stellar binaries. Alternatively if there is a contribution from the dynamical formation channel, hierarchical mergers may fill the upper mass gap and lead to a subpopulation with asymmetric masses from the combination of a second-generation black hole merging with a first-generation black hole [757].

The distribution of black hole spin magnitudes may favor small spins due to the expected efficiency of angular momentum transport in massive stars [387]. When the stellar envelope, which is predicted to carry most of the star's angular momentum, is shed via mass transfer and stellar winds, a slowly-spinning core that collapses to a slowly-spinning compact object is left behind. The secondary star in a field binary can be spun up due to tidal interactions with the first-formed black hole, leading to a more rapidly spinning second black hole [725, 128]. It is possible that systems formed via chemically homogenous evolution may retain higher spins since stellar winds should be insignificant in these systems [596]. Second-generation black holes should also have larger spins due to the conservation of angular momentum, with a characteristic spin magnitude of $|\chi| \equiv \chi \approx 0.7$ [714, 191, 423].

Binaries formed in isolation are expected to have spin tilts preferentially aligned to the orbital angular momentum due to the effect of tides in their stellar progenitors [889, 492, 428, 709, 134, 582, 596, 758, 669, 837]. Some dispersion in the tilts of field binaries is expected due to the velocity dispersion of natal kicks, which can tilt the orbital plane and misalign the compact object spin [373, 299, 385, 966, 591, 402, 133, 410, 586]. Binaries formed dynamically should have isotropically-distributed spin tilts since they pair up randomly without a preferred direction [812, 625, 587, 707, 140, 758]. The merger rate as a function of redshift is expected to follow the star formation rate convolved with a delay time distribution, since the merging compact objects form from stars, but there is a delay between the birth of the stellar progenitors and the eventual merger of the binary [e.g., 670, 671, 592, 351].

The possible binary formation channels may naturally impart correlations between binary parameters. For example, in Chapter 10, I search for a correlation between binary black hole spin and redshift, which may arise because low-metallicity environments where stellar winds are inefficient preferentially occur at higher redshifts, for example. More massive systems may also have higher spins since they form more frequently via chemically homogenous evolution [582, 596] and hierarchical mergers [403]. Similarly, more massive systems may preferentially merge at higher redshifts, since they preferentially form in low-metallicity environments [133, 593, 651].

1.1.2 Electromagnetic counterparts

Postmerger outcomes

Compact-object mergers including a neutron star may be accompanied by a menagerie of electromagnetic counterparts, the properties of which depend on the properties of the binary and merger remnant. If the total mass of the binary is less than the maximum mass of a nonspinning neutron star, M_{TOV} [879, 667], the remnant will be a stable neutron star. The remnant is called a supramassive neutron star if its mass exceeds M_{TOV} but it is supported against collapse by rigid rotation [378], and a hypermassive neutron star if it is supported by differential rotation [118]. For total binary masses above $\sim 1.5 M_{\mathrm{TOV}}$, the remnant collapses directly to a black hole, with the exact collapse mass given by the EoS [120]. Hypermassive neutron stars will collapse to black holes on dynamical timescales of $\mathcal{O}(1)$ s [808], while supramassive neutron stars are longer lived, collapsing on secular timescales of $\mathcal{O}(10^5)$ s [736, 412]. See Ref. [789] for a review of BNS postmerger outcomes.

Binary neutron star mergers are expected to be accompanied by two primary electromagnetic signatures—a gamma-ray burst [319, 630, 485] and a kilonova [552, 853, 854, 340]—although whether a long-lived neutron star remnant can power a gamma-ray burst is an area of active research [967, 641, 413]. In the case of neutron star-black hole (NSBH) mergers discussed in detail in Chapter 7, the neutron star must be disrupted outside the black hole's innermost stable circular orbit in order to leave behind material which could potentially power an electromagnetic counterpart [689, 367, 368]. A schematic of the aftermath of a compact-object merger including potential counterparts is shown in Fig. 1-2.

Kilonovae

The kilonova emission that can accompany compact object mergers including a neutron star is powered by the radioactive decay of the heavy elements synthesized via the rapid neutron-capture process in the material disrupted during the merger. BNS and NSBH mergers have long been considered a promising site of astrophysical heavy-

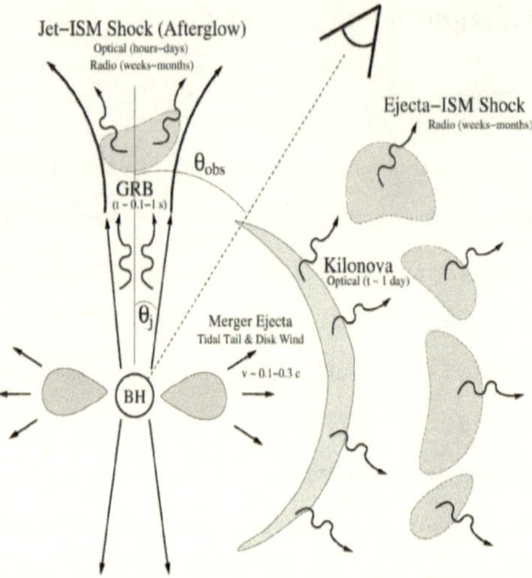

Figure 1-2: Schematic of the aftermath of a compact object merger including a neutron star assuming the merger remnant is a black hole from Ref. [617]. A short gamma-ray burst jet has been launched along the black hole spin axis, while the equatorial merger ejecta is responsible for powering the kilonova emission.

element nucleosynthesis [541, 542, 842] due to the expected low electron fraction of the ejecta, Y_e, the ratio of the proton density to the sum of the proton and neutron densities. This neutron-rich material more readily undergoes the rapid neutron capture process—where the neutron capture timescale is shorter than the β-decay timescale—required to form elements heavier than iron [e.g., 254]. Two main sources of neutron-rich ejecta from compact object mergers are typically considered [616], 1) ejecta that is either tidally disrupted [121] or squeezed from the contact interface of the two merging bodies for BNSs [472, 121] on the dynamical timescale ($\sim$ ms), and 2) disk wind ejecta produced via outflows from the accretion disk that forms around the remnant object on a longer timescale on the order of seconds [341, 489].

The tidal tail ejecta will produce a "red" kilonova with higher opacity and lower Y_e [121, 283] predominantly in the equatorial plane. In the case of NSBH mergers, the tidal ejecta are often additionally azimuthally asymmetric [529], so that an angular dependence is introduced not only relative to the orbital axis of the binary but also relative to the azimuthal orientation of the observer. The contact interface ejecta is emitted along the polar direction and typically consists of lower-opacity, higher Y_e material [733, 928]. The mass of this ejecta component depends on how promptly the merger remnant collapses to a black hole, as this suppresses ejection of material from the contact interface. The disk wind ejecta resides behind the fast-expanding dynamical ejecta, and the composition and hence the color of the kilonova also depend on the lifetime of the remnant. A longer-lived neutron star remnant in the case of a BNS merger can power weak interactions that increase the electron fraction, favoring a bluer kilonova [497, 618].

The complete picture of the kilonova emission from a compact object merger is the sum of all the individual components described above and their interactions. The mass of the ejected material and its composition thus depend sensitively on the properties of the binary such as mass ratio, total mass, neutron star equation of state, and component spins. Fitting formulae have been derived in the literature based on general-relativistic magneto-hydrodynamical merger simulations to map the binary parameters onto the merger ejecta properties [e.g., 368, 250, 294], but there remain large discrepancies in the predicted kilonova lightcurves based on these remnant mass calculations due to differing treatment of microphysics like radiation transport and the geometry of the emission [e.g., 498, 184, 279]. The variation in kilonova signals expected for BNS mergers and optimal observing strategies for their detection with a new instrument, the Wide-field Infrared Transient Explorer (WINTER), are presented in Chapter 6.

Gamma-ray bursts

Gamma-ray bursts (GRBs) are the most energetic electromagnetic explosions observed in the universe, yet the central engine that powers this emission and the geometric properties of the emission itself remain poorly understood. The GRB population appears to be bimodal in duration and hardness (steepness of the spectral slope), with long-soft and short-hard bursts separated at ~ 2 s [526]. The association of long GRBs with Type Ic supernovae [392, 467, 834, 168] and their exclusive location in star-forming galaxies has led to the consensus that they are produced in the deaths of massive stars [167, 927, 571].

For short GRBs, the lack of an associated supernova [e.g., 369, 520, 825, 280] together with the localization of some short GRBs to early-type galaxies [360, 361, 717] provided early evidence in support of the binary neutron star or neutron star-black hole merger progenitor model [319, 649]. The coincident detection [15] of gravitational wave event GW170817 from a binary neutron star merger [17] and the short, hard burst GRB170817A [417, 793] has confirmed the compact binary progenitor model for at least some sGRBs. The association of a kilonova with the long-duration gamma-ray burst GRB211211A, however, indicates that duration and hardness alone cannot be used to definitively reveal the progenitor system [735].

The generic model for GRB emission has four stages. I) A compact, inner, hidden central engine produces a relativistic energy flow. II) The energy is transported outward and III) converted to the observed prompt γ-ray emission. IV) The remaining energy interacts with the circum-burst medium to produce the afterglow radiation at longer wavelengths. The necessary relativistic energy flow can be produced by a fireball—a large concentration of radiation in a small, baryon-poor region of space [424, 678]. In this scenario, the fireball is an opaque lepton-photon plasma whose kinetic energy is much greater than its rest mass, which is produced via electron-positron pair-production [703]. The fireball energy is deposited into the baryonic load, which is accelerated to a relativistic velocity of $\gamma \sim E/M$ [807, 679].

The conversion of the kinetic energy of the baryon load into the observed γ-rays is thought to occur via collisionless shocks due to the interaction of distinct shells of relativistic material. This type of radial stratification can arise due to the turbulence of the ejecta surrounding the central engine or due to the temporal variability of the activity of the central engine itself [741, 680, 519]. The observed GRB γ-ray spectrum is produced by synchrotron radiation from the shocks—ultrarelativistic charged particles gyrating in a magnetic field [612, 504, 787]. Eventually the relativistic load hits the circum-burst medium, producing the multiwavelength afterglow via synchrotron radiation observable from the X-ray band on the timescale of hours-days to the radio band where the emission can persist for several years. See Ref. [143] for a review of short GRB afterglows.

The relativistic material responsible for the observed γ-ray emission is expected to be narrowly beamed into a collimated jet. Evidence for collimation in GRBs comes from the observation of the "jet break" feature in the lightcurves of some afterglows— an achromatic steepening that occurs when the bulk Lorentz factor of the outflow has decreased to $\gamma \approx 1/\theta_j$ [748, 749, 788], where θ_j is the half-opening angle of the jet. Jet geometry is discussed in detail in Chapter 5, along with the presentation of a new method to measure θ_j using coincident gravitational-wave and γ-ray observations.

The GRB central engine: Two primary central engine theories are considered viable candidates for explaining the high luminosity, temporal variability, collimation, and ultrarelativistic outflow speeds observed in GRBs: hyper-accreting black holes [939, 704, 545, 650, 290, 929, 606] and millisecond magnetars [892, 266, 866, 518, 933, 967, 413]. In the hyper-accreting black hole model, the central black hole is surrounded by a thick disk of extremely hot plasma. The two main mechanisms for launching the jet in this scenario are neutrino annihilation along the spin axis of the black hole [319, 772] and a Poynting flux-dominated outflow launched by the Blandford-Znajek mechanism [166]. For neutrino-driven jets, neutrino annihilation generates a hot photon-lepton gas which expands as a fireball, whose final Lorentz factor depends on the efficiency with which the neutrino annihilation energy is deposited into the baryon load [960, 724, 548].

If the jet is alternatively launched by the Blandford-Znajek mechanism, the power of the Poynting flux depends on the spin of the central black hole [545, 551, 929, 606, 548]. The accretion rate should be very high, and neutrino annihilation and neutrino-driven winds will still occur in the disk. Thus, the jet will have a "hot" neutrino component and a "cold" Poynting flux component [548]. The rotation of the black hole creates a funnel along the spin axis where the density of the surrounding material is much lower than in the plane of the accretion disk, since material in this region falls into the black hole on the free-fall timescale instead of being centrifugally supported [613]. The bipolar jets propagate out along this funnel, and the strong magnetic field acts as a barrier for protons drifting into the jet, which reduces the baryon load in agreement with the ultrarelativistic nature of the observed jets.

When the jet is launched into a dense medium (a stellar envelope for long GRBs or merger ejecta for sGRBs), a Kelvin-Helmholtz instability develops at the jet edge where there is significant differential motion with respect to the material. This will induce variability in the jet light curve even if the central engine exhibits smooth behavior [637]. The material also collimates the jet via the production of a hot cocoon, resulting in a mildly relativistic outflow surrounding the central, ultrarelativistic, narrow core [613, 747, 646].

Another possibility for the GRB central engine is a rapidly rotating ($P_0 \sim 1$ ms), highly-magnetized (surface field $B_s \sim 10^{15}$ G) neutron star, called a magnetar. The rotational energy of such a neutron star provides an upper limit for the energy of the jet. Initially, a newborn neutron star is very hot ($T > 10^{11}$ K), which leads to heavy baryon loading of the magnetar wind due to neutrino-driven mass loss from the surface [469]. This early outflow is not fast enough to power a GRB jet. As the neutron star cools down on the timescale of seconds, the neutrino-driven wind begins to disappear and a jet with a large magnetization factor is produced. The complete disappearance of the winds leads to a further increase of the magnetization factor and brings the prompt emission phase to an end. The energy that powers the prompt emission comes from the differential rotation of the neutron star, which generates magnetic energy via a dynamo mechanism [180, 181, 413].

The magnetar continues to spin down after the initial GRB. The Poynting flux from the spin-down can be injected into the blast wave, and if this late-time energy injection exceeds the energy of the prompt phase, the light curve will exhibit a shallow decay [967]. Bursts of magnetic activity can also power late-time X-ray flares [267]. The millisecond magnetar model provides an appealing explanation for the internal plateaus and X-ray flares observed in some short GRBs, while a steep decline immediately after the X-ray plateau is more easily explained by a corresponding sharp decline in the accretion rate of a black hole central engine [528].

Both the millisecond magnetar and hyper-accreting black hole central engine models are compatible with a compact object merger progenitor for the GRB. If the remnant collapses promptly to a black hole, the GRB jets will be launched by accretion. If instead a meta-stable HMNS is formed, this serves as the magnetar central engine [342].

1.1.3 The stochastic gravitational-wave background

The last four paragraphs of this section were previously written by ASB for Ref. [323].

In addition to the mergers of individual compact-object binaries that have been observed by the ground-based gravitational-wave detector network, a random gravitational-wave background is predicted to arise due to the superposition of many individually indistinguishable sources, which may be both astrophysical and cosmological in nature. This stochastic gravitational-wave background is typically characterized by Ω_{GW}, the ratio of the energy density of the Universe contained in gravitational waves to the critical energy density needed to close the Universe [73, 761],

$$\Omega_{\text{GW}} = \frac{1}{\rho_c} \frac{d\rho_{\text{GW}}}{d\ln f}, \quad \rho_c = \frac{3c^2 H_0^2}{8\pi G} \tag{1.2}$$

where H_0 is the Hubble expansion rate. The gravitational wave energy density describes the spectrum of the stochastic background as a function of frequency and is often modeled as a power law,

$$\Omega_{\text{GW}}(f) = \Omega_\alpha (f/f_0)^\alpha. \tag{1.3}$$

The spectrum for the astrophysical background of compact-object mergers is estimated to have an amplitude of $\Omega_{\text{GW}}(f_0 = 25 \text{ Hz}) \approx 7 \times 10^{-10}$ [52] with a power-law index of $\alpha = 2/3$ determined by the frequency evolution of the gravitational-wave signal from these sources [701, 973, 764, 595, 10]. Such a background encodes information about binaries at much larger redshifts than can be probed with individual detections, and its detection would provide information on the star formation history, merger rates, and mass ranges of the progenitor systems of these high-redshift binaries. In addition to the compact binary coalescences discussed in Section 1.1.1, the astrophysical background can include contributions from core-collapse supernovae [192, 974, 256, 257] and from individual neutron stars [765, 943, 538, 229].

A cosmological stochastic background represents the gravitational analog to the cosmic microwave background. This signal is predicted by most cosmological models to produce a flat spectrum with an amplitude several orders of magnitude smaller than the astrophysical background. Standard slow-roll inflationary models predict $\Omega_{\mathrm{GW}} \sim 10^{-17}$ [575, 57], too weak to be directly detected by all but the most ambitious space-based gravitational-wave detectors [259, 506]. However, nonstandard inflationary and cosmological models can produce backgrounds due to processes like preheating, first-order phase transitions, and cosmic strings [210], all with energy densities within the reach of next-generation ground-based gravitational-wave detectors.

Following inflation, the universe must undergo a period of particle production during which the inflaton field couples to other particle species into which it eventually decays. If this process occurs non-perturbatively, it is called *preheating* (see Refs. [68, 81] for reviews). While the amplitude of the background generated during this process of particle production is expected to be independent of the temperature scale at which it occurs, standard inflationary models predict that it will peak at $f \sim 10^7 - 10^8$ Hz, well beyond the frequency band of ground-based gravitational-wave detectors [511, 308]. However, for hybrid inflation occurring around $\sim 10^9$ GeV, the background from preheating peaks in the band of ground-based detectors with an energy density of $\Omega_{\mathrm{gw}} \sim 10^{-11}$ [395, 310, 309].

Cosmic strings are one-dimensional topological defects produced in spontaneous symmetry breaking phase transitions following inflation [512, 913]. When a string folds upon itself, it produces a loop, which oscillates under its tension, emitting gravitational waves in a series of harmonic modes [893, 806, 533]. During the course of these oscillation, a cosmic string loop can produce a cusp at the point where the string reaches relativistic velocities, and the intercommutation of two string segments can produce a kink [71, 72]. These string features emit higher frequency bursts of gravitational radiation [277] whose superposition creates a stochastic gravitational-wave background accessible to ground-based gravitational-wave detectors [278, 811]. The spectrum of the background depends on the cosmic string tension, $G\mu$, and the loop model, among other parameters [6, 456].

Phase transitions in the early universe, such as the decoupling of the electromagnetic and weak forces, can also produce a stochastic gravitational-wave background under some modifications of the Standard Model if they are strongly first order [494, 211, 483, 930], i.e., if there is a discontinuity in the first derivative of the free energy during the transition. In this scenario, gravitational waves are emitted due to the collision of bubbles of the new phase [525, 479] and due to the anisotropic stresses generated by magnetohydrodynamical turbulence and discontinuities in the shocked plasma surrounding the expanding bubbles [208, 209, 464, 465]. A first-order electroweak phase transition also has implications for electroweak baryogenesis, which could provide an explanation for the cosmic baryon asymmetry [635]. The peak frequency of the stochastic background energy density spectrum depends on the energy scale of the transition, with a transition occurring at 10^9 GeV producing a background peaking in the frequency band of ground-based gravitational-wave detectors [575, 490].

1.2 Gravitational-wave detection

1.2.1 The LIGO detectors

The Laser Interferometer Gravitational Wave Observatory (LIGO) is a pair of ground-based interferometers located in Hanford, Washington, and Livingston, Louisiana. Each detector consists of an L-shaped Michelson Interferometer in vacuum with 4 km long arms, with a Fabry–Pérot cavity in each arm [3]. A simplified schematic of the interferometer is shown in Fig. 1-3. Laser light travels to a beam splitter, which in turn redirects the beam down the two arms. Upon reaching the end optics, the light is reflected and recombined at the beam splitter, eventually reaching a photodetec-

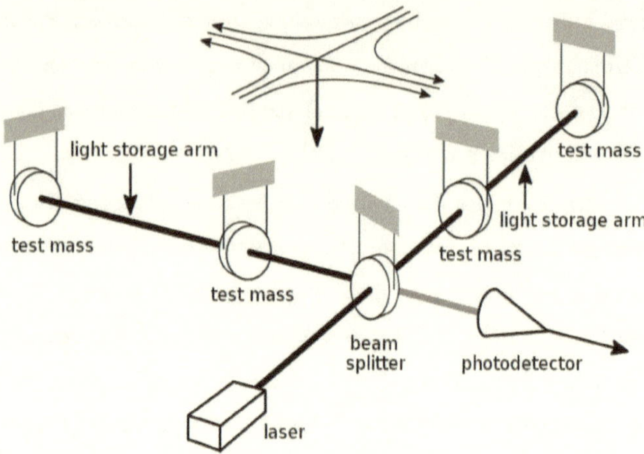

Figure 1-3: Schematic of a Michelson Interferometer like the one used in the two LIGO detectors. Laser light incident on the beam splitter travels down the two arms and is reflected by the end test mass. In the presence of a gravitational wave, the arm length will be compressed in one direction and extended in the other, leading to a phase shift observable as an interference pattern in the recombined light received by the photodetector.

tor. The effect of a gravitational wave is to compress the arm length in one direction while stretching it in the perpendicular direction. This results in a phase shift between the two light beams, producing an interference pattern that is recorded by the photodetector upon recombination. Gravitational-wave interferometers measure the dimensionless strain as a function of time, $h \sim \delta L/L$, where L is the arm length.

The interferometer sensitivity is typically characterized in terms of the one-sided power spectral density (PSD) with units of [1/Hz] or the amplitude spectral density (ASD) with units of $[1/\sqrt{\text{Hz}}]$,

$$\text{PSD}(f) \propto \langle \tilde{h}(f)^*\tilde{h}(f)\rangle, \quad \text{ASD}(f) = \sqrt{\text{PSD}(f)} \tag{1.4}$$

where $\tilde{h}(f)$ is the complex-valued Fourier transform of the real-valued time-domain strain measured by the interferometers, $h(t)$. The simplified noise budget in Fig. 1-4 shows the contributions of individual noise sources to the total interferometer ASD. A lower value of the ASD or PSD indicates that the interferometer is more sensi-

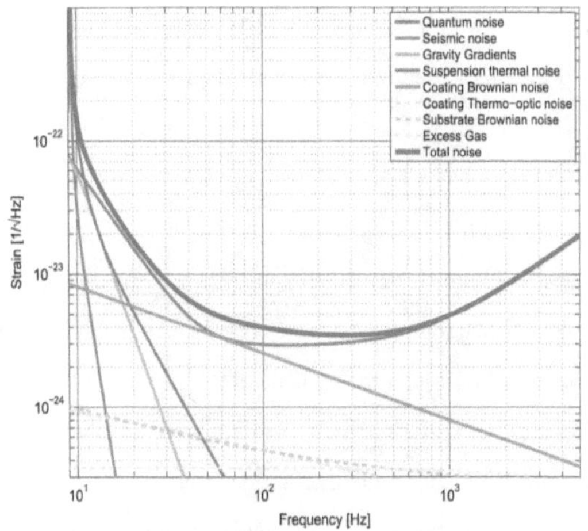

Figure 1-4: Noise budget showing the contributions of individual noise sources to the the amplitude spectral density for Advanced LIGO at design sensitivity from Ref. [3].

tive to gravitational waves at that frequency. The sensitive band of ground-based gravitational-wave detectors spans the range from $\mathcal{O}(10) - \mathcal{O}(10^3)$ Hz.

After a series of updates to the Initial LIGO instrument, the Advanced LIGO (aLIGO) detectors have undertaken three observing runs from September 2015 to March 2020. The fourth observing run, O4, is currently scheduled to begin at the end of May 2023 [22]. The LIGO optical configuration features dual-recycling cavities [607, 455, 839] with a power-recycling mirror at the input to the main interferometer designed to increase the circulating power in the arms [516, 379] and a signal-recycling mirror at the output that broadens the detector bandwidth [185]. Increased laser power in the arms amplifies the effect of the phase shift induced by an arm length change, leading to a stronger detector response to a passing gravitational wave. Increasing both the input laser power and the power circulating in the arms contributed significantly to the improved sensitivity of aLIGO relative to initial LIGO [22, 183].

The diameter and mass of the test masses were correspondingly increased to counteract the effects of *thermal noise* due to Brownian motion of the mirror coatings [174, 550, 471, 951]. Loss in the fused silica fibers that suspend the test masses also contributes to the overall thermal noise [792, 260]. The aLIGO suspension system features a quadruple-pendulum system to improve the seismic isolation of the interferometer [100, 932, 601]. *Seismic noise* due to the effect of ground motion shaking the test masses dominates the noise budget at low frequencies. The suspension system upgrade led to an extension of the lower limit of the frequency band from ~ 40 Hz in initial LIGO down to ~ 20 Hz [3].

Heisenberg uncertainty comprised of shot noise at high frequencies above ~ 100 Hz and radiation pressure noise at low frequencies is called *quantum noise* [185, 214]. Shot noise arises due to variations in the arrival rate of photos at the detection port, which generate laser amplitude fluctuations that cannot be distinguished from the effect of a passing gravitational wave. Similarly, radiation pressure noise due to variations in the number of photons arriving at the mirrors generates a force that pushes on the optics [213]. During the last observing run (O3), squeezed vacuum was injected into the detection port to reduce the Heinsenberg uncertainty along the phase quadrature responsible for shot noise [1, 886, 602]. This induces anti-squeezing along the amplitude quadrature, leading to increased levels of radiation pressure noise [665]. Because the two noise sources dominate at different frequencies, the next observing run will introduce frequency-dependent squeezing to minimize the effect of radiation pressure noise by squeezing along the amplitude quadrature at low frequencies and squeezing along the phase quadrature at high frequencies [956, 307].

1.2.2 Data analysis methods for compact-object mergers

The gravitational-wave signal

The gravitational-wave signal for a compact binary coalescence (CBC) is uniquely predicted by General Relativity for a given set of binary properties. A CBC is completely characterized by 17 parameters, shown in Fig. 1-5. Each object has a mass, a three-dimensional spin vector, and a tidal deformability for neutron stars, introduced in Section 1.1.1. The spin is typical parameterized in terms of the dimensionless spin magnitude χ, the tilt angle relative to the orbital angular momentum θ, and two az-

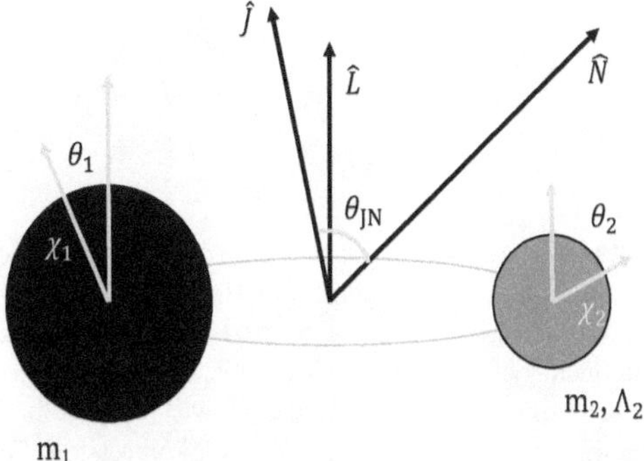

Figure 1-5: Schematic of a compact binary coalescence. $\hat{\mathbf{L}}, \hat{\mathbf{N}}$ and $\hat{\mathbf{J}}$ represent the orbital angular momentum, line of sight, and total angular momentum vectors, respectively.

imuthal angles (see Fig. 8-1). The more massive object is generally called the primary and indexed with 1, although in Chapter 9, I present an alternative parameterization where the binary components are instead sorted by their spin magnitudes.

An example gravitational waveform for a binary black hole merger is shown in Fig. 1-6. The amplitude and frequency both increase as a function of time until the moment of the merger. More massive signals merge at lower frequencies, meaning they spend less time in LIGO's sensitive band, corresponding to a shorter observable waveform [e.g., 163, 576]. The portion of the signal before the merger is called the inspiral. After the merger, the perturbed black hole "rings down" to a steady state, emitting gravitational waves characterized by damped sinusoids at specific frequencies depending on the final mass and spin of the merger remnant [916, 713, 863, 215, 312]. The gravitational-wave emission carries away energy, linear momentum, and angular momentum, meaning that the mass of the remnant black hole is smaller than the total mass of the binary before the merger [162].

The postmerger signal for binary neutron stars is much more theoretically uncertain and depends on the type of remnant. A newborn neutron star is generally expected to emit high-frequency gravitational waves due to a variety of stellar os-

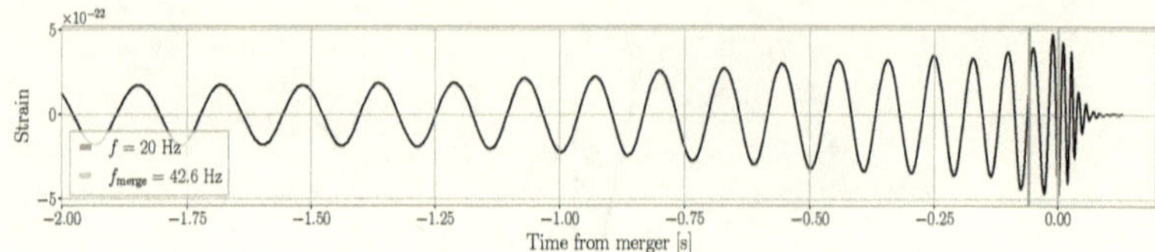

Figure 1-6: Time-domain gravitational waveform, $h_+(t) - ih_\times(t)$, for a binary with the maximum-likelihood parameters inferred for the massive BBH event GW190521. The start of the LIGO sensitive band is indicated with the blue line at $f = 20$ Hz, while the merger frequency of 42.6 Hz, which is typically assumed to correspond to the frequency at the maximum amplitude, is indicated with the red line. The amplitude modulations in the waveform are characteristic of spin precession. For this particular system, LIGO is only sensitive to the last ~ 0.1 s of the signal.

cillations [950, 773], the frequency of which should encode information on the EoS complementary to that encoded in the inspiral due to the effect of the tidal deformability on the phasing of the signal [848, 122, 123, 738, 847, 124]. This high-frequency emission will be damped when the remnant collapses to a black hole.

The set of eight parameters including the masses, spins, and tidal deformabilities are called the *intrinsic parameters*, which affect the amplitude and phase of the gravitational-wave signal. Gravitational waves have two polarizations, called the plus and cross modes. The plus mode squeezes and stretches an extended test mass along the horizontal and vertical directions, while the cross mode acts along the diagonal. The gravitational-wave contribution for each polarization can be decomposed into three terms corresponding to the amplitude $\mathcal{A}_{\mathrm{GW}}$, phase ϕ_{GW}, and viewing angle effect for the dominant (2,2) quadrupolar gravitational-wave emission mode,

$$\tilde{h}_+(f) = \frac{1}{2}\mathcal{A}_{\mathrm{GW}}(f)(1 + \cos^2 \iota) \cos \phi_{\mathrm{GW}}(f), \tag{1.5}$$

$$\tilde{h}_\times(f) = \mathcal{A}_{\mathrm{GW}}(f) \cos \iota \sin \phi_{\mathrm{GW}}(f), \tag{1.6}$$

where ι is the inclination angle between the total angular momentum and line-of-sight vectors. The phase, ϕ_{GW}, includes the effect of the intrinsic parameters on the frequency evolution of the gravitational-wave signal and an offset by the phase at coalescence, ϕ_c [349].

The amplitude depends primarily on the luminosity distance to the source and a particular combination of the component masses called the *chirp mass* [264],

$$\mathcal{A}_{\text{GW}}(f) \propto \frac{\mathcal{M}^{5/6} f^{-7/6}}{d_L}, \quad \mathcal{M} = \frac{(m_1 m_2)^{3/5}}{(m_1 + m_2)^{1/5}}. \tag{1.7}$$

Gravitational-wave detectors are sensitive to the redshifted masses as observed in the detector frame, $m_1 = (1 + z)m_{1,s}$, where z is the merger redshift and $m_{1,s}$ is the primary mass in the source frame. Large spins aligned to the orbital angular momentum cause a delay in the merger relative to an identical system with small spin, a process called the orbital hangup effect [200]. If the spin tilts are misaligned, this causes a phenomenon called General Relativistic spin precession, where the spin and orbital angular momentum vectors precess about the total angular momentum [90, 513]. This introduces characteristic amplitude modulations in the inspiral portion of the waveform as in Fig. 1-6, which should thus carry most of the information about the spin of the binary. In Chapter 8, I conduct a systematic study on the measurability of spin in massive BBHs where most of the inspiral portion of the waveform is inaccessible due to the low merger frequency of these massive systems.

The effect of the various intrinsic parameters along with distance and inclination on the gravitational waveform introduces an observational selection bias since sources with larger amplitudes (i.e., larger masses and smaller distances) and longer signals (i.e., high aligned spins) are easier to detect. We will see later how this *Malmquist bias* can be accounted for due to the deterministic nature of the gravitational-wave signal.

The remaining extrinsic parameters including the sky location parameterized via the right ascension and declination, polarization, and time at coalescence, $(\alpha, \delta, \psi, t_c)$, decouple from the intrinsic parameters in terms of their effect on the signal. The detector's sensitivity to the plus and cross modes of the gravitational wave depends on its orientation and geometry via the *antenna pattern functions*, plotted in Fig. 1-7 [867]. The total gravitational-wave signal is thus the sum of the plus and cross components multiplied by their respective antenna patterns, with a time-shift intro-

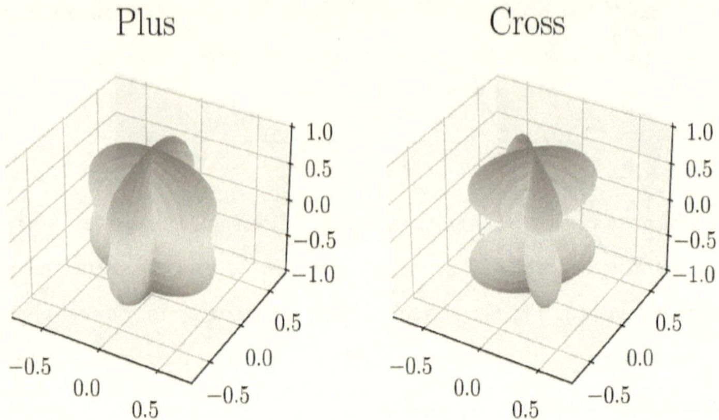

Figure 1-7: Antenna pattern functions for the plus and cross polarizations illustrating the sensitivity of the instrument to a gravitational wave as a function of the position relative to the detector.

duced to account for the time at coalescence,

$$\tilde{h}(f) = \left[F_+(\alpha, \delta, \psi)\tilde{h}_+(f; m_1, m_2, \chi_1, \chi_2, \Lambda_1, \Lambda_2, \iota, d_L, \phi)\right. \tag{1.8}$$

$$\left. + F_\times(\alpha, \delta, \psi)\tilde{h}_\times(f; m_1, m_2, \chi_1, \chi_2, \Lambda_1, \Lambda_2, \iota, d_L, \phi)\right]e^{-2\pi i t_c}. \tag{1.9}$$

Because the gravitational-wave signal predicted for a specific binary is costly to calculate using full numerical relativity, approximations called *waveform approximants* have been developed to enable the rapid evaluation of the waveform for a given set of binary parameters. Different approximant families employ different approximations to the full numerical relativity solution, including inspiral-only waveforms that rely on the post-Newtonian expansion [791, 165, 416, 164, 162, 171, 93, 170, 629, 915], inspiral-merger-ringdown phenomenological waveforms tuned to numerical relativity [445, 482, 509, 710, 711, 397], effective one-body [e.g., 188, 189, 273, 271, 187, 274, 272, 275, 858, 719] or self-force methods [e.g., 628, 727, 109, 110, 799, 898, 897, 86], and numerical relativity surrogate models [160, 159, 161, 903]. These different approximants may or may not include certain physical effects, like spin precession, tides for neutron stars, and gravitational-wave emission at higher-order multipoles.

Source detection and characterization

Detection: The LIGO detectors are sensitive to CBCs at cosmological distances. The optimal signal-to-noise ratio (SNR) of a particular signal depends on both the waveform and the detector PSD, $\rho_{\mathrm{opt}} \equiv \sqrt{\langle h(\boldsymbol{\theta})|h(\boldsymbol{\theta})\rangle}$, where the noise-weighted inner product is defined as

$$\langle a|b\rangle \equiv \frac{4}{T} \sum_j \Re\left(\frac{\tilde{a}(f_i)^* \tilde{b}(f_i)}{S_n(f_i)}\right), \tag{1.10}$$

and T is the signal duration. Because gravitational-wave data are discretely sampled, Eq. 1.10 includes a sum over discrete frequencies rather than a continuous integral over frequency. A typical measure of the detector sensitivity is the distance out to which an equal-mass binary neutron star merger with $m = 1.4\ M_\odot$ can be detected with SNR > 8 averaged over sky location and polarization. During O3, the median *range* was 134 Mpc for the Livingston observatory and 111 Mpc for Hanford [183].

Gravitational-wave detection of CBCs relies on a technique called *matched filtering* [e.g., 70, 202, 270, 891, 659, 658, 716, 610, 56, 235]. A template bank [791, 673, 674, 448, 103, 63, 238] of waveforms covering the binary parameter space of interest is compared to the strain data in real time, and the matched-filter SNR is calculated for each waveform in the bank, $\rho_{\mathrm{MF}} = \frac{\langle d|h(\boldsymbol{\theta})\rangle}{\rho_{\mathrm{opt}}}$. If this detection statistic exceeds a specified threshold, a candidate gravitational-wave event has been identified at that moment in time. The template that maximizes this detection statistic provides an estimate of the binary parameters. In Chapter 3, I investigate the accuracy of this point estimate for the mass parameters of binary neutron stars, in order to determine if it can be used to inform electromagnetic follow-up strategies in low latency.

Noise fluctuations in the detector can produce spurious triggers that may be conflated with astrophysical gravitational-wave events. This is partially mitigated by requiring that the observed signal be coherent in multiple detectors, as the noise should be uncorrelated. In order to determine the significance of a given trigger, an artificial time-shift is introduced in the data streams of multiple detectors, such that any astrophysical signals in the data become incoherent. Matched filtering is performed on the time-shifted data to create a background of known noise triggers. The frequency with which a given value of the detection statistic occurs among the background triggers gives the false alarm rate (FAR) that determines the significance of a trigger in the original data.

Parameter estimation: Following the identification of candidate events using matched filtering, further source characterization is performed using Bayesian inference to obtain posterior probability distributions on the binary parameters,

$$(m_1, m_2, \boldsymbol{\chi}_1, \boldsymbol{\chi}_2, \Lambda_1, \Lambda_2, d_L, \iota, \alpha, \delta, \psi, \phi, t_c) \in \boldsymbol{\theta}, \tag{1.11}$$

rather than just a point estimate. See Refs. [874, 234] for reviews of Bayesian inference in gravitational-wave astronomy. Bayes' theorem states that the posterior on $\boldsymbol{\theta}$ given data d and model H is given by

$$p(\boldsymbol{\theta}|d, H) = \frac{\mathcal{L}(d|\boldsymbol{\theta}, H)\pi(\boldsymbol{\theta}, H)}{\mathcal{Z}(d|H)}, \quad \mathcal{Z}(d|H) = \int \mathcal{L}(d|\boldsymbol{\theta}, H)\pi(\boldsymbol{\theta}, H)d\boldsymbol{\theta}, \tag{1.12}$$

where $\mathcal{L}(d|\boldsymbol{\theta}, H)$ is the likelihood of observing data d given the parameters $\boldsymbol{\theta}$ and model H, $\pi(\boldsymbol{\theta}, H)$ represents the prior probability distribution assumed for $\boldsymbol{\theta}$ given H, and $\mathcal{Z}(d|H)$ is the evidence for model H, or marginalized likelihood. For compact binary parameter estimation, H includes the physical assumptions going into the chosen waveform model and the prior assumptions (i.e., General Relativity, tidal effects, precessing spins, etc.).

Building the likelihood requires incorporating knowledge of the statistical properties of the noise in the detector. For well-behaved noise in the absence of a signal, the real and imaginary parts of the frequency-domain strain each follow a unit Gaussian distribution about the amplitude spectral density, such that the data are *whitened* by the frequency-dependent varaince, $\sigma_i^2 = \frac{TS_n(f_i)}{4}$, where $S_n(f_i)$ is the PSD and the index i indicates the frequency bin. Assuming that the data consist of both a noise component and an astrophysical component, $\tilde{d}(f) = \tilde{n}(f) + \tilde{h}(\boldsymbol{\theta}; f)$, the difference between the data and the signal template should also be Gaussian distributed with variance σ_i^2. This difference is called the *residual*,

$$\tilde{r}(\boldsymbol{\theta}; f_i) = \tilde{d}(f_i) - \tilde{h}(\boldsymbol{\theta}; f_i), \tag{1.13}$$

$$p(\Re\tilde{d}(f_i)|\boldsymbol{\theta}) = \frac{1}{\sqrt{2\pi\sigma_i^2}} \exp\left(-\frac{(\Re\tilde{r}(\boldsymbol{\theta}; f_i))^2}{2\sigma_i^2}\right), \tag{1.14}$$

$$p(\Im\tilde{d}(f_i)|\boldsymbol{\theta}) = \frac{1}{\sqrt{2\pi\sigma_i^2}} \exp\left(-\frac{(\Im\tilde{r}(\boldsymbol{\theta}; f_i))^2}{2\sigma_i^2}\right). \tag{1.15}$$

The total likelihood, known as the *Whittle likelihood*, is the product of the real and imaginary likelihoods [e.g., 761],

$$\mathcal{L}(\tilde{d}(f_i)|\boldsymbol{\theta}) \equiv p(\Re\tilde{d}(f_i)|\boldsymbol{\theta})p(\Im\tilde{d}(f_i)|\boldsymbol{\theta}) \tag{1.16}$$

$$\mathcal{L}(\tilde{d}(f_i)|\boldsymbol{\theta}) = \frac{2}{T\pi S_n(f_i)} \exp\left(-\frac{2|\tilde{d}(f_i) - \tilde{h}(\boldsymbol{\theta}; f_i)|^2}{T\, S_n(f_i)}\right), \tag{1.17}$$

$$\mathcal{L}(d|\boldsymbol{\theta}) = \prod_i \mathcal{L}(\tilde{d}(f_i)|\boldsymbol{\theta}), \tag{1.18}$$

and the likelihood for multiple frequencies is the product of the individual-frequency likelihoods; the same applies for multiple detectors,

$$\mathcal{L}(\{d\}_j|\boldsymbol{\theta}) = \prod_j \mathcal{L}(d_j|\boldsymbol{\theta}). \tag{1.19}$$

The Whittle likelihood assumes that the PSD is fixed and known with arbitrary precision. In reality, the PSD must be calculated from the data and is inherently uncertain. In Chapter 2, I present a method to relax this assumption and marginalize over the uncertainty in the PSD during individual-event parameter estimation.

The typical prior shapes assumed for compact binary parameter estimation are given in Table 1.1 (see also [762]). Because of the correlations in the recovery of the various intrinsic parameters, the choice of prior on one parameter can affect the inference of another. For example, in Chapter 4 I explore the effect of the spin prior choice on the inferred BNS mass distribution. The priors on the tidal deformabilities are usually chosen to be independent and uniform to be astrophysics-agnostic rather than imposing a particular EoS or EoS family. The widths of the priors on the mass

Parameter	Description	Prior
m_1	Detector-frame primary mass	Uniform
m_2	Detector-frame secondary mass	Uniform
χ_1	Primary spin magnitude	Uniform on $[0,0.99]$
χ_1	Primary spin magnitude	Uniform on $[0,0.99]$
θ_1	Primary spin tilt angle	Uniform in cosine on $[-1,1]$
θ_2	Secondary spin tilt angle	Uniform in cosine on $[-1,1]$
ϕ_{12}	Azimuthal inter-spin angle	Uniform on $[0, 2\pi]$
ϕ_{JL}	Azimuthal precession code angle	Uniform on $[0, 2\pi]$
d_L	Luminosity distance	Uniform in V_c and source-frame time
θ_{JN}	Inclination angle between $\hat{\mathbf{J}}$ and $\hat{\mathbf{N}}$	Uniform in cosine on $[0, \pi]$
α	Right ascension	Uniform on $[0, 2\pi]$
δ	Declination	Uniform in sine on $[-1,1]$
ψ	Polarization angle	Uniform on $[0, \pi]$
ϕ_c	Phase at coalescence	Uniform on $[0, 2\pi]$
t_c	Coalescence time	Uniform on $[t_t - 0.1s, t_t + 0.1s]$
Λ_1	Primary tidal deformability	Uniform on $[0, 5000]$
Λ_2	Secondar tidal deformability	Uniform on $[0, 5000]$

Table 1.1: Descriptions and priors on the 17 parameters that completely characterize a CBC. While the mass priors are typically set in terms of the component masses, sampling is often performed in terms of the chirp mass $\mathcal{M}$ and mass ratio $q = m_1/m_2$ to improve convergence. The distance prior is chosen to be uniform in the source frame, where V_c is the comoving volume. The trigger time in the time prior t_t is determined by the matched filtering process.

and distance parameters are informed by the matched filtering point estimate of the binary properties to ensure that the posterior distribution does not rail against one of the prior edges.

The likelihood in Eq. 1.18 is difficult to evaluate on a grid due to the curse of dimensionality; instead, stochastic sampling methods like Markov Chain Monte Carlo [614, 451] and nested sampling [816] are used to generate samples from the posterior distribution in Eq. 1.12 [906]. These sampling methods can also provide an estimate of the evidence, which can be used to perform Bayesian model selection. As an example, the Bayes factor comparing an aligned-spin model to a precessing spin model is defined as $B_P^A \equiv \mathcal{Z}(d|A)/\mathcal{Z}(d|P)$, where $\ln B > 8$ is generally considered significant evidence in favor of one hypothesis over another. I am a developer of the Bayesian inference library BILBY [96, 762], one of the flagship parameter estimation software packages used by the LIGO-Virgo-KAGRA Collaboration (LVK) and the inference tool used to obtain the majority of the results presented in this book.

Population inference: In addition to characterizing the properties of individual sources, we want to obtain estimates of the population-level distributions from which individual events are drawn. Strongly parameterized functional forms are typically assumed for the population distributions of individual parameters, which are generally taken to be independent (although see Chapter 10 for a method to explore correlations). In this case, the population-level inference amounts to obtaining posterior probability distributions for the hyper-parameters governing these *phenomenological* distributions, Λ (not to be confused with the component tidal deformabilities of individual events, $\Lambda_{1,2}$). Recent works have also begun exploring the use of "non-parametric" models like splines [314, 316] and binned Gaussian processes [584] or models directly parameterized in terms of physical theoretical parameters like the common envelope binding energy [963, 937]. Examples of the phenomenological models employed in the literature and in this book include a power law with a Gaussian peak for the BBH primary mass distribution [851], where the Gaussian accommodates the theoretical pile-up of black holes formed via pulsational pair instability supernovae; a Beta distribution for the spin magnitude [946]; and a flat distribution with a Gaussian peaked at $\cos\theta = 1$ for the cosine of the spin tilt [850], accommodating both an isotropic and an aligned-spin component.

The likelihood in terms of the hyper-parameters Λ is obtained by marginalizing over the individual-event parameters $\boldsymbol{\theta}$,

$$\mathcal{L}(d|\Lambda) = \int \mathcal{L}(d|\boldsymbol{\theta})\pi(\boldsymbol{\theta}|\Lambda)d\boldsymbol{\theta}, \tag{1.20}$$

where $\pi(\boldsymbol{\theta}|\Lambda)$ is the population model. Because the individual-event likelihood does not usually have a continuous representation but rather is represented by a discrete set of posterior samples, Eq. 1.20 can be rewritten as

$$\mathcal{L}(d|\boldsymbol{\Lambda}) = \int p(\boldsymbol{\theta}|d, \mathrm{PE})\mathcal{Z}(d|\mathrm{PE})\frac{\pi(\boldsymbol{\theta}|\boldsymbol{\Lambda})}{\pi(\boldsymbol{\theta}|\mathrm{PE})}d\boldsymbol{\theta}, \tag{1.21}$$

$$\mathcal{L}(d|\boldsymbol{\Lambda}) = \mathcal{Z}(d|\mathrm{PE}) \sum_k \frac{\pi(\boldsymbol{\theta}_k|\boldsymbol{\Lambda})}{\pi(\boldsymbol{\theta}_k|\mathrm{PE})}, \tag{1.22}$$

where k represents the index of the individual posterior sample in $\boldsymbol{\theta}$ and the PE hypothesis includes all the assumptions made in the individual-event parameter estimation step.

Selection effects: A single event will not provide an informative posterior on $\boldsymbol{\Lambda}$, so an ensemble of events must be considered. Because we are interested in characterizing the underlying astrophysical distributions but we only have access to a population of detected events, observational selection effects must be accounted for in the inference formalism [566, 874, 585, 919]. By definition, we are only interested in the likelihood of detectable data, where we can choose to threshold on any detection statistic, ρ_{MF} or FAR, for example. Because the likelihood should be a properly normalized probability distribution with respect to the data, including only the detectable data necessitates a renormalization of the likelihood in Eq. 1.18,

$$\mathcal{L}(d|\boldsymbol{\theta}, \det) = \frac{\mathcal{L}(d|\boldsymbol{\theta})}{p_{\det}(\boldsymbol{\theta})} \tag{1.23}$$

where $p_{\det}(\boldsymbol{\theta})$ represents the probability that an event with parameters $\boldsymbol{\theta}$ will be detected, marginalized over all possible data realizations. This function is defined such that

$$\int_{d,\det} \frac{\mathcal{L}(d|\boldsymbol{\theta})}{p_{\det}(\boldsymbol{\theta})} \mathrm{d}d = 1, \tag{1.24}$$

where the integral is over the detectable data realizations.

Imposing a detection threshold also effectively changes the prior, since certain regions of the parameter space will be disfavored due to the limited detectability of those sources. This requires that the prior also be renormalized,

$$\pi(\boldsymbol{\theta}|\boldsymbol{\Lambda}, \text{det}) = \frac{\pi(\boldsymbol{\theta}|\boldsymbol{\Lambda})p_{\text{det}}(\boldsymbol{\theta})}{\int \pi(\boldsymbol{\theta}|\boldsymbol{\Lambda})p_{\text{det}}(\boldsymbol{\theta})d\boldsymbol{\theta}}, \tag{1.25}$$

where we can define

$$\alpha(\boldsymbol{\Lambda}) = \int \pi(\boldsymbol{\theta}|\boldsymbol{\Lambda})p_{\text{det}}(\boldsymbol{\theta})d\boldsymbol{\theta} \tag{1.26}$$

This allows the likelihood of the detectable data given the hyper-parameters $\boldsymbol{\Lambda}$ to be expressed as

$$\mathcal{L}(d|\boldsymbol{\Lambda}, \text{det}) = \int \mathcal{L}(d|\boldsymbol{\theta}, \text{det})\pi(\boldsymbol{\theta}|\boldsymbol{\Lambda}, \text{det})d\boldsymbol{\theta} \tag{1.27}$$

$$= \int \frac{\mathcal{L}(d|\boldsymbol{\theta})}{p_{\text{det}}(\boldsymbol{\theta})} \frac{\pi(\boldsymbol{\theta}|\boldsymbol{\Lambda})p_{\text{det}}(\boldsymbol{\theta})}{\alpha(\boldsymbol{\Lambda})}d\boldsymbol{\theta} \tag{1.28}$$

$$= \frac{1}{\alpha(\boldsymbol{\Lambda})} \int \mathcal{L}(d|\boldsymbol{\theta})\pi(\boldsymbol{\theta}|\boldsymbol{\Lambda})d\boldsymbol{\theta} \tag{1.29}$$

$$= \frac{1}{\alpha(\boldsymbol{\Lambda})}\mathcal{L}(d|\boldsymbol{\Lambda}). \tag{1.30}$$

In some idealized cases, $p_{\text{det}}(\boldsymbol{\theta})$ can be calculated analytically. For example the distribution of ρ_{MF} in a single interferometer marginalized over all possible Gaussian noise realizations is normally distributed about ρ_{opt} with unit variance [874], so

$$p_{\text{det}}(\boldsymbol{\theta}) = \int_{\rho_{\text{min}}}^{\infty} \mathcal{N}(\rho_{\text{MF}}; \mu = \rho_{\text{opt}}, \sigma = 1)d\rho_{\text{MF}}. \tag{1.31}$$

However, in the case of a more complicated detection statistic, like FAR across a realistic detector network whose data includes non-Gaussian noise artifacts, $p_{\text{det}}(\boldsymbol{\theta})$ must be estimated empirically using an injection campaign [e.g., 52]. The binary parameters for simulated signals that will be "injected" into realistic data are drawn from some chosen "true" distribution, $p_{\text{true}}(\boldsymbol{\theta})$. The data with the injected simulated signals is subjected to the same matched filtering search process as real data, and the resulting distribution of found injections that pass the detection threshold is given by

$$p_{\text{found}}(\boldsymbol{\theta}) = \frac{p_{\text{true}}(\boldsymbol{\theta})p_{\text{det}}(\boldsymbol{\theta})}{\int p_{\text{true}}(\boldsymbol{\theta})p_{\text{det}}(\boldsymbol{\theta})d\boldsymbol{\theta}}. \tag{1.32}$$

The denominator of Eq. 1.32 has the same form as $\alpha(\Lambda)$ introduced in Eq. 1.26, so we can define

$$\alpha_{\text{true}} = \int p_{\text{true}}(\boldsymbol{\theta})p_{\text{det}}(\boldsymbol{\theta})d\boldsymbol{\theta}, \tag{1.33}$$

which is a constant. Because the injection campaign gives us samples from $p_{\text{found}}(\boldsymbol{\theta})$, we can rewrite Eq. 1.26 as

$$\alpha(\Lambda) = \int \frac{p_{\text{found}}(\boldsymbol{\theta})\alpha_{\text{true}}}{p_{\text{true}}(\boldsymbol{\theta})}\pi(\boldsymbol{\theta}|\Lambda)d\boldsymbol{\theta}, \tag{1.34}$$

$$= \frac{\alpha_{\text{true}}}{N_{\text{found}}} \sum_j \frac{\pi(\boldsymbol{\theta}_j|\Lambda)}{p_{\text{true}}(\boldsymbol{\theta}_j)}, \tag{1.35}$$

where the index j indicates the sample from $p_{\text{found}}(\boldsymbol{\theta})$ and N_{found} is the total number of found injections. Since $\alpha(\Lambda)$ represents the fraction of systems drawn from a population distribution with hyper-parameters Λ that will be detected, $\alpha_{\text{true}} = N_{\text{found}}/N_{\text{true}}$, where N_{true} is the total number of injections drawn from the true distribution. For a population of N_{det} detected sources, the total likelihood is the product of the individual likelihoods in Eq. 1.30,

$$\mathcal{L}(\{d\}|\Lambda, \text{det}) = \frac{1}{\alpha(\Lambda)^{N_{\text{det}}}} \prod_n^{N_{\text{det}}} \mathcal{Z}(d_n|\text{PE}) \sum_k^{N_n} \frac{\pi(\boldsymbol{\theta}_k|\Lambda)}{\pi(\boldsymbol{\theta}_k|\text{PE})}, \quad \alpha(\Lambda) = \frac{1}{N_{\text{true}}} \sum_j^{N_{\text{found}}} \frac{\pi(\boldsymbol{\theta}_j|\Lambda)}{p_{\text{true}}(\boldsymbol{\theta}_j)}, \tag{1.36}$$

where N_n is the total number of posterior samples from $p(\boldsymbol{\theta}|d_n)$ for each individual event [334].

Current observations: The catalog of gravitational-wave transient events released following O3, GWTC-3, includes 90 CBCs detected with a probability of astrophysical origin larger than 50% [44], where p_{astro} is another detection statistic along with ρ_{MF} and FAR. Two of these 90 events are binary neutron stars, GW190425 [32] and GW170817 [17], the latter of which was the first multimessenger event and was accompanied by the detection of a short gamma-ray burst and a kilonova counterpart. GWTC-3 also includes two NSBH events detected with FAR < 0.25 yr^{-1}, the first observations of this category of binary [47]. GW190412 represents the first BBH for which gravitational radiation in higher-order multipoles was detected [36], and the mass of the secondary of GW190814 falls in the lower mass gap described in Section 1.1.1, making it either the most massive neutron star or least massive black hole ever detected [39]. The primary of GW190521 with a mass of $\sim 85\ M_\odot$ falls in the putative upper mass gap [37]. The mass of this system along with some evidence for spin precession suggests that it may be of dynamical origin.

On the population level, the BBH primary mass distribution roughly follows a power law with a global maximum at about 10 $M_\odot$ along with another peak at $\sim$ 35 $M_\odot$, likely too small to be explained by a pile-up of black holes formed via pulsation pair instability supernovae. Analyses assuming more flexible population models have found additional sub-structure in the mass distribution, suggesting that the power-law component actually peaks at $\sim 8\ M_\odot$ with the global maximum at 10 $M_\odot$ representing another sub-population [877, 314, 329]. The mass ratio distribution favors equal masses [52].

The spin magnitude distribution favors small spins, with very little support for $\chi \gtrsim 0.7$. The spin tilt distribution is consistent with isotropy, although there is tentative evidence for a peak at positive values of $\cos\theta$ [917, 420, 197]. The binary merger rate increases with redshift out to the maximum redshift probed with current detectors, $z \sim 2$, at a rate that is consistent with the star formation rate [52].

The properties of the black holes and neutron stars observed in gravitational waves differ from those detected electromagnetically in several key ways. The gravitational-wave black holes are more massive and more slowly spinning than those in X-ray binaries, with masses $\lesssim 25\ M_\odot$ and nearly maximal spins [241, 726, 745, 301]. Whether this discrepancy is an astrophysical selection effect or observational selection effect is an area of active research. The neutron star mass distribution inferred in gravitational waves also tentatively has more support for more massive systems, which may be indicative of a difference in the population of extragalactic neutron stars and those observable as pulsars within our own galaxy [536].

1.2.3 Stochastic background searches

Unlike the templated searches for gravitational-wave signals from individual CBCs, standard search methods for the stochastic background assume that it is stationary, isotropic, unpolarized, and Gaussian. For a signal to be stationary, it must depend only on the differences between observation times and not on the absolute times themselves. Because the age of the Universe is at least 20 orders of magnitude larger than the period of gravitational waves detected by ground–based interferometers, no time dependence is anticipated [73].

If the stochastic background is analogous to the Cosmic Microwave Background, it is justified to assume that it would be isotropic on large scales [142, 466]. However, the directional dependence of the background will ultimately depend on its sources. For example, the astrophysical stochastic background from unresolved merging binaries should have some level of anisotropy inherited from the galaxy over-density that it traces [261, 488, 486, 262, 487]. If the signal is unpolarized, that means it should have roughly equal contributions from the plus and cross polarizations.

The final assumption comes from the Central Limit Theorem, which states that any random process created by the superposition of independent random variables will be Gaussian. As long as the stochastic background is a result of many overlapping and independent gravitational wave signals whose individual duration is much larger than the time between events, its statistical properties will be entirely determined by the first and second moments of the strain recorded in a detector [743]. However, the Gaussianity of the signal is also source-dependent. A black hole background, for example, is expected to be populated by events whose duration is shorter than the time separating them, which would result in a non-Gaussian, Poisson background where waveforms no longer overlap and the amplitude at the detector is unpredictable at a given time [743, 20]. In Chapter 11, I develop a method that exploits the non-Gaussianity of the astrophysical foreground to simultaneously search for this signal along with a much weaker cosmological background.

Standard stochastic searches look for excess correlated power in pairs of detectors. See Refs. [73, 761] for detailed reviews of standard stochastic search methods. The noise component of the data should be uncorrelated for detectors with significant physical separation,

$$\langle \tilde{d}(f_i)_I^* \tilde{d}(f_j)_J \rangle = \langle \tilde{h}(f_i)^* \tilde{h}(f_j) \rangle = \frac{T}{10} \delta_{ij} \gamma_{IJ}(f) S_h(f_i), \tag{1.37}$$

where the indices I, J indicate the detector pair and $S_h(f_i)$ is the astrophysical power spectral density, related to Ω_{GW} via

$$S_h(f) = \frac{3H_0^2}{2\pi^2 f^3} \Omega_{\mathrm{GW}}(f). \tag{1.38}$$

The *overlap reduction function* $\gamma(f)$ [232, 358] encodes the sensitivity of the detector pair to the stochastic background as a function of its geometry,

$$\gamma_{IJ} \equiv \frac{5}{8\pi} \int_{S^2} e^{2\pi i f \hat{\Omega} \cdot \Delta \vec{X}/c} (F_I^+ F_J^+ + F_I^\times F_J^\times) d\hat{\Omega}, \tag{1.39}$$

where $\hat{\Omega}$ indicates the direction on the sky, $\Delta \vec{X}$ is the distance vector between the two detectors in the Earth-based coordinate system, and the integral is over the two-sphere. $\gamma(f)$ is a highly oscillatory function that asymptotes to $\gamma(f) = 0$ at high frequencies, meaning that most of the sensitivity to the stochastic background comes from low frequencies.

A cross-correlation statistic can be defined so that its expectation value is proportional to the gravitational-wave energy density,

$$\langle \hat{Y}(f) \rangle = \gamma(f) \Omega_{\mathrm{GW}}(f), \tag{1.40}$$

$$\hat{Y}(f) = \tilde{d}_I(f) \tilde{d}_J(f) \frac{20\pi^2 f^3}{3T H_0^2}. \tag{1.41}$$

By calculating the SNR of the cross-correlation statistic, the significance of a stochastic background detection can be determined. The measurement of $\langle \hat{Y}(f) \rangle$ can also be used to obtain a posterior probability distribution for the energy density amplitude and slope characterizing the signal in Eq. 1.3. A detection of the stochastic background by ground-based gravitational-wave interferometers has not been made, but the O3 data resulted in an upper limit of $\Omega_{\mathrm{GW}}(f_0 = 25 \text{ Hz}) \leq 3.4 \times 10^{-9}$ for an assumed spectral index of $\alpha = 2/3$ consistent with the expectation for the CBC background [50]. The O3 data also placed the most stringent upper limit on the tension of cosmic strings to date, $G\mu \lesssim 4 \times 10^{-15}$ [40], using a loop model based on Ref. [102].

1.3 Outline

In this book, I present new methods for the analysis and astrophysical interpretation ofd-based gravitational-wave detector data focusing on compact-object binaries, their electromagnetic counterparts, and the stochastic background. I begin by presenting an extension of the methods for parameter estimation of individual compact-object binaries summarized in Section 1.2.2 to include the uncertainty in the estimate of the detector noise power spectral density in Chapter 2. This introduces the Bayesian analysis methods that will be used throughout the rest of the book, which is divided into two parts.

Part I covers multimessenger sources including a neutron star. With the improvement in sensitivity expected during the next LVK observing run and beyond, electromagnetic observers may need to prioritize which BNS triggers to follow up. In Chapter 3, I determine that the point estimate of the BNS mass parameters obtained using matched filtering in low latency is accurate enough to inform this prioritization strategy. In Chapter 4, I find that care must be taken when choosing the spin priors for BNS parameter estimation, as assuming low-spin priors that exclude the true spin of some sources in the population leads to misestimating the neutron star mass distribution at the population level. Chapters 5-7 introduce methods for facilitating multimessenger detections and inference, focusing on short gamma-ray bursts and kilonovae from binary neutron stars and the possibility to constrain the EoS using counterpart nondetections for NSBH mergers, respectively.

In Part II, I transition to binary black hole sources, investigating the measurability of spin for both individual mergers and on the population level. In Chapter 8, I find that spin can be measured even for the most massive BBH sources whose signals in the LVK sensitive band are dominated by the merger and ringdown components rather than the inspiral. In Chapter 9, I present an alternative spin parameterization that sorts the binary components by their spin magnitudes rather than their masses, providing an improved constraint on the component spin parameters for individual near-equal-mass systems and enabling the first measurement of the population-level distributions of these new parameters. Chapter 10 presents evidence for a correlation between BBH spin and redshift, with implications for the formation channels of these systems. Finally, in Chapter 11, I present the statistically optimal method for the detection of the cosmological stochastic gravitational-wave background in the presence of a much louder foreground of binary black hole mergers.

Chapter 2

Quantifying the Effect of Power Spectral Density Uncertainty on Gravitational-Wave Parameter Estimation for Compact Binary Sources

The content of this chapter was previously published in Physical Review D as Ref. [151] in July 2020. ASB performed the analysis, wrote the manuscript, and contributed to project development.

Abstract

In order to perform Bayesian parameter estimation to infer the source properties of gravitational waves from compact binary coalescences (CBCs), the noise characteristics of the detector must be understood. It is typically assumed that the detector noise is stationary and Gaussian, characterized by a power spectral density (PSD) that is measured with infinite precision. We present a new method to incorporate the uncertainty in the power spectral density estimation into the Bayesian inference of the binary source parameters and apply it to the first 11 CBC detections reported by the LIGO-Virgo Collaboration. We find that incorporating the PSD uncertainty only leads to variations in the positions and widths of the binary parameter posteriors on the order of a few percent. Our results are publicly available for download on git [150].

2.1   Introduction

A gravitational wave detector, such as the ground-based laser interferometers LIGO and Virgo [3, 53], is assumed to generate an output consisting of background Gaussian noise. Gravitational-wave signals, as well as non-Gaussian noise transients, will introduce a deviation of the detector output from the baseline noise behavior. As the sensitivity of the network of ground-based laser interferometers searching for gravitational waves improves [22], so too must our understanding of their noise properties. These noise properties are usually characterized by the power spectral density (PSD) in each detector, which is a required input for both low latency searches for gravitational waves and further source characterization via Bayesian parameter estimation [610, 776, 658, 891, 56, 235, 906, 96, 478]. Both of these types of analyses typically require that the PSD is measured with infinite precision, and so far LIGO/Virgo template-based results have not accounted for the uncertainty in the PSD estimation since a single point estimate has been used for each analysis segment [26]. Unmodeled searches and follow-up analyses, including CBC waveform reconstruction [26] and short-duration gravitational-wave transient searches [24] have, however, included marginalization over the uncertainty in the PSD estimation. While the low-latency searches employed by LIGO account for the variability of the PSD on longer timescales by recalculating it periodically [658, 776], all PSD estimation methods formally assume that the detector noise is Gaussian and stationary, such that the noise properties do not change over the course of the data segment used in the calculation [221]. However, these assumptions are not generally true and can impact the sensitivity of the searches when the noise is mis-characterized [19].

The properties of the noise do vary in time, though usually the stationarity timescale is much longer than the analysis segment for transient gravitational-wave signals [2, 221]. Recently, a new method for relaxing the stationarity assumption was proposed and applied in [959] and [910], respectively, where the non-stationarity of the data over the duration of the segment used to calculate the PSD is accounted for by applying a "drift" correction to the PSD obtained by tracking the time-dependent variance of the overlaps between the data and signal templates used in low-latency searches. They demonstrated that this had a significant impact on the recovered trigger distribution for the search pipeline. A similar method to account for the change of the detector noise properties over time via the dynamic renormalization of the search trigger ranking statistic was developed in [643] and applied in [660], leading to an improvement in the search sensitivity for low-mass compact binary systems. In addition to these slow variations in the PSD, the data is often plagued by short (ms) transient non-Gaussian excursions, known as glitches [663, 965, 8]. In low latency, glitches are typically excised from the data using a procedure known as "gating" [19], while in higher latency they can be modeled and subtracted from the data, as was the case for the glitch present in the Livingston detector during the first BNS merger, GW170817 [686].

Finally, the noise power spectral density cannot be measured with infinite precision and must be estimated from the data itself in one of two ways. The first is a modification of the standard Welch's method [931], in which a longer stretch of data either before or after, but always excluding, the analysis segment is divided into smaller segments with the same duration as the analysis segment, and the resulting PSD is the mean of the periodogram for each of these sub-segments. This is known as an "off-source" method since the data used to compute the PSD excludes the analysis segment. This requires the noise to be stationary over the entire segment used for the PSD estimation, which can be on the order of 1000 s. Because glitches can bias the mean, the median periodogram is generally used in gravitational-wave data analysis instead [70, 906].

The second, or "on-source" method only uses the data from the specific segment under analysis, therefore assumed to contain both Gaussian noise and a non-Gaussian signal component. The Gaussian contribution is inferred from the data and represented as a frequency-dependent noise variance parameterized in terms of a phenomenological model with two separate components, a cubic spline describing the broadband Gaussian process assumed to generate the noise itself combined with a set of Lorenztians describing narrow-band features[1]. The number and position of both the spline points and the Lorenztians, as well as the line widths and amplitudes, are themselves free parameters in the models that are explored through a trans-dimensional Markov Chain Monte Carlo (MCMC) algorithm [558]. The non-Gaussianity due to the presence of the astrophysical signal is modeled using sine-Gaussian wavelets [558, 240]. Both the on-source and off-source methods assume the noise to be stationary and Gaussian, but as the off-source method requires stationarity over a duration more than an order of magnitude longer than the on-source method, this assumption is more likely to hold true for the on-source method [221].

In this chapter, we demonstrate a new method to relax the assumption that the PSD measurement is infinitely precise as applied to Bayesian parameter estimation for gravitational waves from compact binaries, where instead of using a single point estimate for the PSD, we marginalize over the uncertainty in the PSD estimation. Other studies have previously looked at the effects of the methods and uncertainty associated with modeling the noise power spectral density on compact binary parameter estimation. [2] found that using different noise realizations and hence different PSDs has a similar impact on the variation in the recovered source parameters as using different waveform models. The idea of marginalizing over the uncertainty in the PSD was first proposed in [771] and [770] by analytically marginalizing the standard Gaussian likelihood for gravitational wave data over the uncertainty in the PSD and arriving at the Student's T likelihood. This technique was employed in [107] to obtain unbiased parameter estimates for the properties of neutron star postmerger remnants in the context of the millisecond magnetar model. [822] proposed a similar method using a Gaussian prior on the PSD instead of a scaled inverse χ^2-distribution for compact binary parameter estimation in the presence of a Gaussian stochastic

---

[1]The narrowband features can typically be attributed either to the resonances of the cables suspending the test masses, known as "violin modes", the AC electrical supply "power line", or the "calibration lines" which are injected into the data by driving the test masses at known frequencies [558, 302, 281, 894].

background. In [906] and [559] the uncertainty in the PSD was parameterized as a scale factor that modifies the point estimate for a fixed number of frequency segments, which led to significant improvements of the consistency of compact binary parameter estimation results in real LIGO data. Methods for simultaneously measuring the PSD using a different parameterization in the presence of an astrophysical signal were developed in [558] and [240], although their signal model is a sum of wavelets and not a 17-dimensional compact binary waveform. Another method to simultaneously estimate the PSD and astrophysical signal parameters was presented in [317], which used a nonparametric approach to model the PSD in the presence of a gravitational-wave burst from core-collapse supernovae. More recently, [221] investigated the differences between the two methods for computing the PSD outlined above and found that the on-source method provides a better agreement with the statistical assumptions about the data described previously.

With this in mind, our method differs from the one proposed in [771] because the analytic marginalization still requires the PSD point estimate to be computed via the off-source method, while our method uses the full PSD posterior calculated via the on-source method, recovering many of the same advantages that are detailed in [221]. Additionally, while the parameterization in [906] and [559] models the scale factor as constant over some range of frequencies, the posteriors we obtain for the PSD using the on-source method allow for variation at much higher frequency resolution.

We note that a similar method was proposed in [97] in the context of marginalizing over the uncertainty due to the choice of waveform model, although the implementation differs from the method presented in this work since the waveform model is not an independent parameter for which posteriors are obtained, unlike the PSD.

The rest of this chapter is organized as follows. In Section 2.2 we present our method in detail and provide a primer in gravitational-wave parameter estimation. We then apply our method to the first 11 compact binary merger gravitational-wave signals detected, presenting the results in Section 2.3. We conclude with a summary and discussion of some caveats to the method we have described.

2.2 Marginalizing over PSD uncertainty

The time-varying data in a gravitational-wave interferometer can be written in terms of an astrophysical signal, $h(\theta)$, and a noise term n:

$$d = h(\theta) + n. \tag{2.1}$$

For signals from CBCs with quasi-circular orbits, θ represents the 17 parameters describing the binary including the masses, tidal deformabilities and vector spins of the components, the sky location, the distance and inclination angle relative to the source, a polarization angle, and the time and phase at coalescence. The noise in each detector is typically assumed to be Gaussian and stationary [29], such that the noise covariance matrix is diagonal in the frequency domain [761]:

$$\langle \tilde{n}_i^* \tilde{n}_j \rangle = \frac{T}{2} S_n(f) \delta_{ij}, \tag{2.2}$$

where $\tilde{n}$ denotes the Fourier transform of the noise contribution, the indices correspond to different frequency bins, δ_{ij} is the Kronecker delta, and $S_n(f)$ is the noise PSD for that detector.

Under the assumption of stationary, Gaussian noise, the likelihood of observing data d in one detector given the signal $h(\theta)$ and the power spectral density $S_n(f)$ is [906, 761]:

$$p(d|\theta, S_n) = \prod_i \frac{2}{\pi T S_n(f_i)} \exp\left[-\frac{2|\tilde{d}(f_i) - \tilde{h}(f_i; \theta)|^2}{T S_n(f_i)} \right], \tag{2.3}$$

where T is the duration of the analyzed segment. When using data from multiple detectors, the joint likelihood is obtained by multiplying the individual likelihoods for each detector, while requiring the signal $h(\theta)$ to be coherent across the detector network:

$$p(\{d\}|\theta, \{S_n\}) = \prod_{j}^{N_{\mathrm{IFO}}} p(d_j|\theta, S_{n,j}), \tag{2.4}$$

where the index j indicates the interferometer, N_{IFO} is the total number of interferometers in the network, and the signal parameters θ are assumed to be the same in all detectors. As mentioned in the previous section, the PSD is usually assumed to be measured with infinite precision and is typically computed in one of two ways, although we will focus on the on-source method for the rest of this chapter.

This method is implemented in the `BayesWave` package, which uses the `BayesLine` algorithm to fit the PSD [558, 240]. This algorithm uses the same likelihood defined in Eq. 9.1, with the exception that the astrophysical contribution h is no longer a waveform that depends on the 17 binary parameters θ but rather a sum of wavelets. When `BayesWave` is used to characterize only the properties of the noise without assuming the presence of an astrophysical signal, it is run independently for each interferometer and includes a glitch model that allows for the presence of independent non-Gaussian noise excursions in each detector. These glitches are modeled separately from the background Gaussian noise using wavelets without the requirement that the reconstructed glitch signal be coherent across the different detectors[2]. `BayesWave` samples over the properties of the background Gaussian noise model, again characterized by a variable number of spline points and Lorenztians, recording the parameters of those components as a posterior sample. These components can also be represented as an instance of a posterior set of PSDs describing the inferred variance of the Gaussian noise in the analyzed data. Recent parameter estimation (PE) analyses of CBC GW events [26] have only considered describing the PSD, the S_n term in the likelihood of Eq. 9.1, through a fixed point estimate of the overall PSD posterior distribution inferred by `BayesWave`. As shown by [221], the preferred point estimate in that case is the median PSD. It is obtained by evaluating the median value of $S_n(f)$ in each

---

[2]We note that in this configuration, the "glitch model" is also expected to capture any gravitational wave (GW) signal present, such that the noise model only describes the Gaussian contribution to the data

frequency bin from among the individual PSDs computed for each of the posterior samples `BayesWave` inferred for the splines and Lorentzians. This means that the median PSD by construction does not correspond to any individual PSD posterior sample, and that it formally is not required to be smooth over adjacent bins, something that is enforced by construction by the spline instance in each posterior PSD.

Ideally, the binary parameters θ and the PSD would be estimated simultaneously using the likelihood above, but this is at present prohibitively difficult since fitting the spline and Lorentzian parameters describing the PSD requires a trans-dimensional MCMC algorithm [558], and current frameworks for CBC parameter estimation depend on using fixed-dimensional models [906, 96]. We thus write the combined posterior for the PSD and the binary parameters as the product of two separate posterior probabilities, under the assumption that the binary signal parameters and the PSD are uncorrelated:

$$p(\theta, S_n|d) = p(\theta|S_n, d)p(S_n|d) \tag{2.5}$$

To obtain the posterior on θ we marginalize the above expression over the PSD:

$$p(\theta|d) = \int \mathrm{d}S_n \, p(\theta, S_n|d) = \int \mathrm{d}S_n \, p(\theta|S_n, d)p(S_n|d), \tag{2.6}$$

which is the expectation value of the posterior on the binary parameters averaged over the PSD uncertainty.

In practice, we first obtain a discrete set of N posterior samples for the PSD using `BayesWave`. For each PSD posterior sample k, we run Bayesian PE and obtain a posterior on the binary parameters via:

$$p(\theta|S_{n,k}, d) = \frac{\pi(\theta)}{\mathcal{Z}_k}p(d|\theta, S_{n,k}), \tag{2.7}$$

where $\pi(\theta)$ is the prior, and the denominator is the evidence, or marginalized likelihood, for a particular PSD posterior sample:

$$\mathcal{Z}_k = p(d|S_{n,k}) = \int p(d|\theta, S_{n,k})\pi(\theta)\mathrm{d}\theta. \tag{2.8}$$

Because Eq. 2.6 is just the expectation value of the binary posteriors obtained with each of the PSD posterior samples, it can be rewritten as a sum of the individual posteriors:

$$p(\theta|d) = \frac{1}{N}\sum_k p(\theta|S_{n,k}, d). \tag{2.9}$$

2.3 Application to current gravitational wave detections

We apply the method described in the previous section to each of the 10 BBHs in the first gravitational wave transient catalog, along with the BNS merger, GW170817 [26, 17]. For the BBHs we first run `BayesWave` to generate 200 fair draws from the PSD posterior for each event, while for the BNS we only use 141 fair draws to restrict the computational cost. We discuss the choice of the number of PSD posterior samples used for marginalization further in Section 2.4. To contain the computational cost, we pair the first PSD posterior sample for one detector with the first for the other detectors, so we only generate 200 and 141 total PSD posterior sample pairs for BBH and BNS respectively. For each of the PSD sample pairs, we use the `Bilby` PE package [96, 95] with the `dynesty` nested sampler [828] to obtain a posterior distribution for θ, the 17 binary parameters[3]. For BBH sources, we use the `IMRPhenomPv2` waveform [482, 509, 445], while for the BNS we use the `IMRPhenomPv2_NRTidal` waveform [293, 295] implemented via a reduced order quadrature likelihood [820] to reduce the computational cost.

---

[3]For assumed BBH sources, we fix the tidal deformability parameters to 0.

We use priors that are uniform in the chirp mass,

$$\mathcal{M} \equiv \frac{(m_1 m_2)^{3/5}}{(m_1 + m_2)^{1/5}}, \tag{2.10}$$

asymmetric mass ratio,

$$q \equiv m_2/m_1 \text{ with } m_2 \leq m_1 \tag{2.11}$$

and dimensionless spin magnitudes, and proportional to the square of the luminosity distance, d_L. The mass ratio prior ranges from $q = 1/8$ to 1 for all events except GW151012, where the lower bound of the mass ratio prior was extended to $1/17.95$ because of posterior support at lower mass ratios. The spin magnitude prior covers the range $a \in [0, 0.99]$ for BBHs, while for the BNS we use a restricted spin prior covering the range $a \in [0, 0.05]$ motivated by the component spins of observed galactic double neutron star systems [193, 17]. The priors on the tidal parameters for the BNS, Λ_1 and Λ_2, are independent and uniform from 0 to 5000. The marginalized posteriors are obtained by combining the samples from all 200 runs, choosing 5000 samples from each run since they all have equal weights according to Eq. 2.9, and the result files are available for download on git [150]. We also perform an analysis with the median PSD computed by BayesWave in order to emulate the analyses typically performed by LIGO/Virgo and to compare with the marginalized posteriors. The run settings are described in Table 2.1.

Fig. 2-1 shows the PSD-marginalized posteriors for the detector frame chirp mass, mass ratio, effective spin (χ_{eff}), and luminosity distance compared to the posteriors obtained using the median PSD as a point estimate for the detected BBHs, and Fig. 2-2 shows the same results for the BNS, in addition to the comparison of the mass-weighted average tidal deformability, $\tilde{\Lambda}$. The effective spin is the mass-weighted projection of the component spins along the direction of the orbital angular momentum [200, 729] and is the best measured spin parameter with gravitational wave data [921, 655]. $\tilde{\Lambda}$ determines the gravitational-wave phase to leading order in the

Event	Duration [s]	f_{max} [Hz]	$\mathcal{M}_{min}$ [$M_\odot$]	$\mathcal{M}_{max}$ [$M_\odot$]	$d_{L,min}$ [Mpc]	$d_{L,max}$ [Mpc]
GW150914	4	1024	9	69.9	500	2000
GW151012	4	512	10	30	223	3000
GW151226	8	512	5	12.3	20	1500
GW170104	4	1024	12.3	45	40	3260
GW170608	16	1024	5	12.3	90	2000
GW170729	4	1024	9	173	226	7000
GW170809	4	1024	12.3	43.5	168	4000
GW170814	4	1024	12.3	45	217	3000
GW170817	128	2048	1.18	1.21	1	75
GW170818	4	1024	12.3	45	100	2000
GW170823	4	1024	12.3	55	286	6000

Table 2.1: Run settings for each compact binary signal analyzed. The prior limits on the chirp mass are specified in the detector frame. The starting frequency for the overlap integral is 20 Hz for all BBH events with the exception of GW170608, which suffered from low-frequency noise in the Hanford interferometer, so the starting frequency was chosen to be 30 Hz [16]. The starting frequency for the BNS analysis was 23 Hz, in accordance with the range of validity of the reduced order quandrature used.

individual tidal deformabilities, Λ_1 and Λ_2, and is defined as [359, 338]:

$$\tilde{\Lambda} = \frac{16}{13} \frac{(m_1 + 12m_2)m_1^4\Lambda_1 + (m_2 + 12m_1)m_2^4\Lambda_2}{(m_1 + m_2)^5}. \tag{2.12}$$

While small variations in posterior shape and position are observed, marginalizing over the PSD uncertainty generally appears to lead to only a minor increase in the width of the binary parameter posteriors.

To quantify this effect, we show the difference in the width of the 90% and 50% confidence intervals between the PSD-marginalized and median PSD posteriors in Figs. 2-5 and 2-6 respectively for each event as a function of the network matched filter SNR calculated at the maximum likelihood point. The network SNR is obtained by adding the individual-detector SNRs in quadrature, where the matched filter SNR in a single detector is given by[70]:

$$\rho_{mf} = \frac{\langle d, h \rangle}{\sqrt{\langle h, h \rangle}}, \tag{2.13}$$

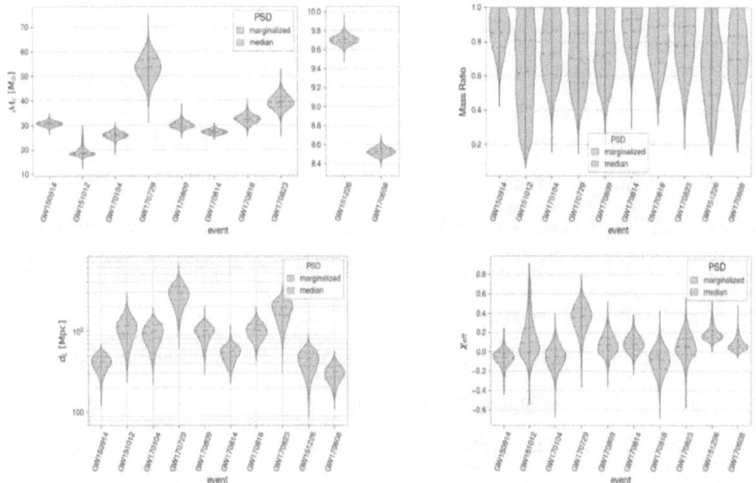

Figure 2-1: Violin plots showing the probability density for the PSD-marginalized posterior samples (blue) compared to the posterior samples obtained using the median PSD (orange) for the chirp mass, mass ratio, luminosity distance, and effective spin for each of the BBH detections. The horizontal lines represent the median (dashed) and 1σ confidence intervals (dotted) for each event.

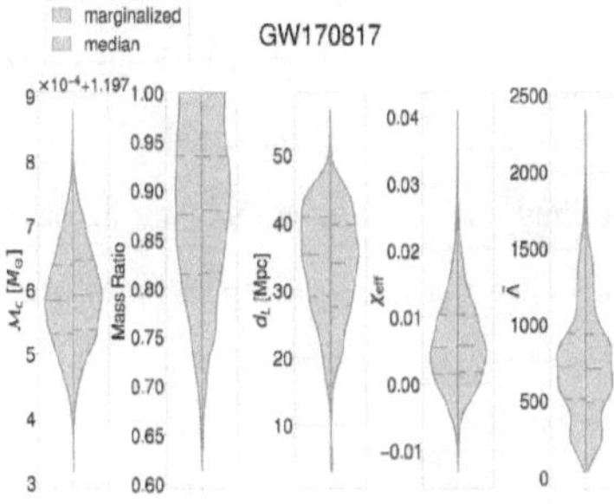

Figure 2-2: Violin plots showing the probability density for the PSD-marginalized posterior samples (blue) compared to the posterior samples obtained using the median PSD (orange) for the chirp mass, mass ratio, luminosity distance, effective spin, and mass-weighted average tidal deformability for the BNS, GW170817. The horizontal lines represent the median (dashed) and 1σ confidence intervals (dotted) for each parameter.

for the inner product defined as:

$$\langle a, b \rangle = 4\Re \int_0^\infty \frac{\tilde{a}^*(f)\tilde{b}(f)}{S_n(f)} \mathrm{d}f. \tag{2.14}$$

These results indicate that marginalizing over the uncertainty in the PSD produces posteriors that are wider than those obtained with the median PSD as a point estimate for about half of the events. The fractional change of the posterior width for both the 90% and 50% confidence intervals is of the order of a few percent, although a few larger excursions are observed for both confidence intervals. No significant trend is observed in the change in the confidence interval width as a function of the SNR, which indicates that the properties of the noise and the subsequent variation in the PSD are independent of the strength of the signal.

The largest deviation occurs in the 90% confidence interval of the chirp mass posterior of GW151012. This behavior can be explained by the variability in the Hanford PSD posterior at low frequencies shown in Fig. 2-3 that the median PSD cannot account for. This variability translates into significant posterior support for higher chirp masses for some of the individual PSD posterior samples, as shown in Fig. 2-4. This effect is minimized when averaging over the full PSD posterior, but not for the run with the median PSD alone, which also has increased support for higher chirp masses. Because the posterior obtained with the median PSD has wider tails, it has a correspondingly wider 90% confidence interval without affecting the 50% confidence interval, indicating good agreement between the two results for the bulk of their posterior distributions.

Table 2.2 shows the fractional change in the sky area Ω, in square degrees, contained within the 50% and 90% confidence intervals between the PSD-marginalized and median PSD posteriors, $\Delta\Omega = (\Omega_{\mathrm{marg}} - \Omega_{\mathrm{med}})/\Omega_{\mathrm{marg}}$, as well as the absolute difference in the sky area contained in the 90% confidence interval. The change in the confidence intervals for the sky area exhibits similar variation to the other parameters shown in Figs. 2-5 and 2-6, on the order of $\sim 10\%$. The biggest deviations occur for

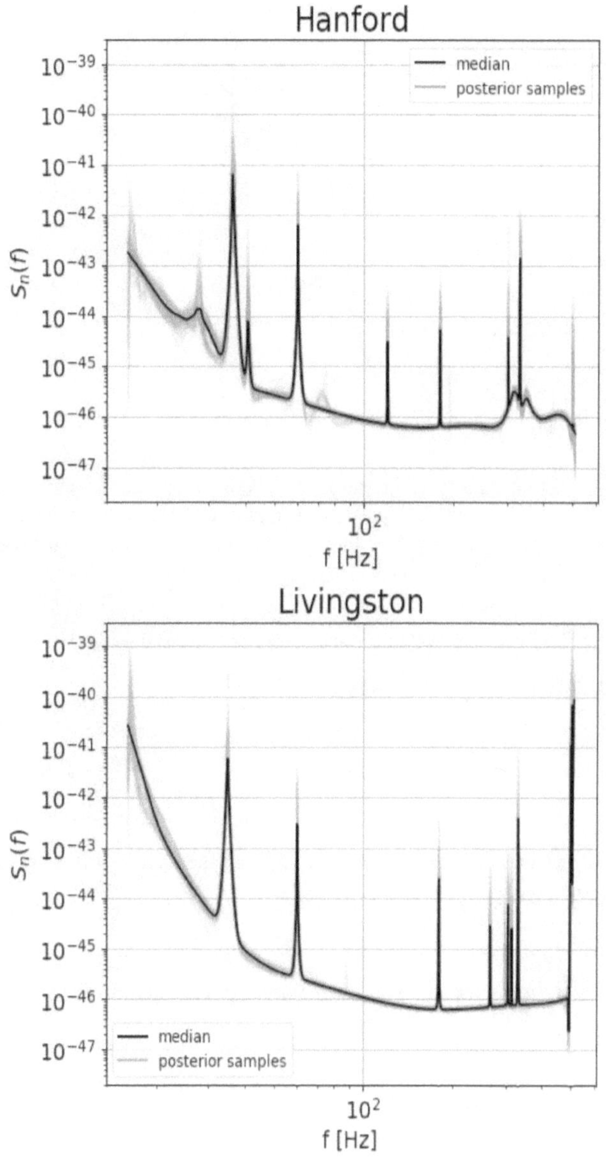

Figure 2-3: The 200 analyzed PSD posterior samples for the Hanford and Livingston LIGO detectors for GW151012 along with the median PSD in black

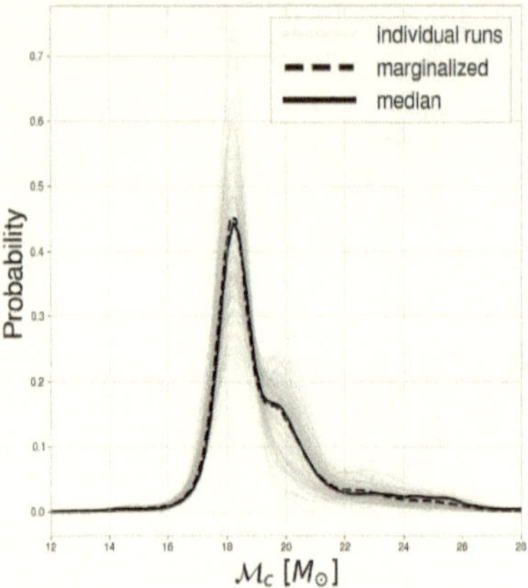

Figure 2-4: Kernel density estimate of the posterior on the detector-frame chirp mass for all 200 individual PSD posterior samples with the total PSD-marginalized posterior shown in the dotted black line and the posterior obtained with the median PSD in the solid black line for GW151012.

the best-localized event, GW170817, although the total change is only a few square degrees for both confidence intervals.

In order to compare the posterior variations due to marginalizing over the PSD uncertainty to those expected due to statistical fluctuations, we use bootstrapping to generate different sets of samples from the median PSD posterior for different parameters and events. We find that on average the change in the width of the 90% confidence interval among the bootstrapped samples is of the order of $\sim 0.1\%$ across different events and parameters, about an order of magnitude smaller than the deviations we observe due to marginalizing over the PSD uncertainty.

2.4   Discussion

In this chapter, we demonstrate a new method for marginalizing over the uncertainty in the noise power spectral density when performing gravitational wave parameter estimation for compact binary sources. We first obtain posterior samples for the PSD

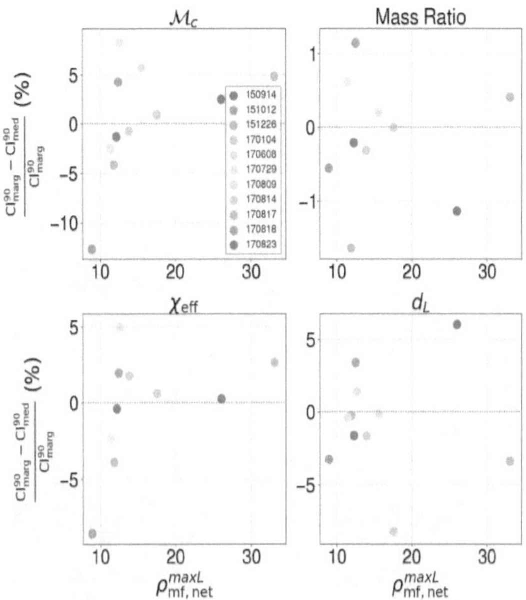

Figure 2-5: Normalized difference between the width of the 90% confidence interval of the PSD-marginalized and median PSD posteriors for chirp mass, mass ratio, effective spin, and luminosity distance for each event as a function of the maximum-likelihood network matched filter SNR calculated using the median PSD.

Event	$\Delta\Omega_{50}$ (%)	$\Delta\Omega_{90}$ (%)	$\Delta\Omega_{90}$ (deg^2)
GW150914	13.7	12.3	21
GW151012	1.5	-1.1	-20
GW151226	-9.6	-6.7	-93
GW170104	15.3	7.5	77
GW170608	-12.0	1.9	8
GW170729	19.0	10.4	136
GW170809	-12.1	2.9	9
GW170814	0	-18.6	-24
GW170817	28.6	25.9	7
GW170818	11.1	6.5	2
GW170823	2.9	-0.9	-14

Table 2.2: Fractional change in the 50% and 90% confidence intervals for the sky area in square degrees between the PSD-marginalized and median-PSD posteriors for each event, defined as $\Delta\Omega = (\Omega_{\mathrm{marg}} - \Omega_{\mathrm{med}})/\Omega_{\mathrm{marg}}$, and the absolute change in square degrees for the 90% confidence interval, which is $\Omega_{\mathrm{marg}} - \Omega_{\mathrm{med}}$.

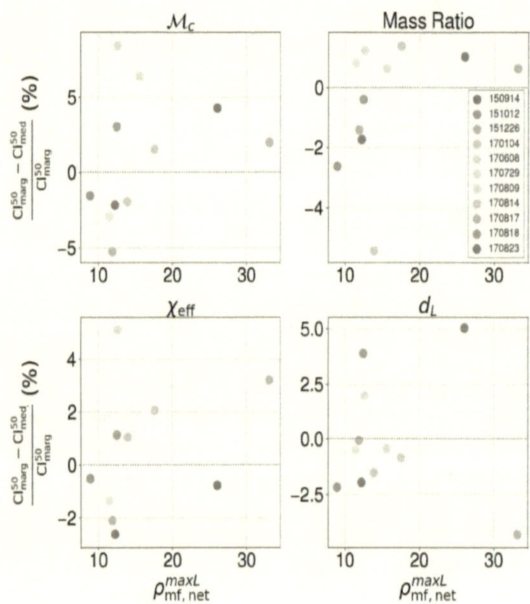

Figure 2-6: Normalized difference between the width of the 50% confidence interval of the PSD-marginalized and median PSD posteriors for chirp mass, mass ratio, effective spin, and luminosity distance for each event as a function of the maximum-likelihood network matched filter SNR calculated using the median PSD.

itself using the BayesWave software and then perform parameter estimation using 200 fair draws from the PSD posterior, combining the samples with equal weights to obtain the PSD-marginalized posterior. While no critical difference is observed in the posterior peak and shape for the binary parameters between the posteriors obtained using the median PSD as a point estimate and the PSD-marginalized posteriors, the posterior widths including the sky area vary on the order of a few percent, with the PSD-marginalized posterior being broader than that obtained using the median PSD for about half of the events. We do find more significant variation in the binary parameter posteriors obtained using individual PSD posterior samples for each gravitational-wave event, which can pick up secondary peaks or stronger support in the tail of the posterior, as was the case for GW151012. Based on these results, we conclude that the median PSD provides posteriors whose position and width are similar to those obtained when marginalizing over the PSD uncertainty to within a few percent for gravitational-wave signals of these SNRs observed in detectors with the current noise properties. The variations between the median and PSD-marginalized posteriors are an order of magnitude larger than those expected due to statistical fluctuations.

We close with a discussion of some caveats to our analysis. While we have shown that there is only minimal variation between the binary parameter posteriors obtained by marginalizing over the uncertainty in the PSD measurement and those obtained using the median PSD as a point estimate, we have not determined which of the two PSD models is preferred by the data. In our case, the two models we wish to compare are the "varied spline and Lorentzian model", where the PSD is allowed to vary and is parameterized in terms of a broadband spline and a series of Lorentzians to fit narrowband features, and the "median PSD model", where the PSD is fixed to the median of the full posterior computed with BayesWave. In both cases, the model also includes the presence of a CBC signal in the data, so we denote the varied spline and Lorentzian model as CBC+SL and the median PSD model as CBC+Me. This question of which model is statistically preferred is typically answered via Bayesian model selection and quantified using a Bayes factor between the two models being compared:

$$\text{BF}^{\text{CBC+SL}}_{\text{CBC+Me}} = \frac{\mathcal{Z}^{\text{CBC+SL}}}{\mathcal{Z}^{\text{CBC+Me}}}, \tag{2.15}$$

where the evidence, $\mathcal{Z}$, is defined as the normalization factor of the posterior obtained using a particular model. In addition to comparing the CBC+SL and CBC+Me models, obtaining the evidence under the CBC+SL model could also be used for comparing different variations of the CBC+SL model, for example precessing versus aligned compact binary component spins, the presence of higher order modes in the CBC waveform, or deviations from general relativity, which would all be represented as $\mathcal{Z}^{\text{CBC+X+SL}}$.

Using the method described in Eq. 2.9, we obtain posterior samples for for the CBC+SL model but not an evidence:

$$\mathcal{Z}^{\text{CBC+SL}} = p(d|\text{CBC} + \text{SL}) = \int p(d|\theta, \text{CBC} + \text{SL}) \tag{2.16}$$
$$\times\, \pi(\theta|\text{CBC} + \text{SL})\text{d}\theta,$$

where $\pi(\theta|\text{CBC} + \text{SL})$ is the prior defined in Eq. 2.7 where the explicit dependence on the $\text{CBC} + \text{SL}$ model had previously been suppressed. $p(d|\text{CBC} + \text{SL})$ is the likelihood of the data given the $\text{CBC} + \text{SL}$ model, which we need to define in order to calculate the evidence above.

The evidence can be calculated during sampling if instead of marginalizing over the PSD uncertainty by combining the binary parameter posteriors obtained using different PSD posterior samples, the likelihood is modified to account for the PSD uncertainty, and this modified likelihood is used to estimate the PSD-marginalized binary parameters directly. The marginalized likelihood for the $\text{CBC} + \text{SL}$ model is given by:

$$p(d|\theta, \text{CBC} + \text{SL}) = \int p(d|\theta, S_n, \text{CBC} + \text{SL}) \tag{2.17}$$
$$\times\, \pi(S_n|\text{CBC} + \text{SL})\text{d}S_n.$$

However, the likelihood $p(d|\theta, S_n, \mathrm{CBC + SL})$ doesn't depend on whether the PSD uncertainty is being included; it is equivalent to the likelihood for CBC signals defined in Eq. 9.1, so we drop the explicit dependence on SL. Similarly, the prior on the PSD, $\pi(S_n|\mathrm{CBC + SL})$ doesn't depend on the presence of a CBC signal, so we drop the dependence on CBC:

$$p(d|\theta, \mathrm{CBC + SL}) = \int p(d|\theta, S_n, \mathrm{CBC})$$
$$\times \, \pi(S_n|\mathrm{SL})\mathrm{d}S_n. \tag{2.18}$$

The prior on the PSD under the varied spline and Lorentzian model is the prior used by `BayesWave`, which is actually a complicated function of the spline and Lorentzian parameters and cannot be straightforwardly expressed in terms of S_n directly. Unfortunately, it also cannot be extracted from the `BayesWave` sampler products. Since `BayesWave` must be run in the configuration where it also models the non-Gaussian data component from the astrophysical CBC signal in order to obtain an unbiased estimate of the PSD, the prior that can be constructed from the sampler products is a joint prior on the spline and Lorentzian parameters and the wavelet parameters used to model the non-Gaussian component. Thus, the marginalized likelihood in Eq.2.18 cannot be obtained using the two-step system we have employed in the rest of the analysis where the PSD is estimated first, separately from the estimation of the binary parameters. One possible way to obtain the evidence for the CBC+SL model would be to simultaneously estimate both the binary parameters and the PSD using the spline and Lorentzian parameterization.

We note that the likelihood under the CBC+Me model can be recovered from the form of the likelihood in Eq. 2.18 by substituting a Dirac delta for the prior on S_n:

$$\pi(S_n|\mathrm{SL}) \to \pi(S_n|\mathrm{Me}) = \delta(S_n - S_{n,\mathrm{Me}}). \tag{2.19}$$

The evidence for this model is hence obtained during the sampling of the binary parameters using the median PSD.

While our method is embarrassingly parallel, so that each of the N parameter estimation analyses with different PSDs can be launched simultaneously, using the modified likelihood above requires that the CBC signal likelihood in Eq. 9.1 is evaluated N times in series for each binary parameter sample. This computation could in principle be accelerated through the use of likelihood reweighting [695] or other parallel processing techniques such as graphical processing units (GPUs) [849], multiprocessing [819], or the Message Passing Interface (MPI) [823]. Furthermore, our method still requires N times the computational resources compared to using the median PSD as a point estimate. We have chosen N=200 somewhat arbitrarily, balancing the computational cost against the number of samples needed to adequately represent the complete PSD posterior. We note, however, that producing the 200 draws from the PSD posterior comes at no extra computational cost compared to the production of the median PSD, so it would be possible to inspect the variability of the PSD posterior before starting the follow-up parameter estimation with each posterior draw. While we did find in the case of GW151012 that increased variability in the PSD posterior led to a larger change in the width of the binary parameter posteriors, there were also changes on the order of $\sim 5\%$ for events whose PSD posteriors seemed "smooth" like for the mass ratio of GW170104. We leave the systematic determination of the optimal N to future work, along with the investigation of the applicability of the likelihood reweighting method to marginalization over PSD uncertainty.

We emphasize that while BayesWave can account for the presence of non-Gaussian "glitches" in the data such that they do not affect the inference of the PSD parameters, it still assumes that the noise is stationary and Gaussian over the duration of the analysis segment. Thus, the method we have described here does not include marginalizing over PSD uncertainty due to the variation of the noise properties over time. This should not have a significant effect for short signals like BBHs, but can be more pronounced for BNS signals requiring longer analysis segments [221, 959].

2.5   Supplementary material

2.5.1   PSD Posteriors

In Figs. 2-7-6-2, we present the posteriors for the PSDs for each detector, LIGO–
Hanford and LIGO–Livingston as well as Virgo where such data were available, for
each event as well as the resulting median PSD.

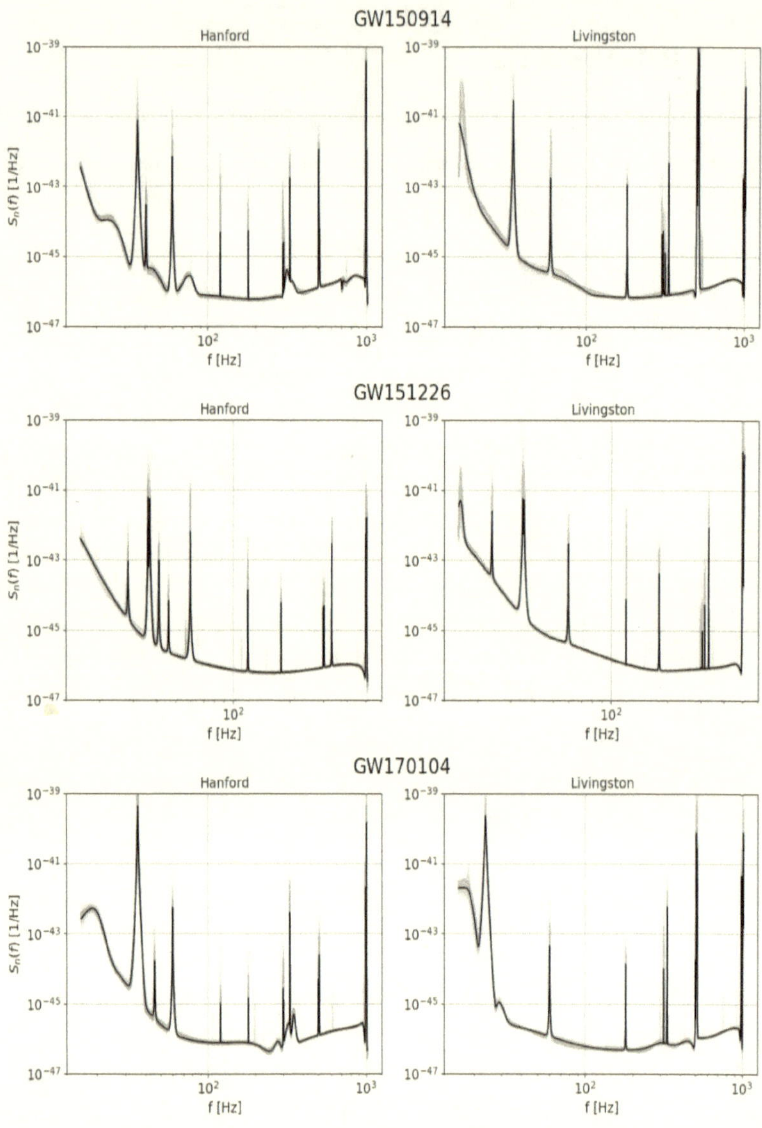

Figure 2-7: PSD posterior samples (colored lines) and the median PSD (black line) for each of the 11 events in GWTC-1. The PSDs start at 16 Hz, and the maximum frequency for each event is given in Table. 2.1.

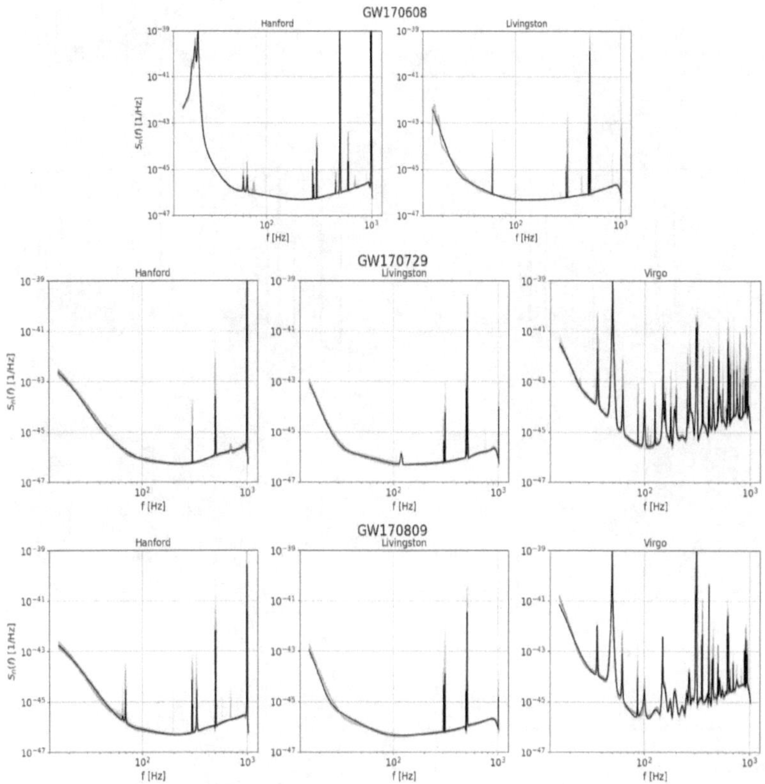

Figure 2-8: PSD posterior samples (colored lines) and the median PSD (black line) for each of the 11 events in GWTC-1. The PSDs start at 16 Hz, and the maximum frequency for each event is given in Table. 2.1.

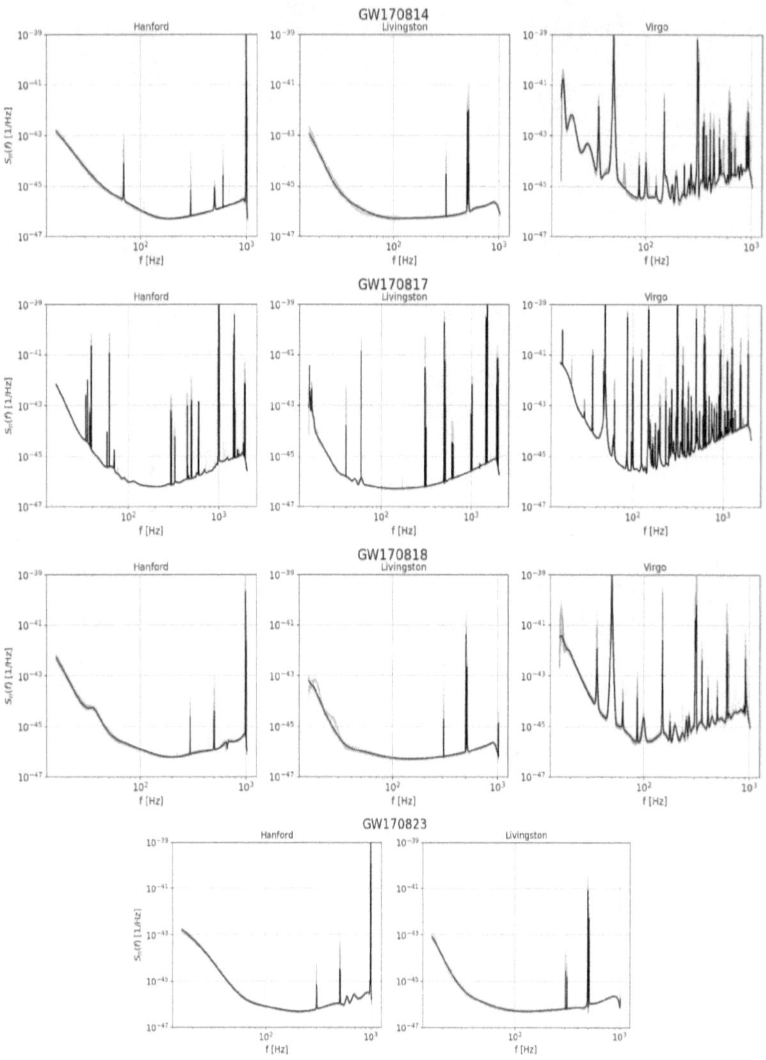

Figure 2-9: PSD posterior samples (colored lines) and the median PSD (black line) for each of the 11 events in GWTC-1. The PSDs start at 16 Hz, and the maximum frequency for each event is given in Table. 2.1.

Part I

Multimessenger sources

Chapter 3

The reliability of the low-latency estimation of binary neutron star chirp mass

The content of this chapter was previously published in the Astrophysical Journal Letters as Ref. [158] in Oct. 2019. ASB performed the analysis and contributed to project development and manuscript writing.

Abstract

The LIGO and Virgo Collaborations currently conduct searches for gravitational waves from compact binary coalescences in real-time. For promising candidate events, a sky map and distance estimation are released in low-latency, to facilitate their electromagnetic follow-up. Currently, no information is released about the masses of the compact objects. Recently, [598] have suggested that knowledge of the chirp mass of the detected binary neutron stars could be useful to prioritize the electromagnetic follow-up effort, and have urged the LIGO-Virgo collaboration to release chirp mass information in low-latency. One might worry that low-latency searches for compact binaries make simplifying assumptions that could introduce biases in the mass parameters: neutron stars are treated as point particles with dimensionless spins below 0.05 and perfectly aligned with the orbital angular momentum. Furthermore, the template bank used to search for them has a finite resolution. In this chapter we show that none of these limitations can introduce chirp mass biases larger than $\sim 10^{-3}\ M_\odot$. Even the total mass is usually accurately estimated, with biases smaller than 6%. The mass ratio and effective inspiral spins, on the other hand, can suffer from more severe biases.

3.1 Introduction

The discovery of the binary neutron star (BNS) merger GW170817 [17] in both the gravitational-wave (GW) and electromagnetic (EM) bands [18] has shown the great potential of multi-messenger astrophysics. Besides proving that BNS mergers are the progenitors of at least some short gamma-ray bursts (sGRBs) (e.g. [319, 15, 417, 793]), that event has given a glimpse at the energetics of BNS mergers at all frequencies and at the formation of a kilonova (e.g. [552, 92, 253, 702, 824, 857]) and allowed for a standard-siren measurement of the Hubble constant [14]. In the hours and weeks following the detection of the GWs and the sGRB, virtually all telescopes in the world were observing that part of the sky.

As the number of interesting GW sources to follow-up in the EM band increases, it might become necessary to prioritize and decide which ones are especially worth following up. During their third observing run, which began in April 2019 and is scheduled to last one year, the GW detectors LIGO [3, 447] and Virgo [53] might detect up to 10 BNS systems, a number that will become even higher as further upgrades are made to the detectors [5].

The LIGO-Virgo Collaborations (LVC) search for compact binary coalescences, including BNSs, in real time [26, 56, 891, 235, 610, 658, 776]. For candidate events below some pre-determined false-alarm threshold, a public alert is released in low-latency, together with a skymap and a distance estimation, to enable their EM follow-up [5]. At the time of writing, no estimation of the component masses (or derived quantities, such as the chirp mass or total mass) is released. Instead, the alert contains the probability that the source belongs to a specific category (BNS, binary black hole, neutron star–black hole, mass gap, and terrestrial) [5] and the probability that it will emit light, according to published models [368, 495].

Recently, [598] have argued that releasing an estimate of the chirp mass (defined below) of BNSs in low latency will allow EM astronomers to better allocate follow-up resources, based on which gravitational-wave events have the largest potential for enriching our understanding of merger outcomes and providing equation-of-state constraints with a multi-messenger observation. Kilonova modeling for GW170817 revealed that most of the ejecta came from accretion disk outflows in the post-merger phase rather than from dynamical processes during the merger [620, 341, 697, 810, 343]. The accretion disk properties depend most sensitively on the total mass of the binary, rather than the mass ratio, which is not well-constrained by GW observations [250, 732]. [598] then argue that the total mass can be inferred from the chirp mass of the binary given a prior on the mass ratio informed by galactic neutron star binaries. The total mass can then be used to predict the properties of the accompanying EM signal, allowing observers to prioritize binaries with the largest potential for placing constraints on the neutron star equation of state according to their "multi-messenger matrix".

In this chapter we investigate whether the estimation of BNS mass parameters that can be obtained in low-latency by the search algorithms is accurate. There are several reasons why it might not be. Compact binary coalescence (CBC) search algorithms are currently based on matched filtering (see e.g. [70, 203]). While the details of its implementation depend on the algorithm and involve subtle points (see Sec. 3.3 for a discussion of some of them), the concept is quite simple: a bank of waveforms is pre-built and used to filter the data [673, 674, 448, 63]. The waveform template that yields the highest signal-to-noise (SNR) ratio, or an equivalent derived quantity (e.g. a chi-square weighted SNR, or a likelihood ratio [204]) [56, 891, 235, 610, 658, 776], is selected and provides a point-estimate of some of the source's intrinsic parameters: masses and spins. We stress that detection algorithms are sensitive to the detector-frame mass, which is larger than the astrophysically relevant source-frame mass by a factor of $(1 + z)$, with z the redshift of the source. Unless otherwise indicated, we will only deal with the detector-frame chirp (or total) mass in this chapter, and go back to the difference between these two quantities in Sec. 3.3. Size and placement (i.e. the values of the intrinsic parameters of the templates) of the bank are chosen to guarantee a minimal overlap between expected GW signals and at least one waveform in the bank. Usually, one requires a minimal overlap of 97% percent [1], where the overlap is defined as the inner product between the source waveform and the template waveform, weighted by the instrument noise spectral density and maximized over the extrinsic parameters (sky position, distance, etc.) [644].

However, recovering most of the SNR of the source signal does not necessarily imply that the point-estimate of its masses and spins is accurate. While this possible inaccuracy is especially relevant for high-mass systems, such as binary black holes, due to the smaller numbers of templates in that region [9, 26], it might also be an issue for BNSs. In addition, biases might arise due to missing physics in the template bank. To keep the size of the template bank manageable, the spins of the compact objects are assumed to be perfectly aligned with the orbital angular momentum [2], and at least in the part of the bank that targets BNSs, the magnitude of the dimensionless spin is limited to ≤ 0.05. [716, 205]. Furthermore, corrections to the point-particle approximation, such as the presence of tidal effects in neutron stars, are entirely neglected.

All of these factors might introduce biases in the point-estimate of the mass and aligned-spin parameters obtained from the template bank in low-latency. A last reason is that the region of the template bank used to search for BNSs uses a very simple waveform model (`TaylorF2` [190]) which does not include any post-inspiral GW signal. While the choice of waveform model should have a minimal effect on the recovery of BNS parameters, since these systems merge at high frequencies where the sensitivity of the detector is poor, this needs to be explicitly checked.

It is worth stressing that these limitations can be lifted when high-latency dedicated source characterization algorithms are run on candidate events. These are usually based on stochastic sampling across a continuous parameter space and provide posterior distributions on the source parameters [906, 96], while also being able

---

[1]As the availability of computational resources improves, the minimum match is increasing, particularly in the binary black hole part of the bank where detections are more frequent, and increasing the minimum match only adds a small number of templates compared to the overall size of the bank [268].

[2]Allowing for generic spin tilt angles would double the dimensionality of the template bank, from 4 to 8 parameters, and result in a loss of sensitivity for sources with aligned spins [449].

to use more sophisticated waveform families that can account for larger spins with arbitrary orientation, tidal effects, and post-inspiral signal. The extra complexity is compensated by a longer run time, usually hours or days depending on the mass of the source (lower mass systems take longer to analyze due to their longer in-band duration). Alternative approaches also exist if only the intrinsic parameters (detector-frame masses, spins, tidal deformability) are of interest and a measurement of the extrinsic parameters is not necessary [537].

In this chapter we show that the estimation of the (detector frame) chirp mass of BNSs obtained by low-latency search algorithms is a) not significantly biased even when the true signal has features which are not included in the template bank, and b) usually consistent with what would be obtained by more sophisticated follow-up parameter-estimation codes, in spite of the missing physics and the finite size of the template bank. For a GW170817-like source, the difference between the source-frame and the detector-frame chirp mass would be much larger than the biases we find, making that the dominant error if the chirp mass measured by the search is inter-preted directly as the astrophysical chirp mass. Depending on the level of systematic uncertainty that can be tolerated, the total mass could also be used, as we find biases smaller than 6% in all cases. Conversely, the mass ratio can suffer from large offsets, although these are not much larger than the statistical uncertainty obtained from analyses using more sophisticated higher-latency source characterization codes.

Our study suggests that the low-latency chirp mass point estimate for BNSs pro-duced by current algorithms is adequate to inform EM follow-up strategies.

3.2   Method and results

3.2.1   Biases from the template bank

We first create a set of simulated BNS signals with features that are not accounted for in the template bank: large spin magnitude, spin misalignment and possibly tidal effects. To isolate the contribution of various terms, we proceed by gradually increasing the complexity of the simulated signals.

To verify if and how much the results depend on the region of the parameter space that is being probed, we consider two different values of the mass ratio

$$q \equiv m_2/m_1 \text{ with } m_2 \leq m_1$$

and three values of chirp mass

$$\mathcal{M} \equiv \frac{(m_1 m_2)^{3/5}}{(m_1 + m_2)^{1/5}}$$

where m_1 and m_2 are the component masses. Specifically, the signals we simulate have all possible combinations of

$$q \in [0.8, 1.0]; \quad \mathcal{M} \in [1.0, 1.15, 1.35] \, M_\odot.$$

These ranges are consistent with what is used in [598].

For each value of mass ratio and chirp mass, we create five BNS signals that differ by the value of the NS spin tilt angle, i.e. the angle between the spins and the orbital angular momentum, from $0°$ (spins aligned with the orbital angular momentum vector) to $90°$ (spins in the orbital plane). The inclination angle, i.e. the angle between the orbital angular momentum and the line of sight, is $35°$ for all sources, which is near the peak of the distribution of orientations for detectable BNS systems [797, 228].

For the first set of simulations, we assign to all the sources a spin magnitude $a = 0.05$, equal to the maximum allowed by the template bank, and we do not include tidal effects. We use the `IMRPhenomPv2_NRTidal` waveform [445, 293, 296]. The distance of the sources is chosen such that they have an optimal network SNR of 35 [13] (in a network made of the two LIGO detectors and Virgo, all at design sensitivity [3, 447, 53], comparable to GW170817. We focus on loud events as for these the effects of potential systematic uncertainties and/or biases should be more prominent. This is because the bias is an inherent property of the template bank for a given set of intrinsic parameters, and the statistical error shrinks with increasing SNR, so any bias will be more statistically significant for louder events.

The simulated signals are then filtered with the same template bank that was used by the `pycbc` search algorithm during the second observing run of LIGO and Virgo, which is publicly available online [722]. In the BNS region, the template bank uses the `TaylorF2` waveform model [190], and component masses in the range $1M_\odot < m < 2M_\odot$, which results in a chirp mass range $0.87M_\odot < \mathcal{M} < 1.73M_\odot$. As mentioned above, spins are assumed to be aligned with the orbital angular momentum and in the range $-0.05 \leq a \leq 0.05$. This parameter space is covered with $14,975$ templates.

For each simulated BNS, we use the `banksim` routine of the `pycbc` library [661] to select the template in the bank that yields the best match, and we use the mass and spins of that template as the point-estimate that would be produced in low-latency (some caveats associated with our choices are listed in Sec. 3.3). The starting frequency of the overlap integrals needed to calculate the matched filter SNR is 27 Hz, following the convention used in the construction of the bank [268]. Simulated detector noise is not added to the signals before computing the overlap.

In the top panel of Fig. 3-1 we show the difference between the point-estimate and the true value of the chirp mass, plotted against the true value of the spins tilt angle for the simulated BNSs. Triangles refer to equal-mass BNSs, whereas circles are used for the $q = 0.8$ sources. Different colors refer to the true value of the chirp mass.

We see that no matter the true value of the chirp mass, mass ratio, and spin orientation, the difference between the measured and the true chirp mass is smaller than 10^{-3} $M_\odot$, and usually smaller than 5×10^{-4} $M_\odot$. The offsets are larger for the total mass (Fig. 3-1 middle panel) though they are still smaller than 6% of the true value for all sources.

Finally, in the bottom panel of Fig. 3-1 we show the offset for the mass ratio q. For this parameter larger biases are visible when the spin tilt angles are larger than $20°$. This is not unexpected since the mass ratio and the spins are correlated in the inspiral phase of CBC signals (e.g. [90, 264, 186, 106, 444, 655]). The biases in q will

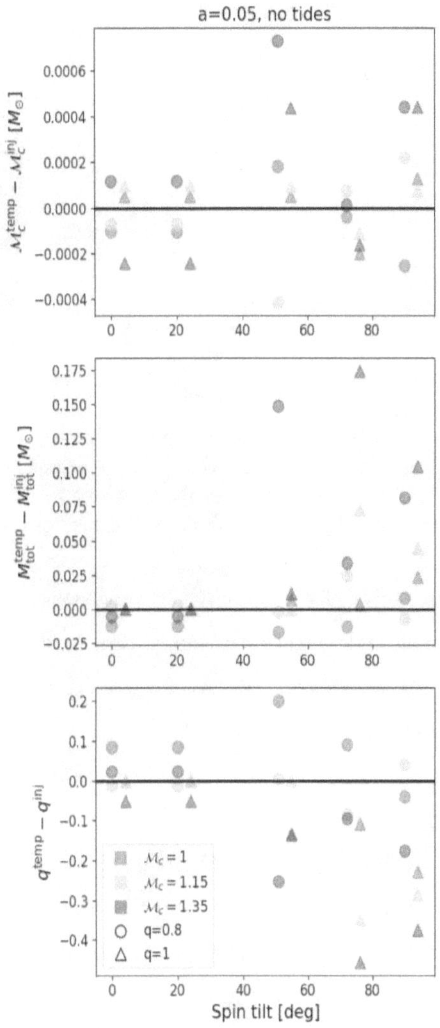

Figure 3-1: Difference between the value of parameters measured by the template bank and the true value, as a function of the spin tilt angle. An artificial offset in tilt angle is introduced for the equal-mass sources for clarity. The simulated BNSs have no tides and dimensionless spin $a = 0.05$. The values of the true chirp mass and mass ratio are given in the legend. These results can be seen as the bias due to missing physics in the template bank.

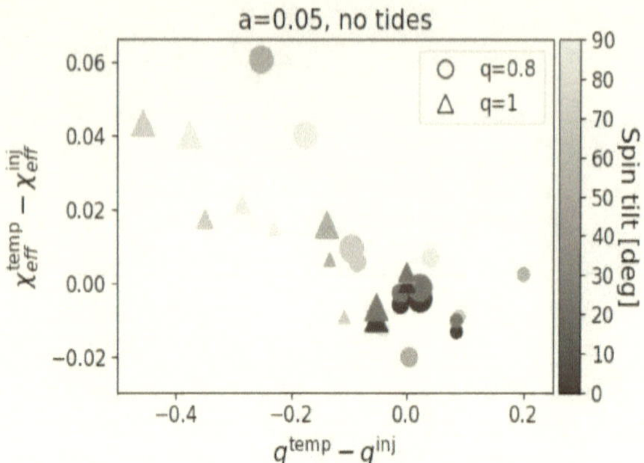

Figure 3-2: Correlation between the mass ratio and χ_{eff} biases. The true value of the spin tilt angle is given in the color bar. The size of the markers is representative of the chirp mass, from $\mathcal{M} = 1\ M_\odot$ for the smallest markers to $\mathcal{M} = 1.35\ M_\odot$ for the largest ones. The simulated BNSs have no tides and dimensionless spin $a = 0.05$.

usually be correlated with similar offsets in χ_{eff}, the mass weighted component of the total spin projected along the angular momentum [729]. As shown in Fig. 3-2, when the spin tilt angles are larger than a few tens of degrees, significant biases appear in both the mass ratio and χ_{eff}.

Next we modify our simulated signals to include tidal effects in the neutron stars. For all sources, we use the APR4 [66] equation of state, which results in a dimensionless tidal deformability, Λ, [359, 463, 915, 222] in the range $[125, 809]$, depending on the neutron star masses. We do not find that the results are significantly different than what we obtained in the absence of tides in the simulated signals, which is not surprising since tidal effects modify the phase evolution in the late inspiral (see e.g. [148, 521, 740, 276]), at frequencies where LIGO and Virgo are not very sensitive. Given that the results are basically identical, we do not show plots for this setting.

We then proceed to verify the effect of neutron star dimensionless spins larger than the maximum value allowed in the template bank, i.e. 0.05. In fact, while 0.05 is a value informed by the maximum spin of observed galactic double neutron star systems, pulsars have been observed with larger spins, the exact maximum depending on the (unknown) equation of state [567, 563]. For these runs we only analyze sources with $\mathcal{M} = 1.15 \ M_\odot$. In Fig. 3-3 we show the offsets for the chirp mass for systems with true spin magnitudes of 0.05 (gold), 0.1 (purple) and 0.2 (green), plotted against the true value of the spin tilt angle,for both mass ratios $q = 0.8$ (circles) and $q = 1$ (triangles). We stress again that the template bank does *not* extend to spins larger than 0.05 in the BNS region, although the best match could be provided by a template in the NSBH region with larger spins for the same chirp mass but more unequal mass ratios. For these runs there thus are two different sources of potential bias: the spin magnitudes are too large for the template bank and the spin vectors can be misaligned. The results reported in Fig. 3-3 might be surprising at first, since they show that the chirp mass bias *decreases* as the tilt angles of the spins go to 90°, i.e. to a configuration that yields the largest amount of orbital precession, which is a feature missing from the template bank. These results can be interpreted by remembering that χ_{eff} is the spin parameter that is best measured with GW data. Conversely, the component of the spins perpendicular to the angular momentum is usually poorly measured [923, 332, 721, 26, 25]. χ_{eff} modifies the duration of the signal (keeping everything else the same, larger and positive χ_{eff} yields longer GWs as the system radiates less efficiently [200]) and is conserved up to the second post-Newtonian order [729, 162]. Since it affects the waveform evolution so much, χ_{eff} is usually well measured [921, 26, 25, 655]. On the other hand, the effects of spin precession in the detector frame are suppressed in systems for which: the spin magnitude is small; the orbital inclination angle is not close to 90°; the mass ratio is not very different from 1. Neutron star binaries suffer from all these limitations.

The results in Fig. 3-3 can thus be explained as follows. When the spin vectors are aligned with the angular momentum (i.e. the spin tilts are 0) then the entirety of the spin vectors contribute to χ_{eff}, resulting in a value that cannot be matched by the template bank, unless $a = 0.05$. Physically, the source signal is longer than any waveform in the template bank with the same masses, due to its χ_{eff}. To compensate for the missing waveform length, a template with a smaller chirp mass is chosen, since these also yield longer signals [576]. This explains why for small tilt angles the template bank yields a point estimate of the chirp mass that is smaller than the true value. As the true value of the tilt angle increases, the spin vectors contribute less and less to χ_{eff}, leaving the waveform duration and phase evolution (at the 2nd post-Netwonian order) unaffected and removing the need to bias the chirp mass. While the trends in Fig. 3-3 are interesting, the main conclusion remains the same: no matter the value of the tilt angle, the chirp mass is not biased to a level larger than 2×10^{-3} $M_\odot$. For the total mass offset, we obtain results which are similar to what was shown in Fig. 3-1 for sources with $\mathcal{M} = 1.15$ $M_\odot$. We find that the mass ratio is overestimated for small tilts for the BNSs with $a = 0.1$ and $a = 0.2$. While the chirp mass bias can modify the waveform duration, the part of χ_{eff} that cannot be matched by the template bank will also create a dephasing. At the lowest order in the post-Netwonian phase, that can be at least partially compensated for by overestimating the mass ratio (e.g. [106] and [655] eq. A2).

To assess the possible impact of the choice of the waveform model on the results we have presented, we have repeated a subset of the tests described above using the `TaylorF2` waveform model for both the simulated BNS signals and the template bank. Specifically, we re-analyzed all combinations of the $q = 0.8$ sources and the $q = 1$ sources with $a = 0.05$, no tides, and aligned spin (due to the limitation of the waveform) and find that this does not result in any significant difference with respect to the results we have presented above.

3.2.2 Results from full source characterization

The results we have presented in the previous section answer the following question: how much bias can there be in the point estimate provided by CBC searches based on current template banks? We have found numbers that are small for the chirp mass and the total mass, and potentially larger for the mass ratio.

We now want to address a different question: how do the biases we found compare with the *statistical* uncertainty on the measurement of the same parameters that

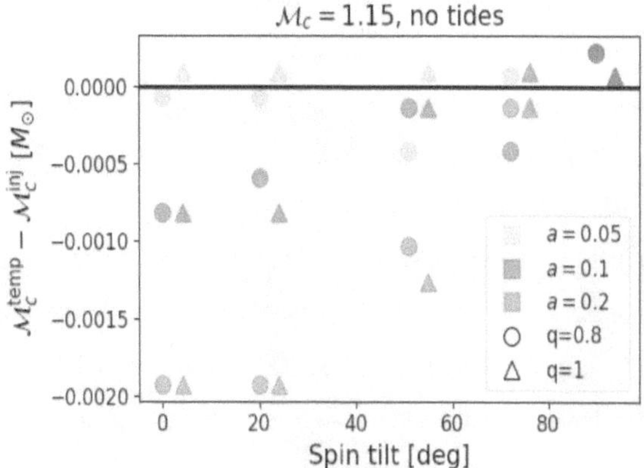

Figure 3-3: Same as in Fig. 3-1, for BNS sources with spin magnitudes from 0.05 to 0.2. The template bank only allows for spins up to 0.05. These simulated signals have $\mathcal{M} = 1.15\ M_\odot$ and no tidal effects. The values of the true dimensionless spin and mass ratio are given in the legend.

one would obtain using the full posterior distribution from higher-latency parameter estimation (PE) codes with waveforms that include all the relevant physics [3]?

To answer this question we have run the parameter estimation algorithm `bilby` [96] using the `dynesty` sampler [828] and produced posterior distributions for all the extrinsic and intrinsic parameters on which BNS systems depend. These simulations are carried out by using the reduced order quadrature likelihood (ROQ) [820] for the `IMRPhenomPv2_NRTidal` waveform family for *both* the source signal and the template signal used in the parameter estimation algorithm. We do not add simulated detector noise to the signals to be consistent with the `banksim` calculation, and start the integration at 32 Hz, which is the lowest frequency allowed by the ROQ likelihood [4]. As we have seen above, the choice of the waveform family is not a significant source

---

[3]Most waveform models available for general use implement significant simplifications in their description of the physical processes governing the binary evolution, especially regarding the detailed microphysics of the compact objects near merger. See [783, 784] for a discussion of BNS waveform systematics, and [809, 292, 297, 306] for a current overview of the status of BNS numerical relativity simulations.

[4]If the parameter estimation analysis started at lower frequency, the statistical uncertainties would decrease, owing to the larger number of waveform cycles that can matched. In the context of our study, that would only result in the biases found in Sec. 3.2.1 to be statistically more significant.

of systematic errors for the BNSs we are working with. We thus do not rerun all of the PE analyses with different waveform models.

To keep the computational cost contained, we follow the same strategy outlined above and run all combinations of mass ratio, chirp mass, and spin tilt angle only for the set of BNS for which $a = 0.05$ and no tidal effects are added. When adding tides or simulating BNS systems with spins of 0.1 and 0.2, we only analyze sources with $q = 0.8$ and the same combination of chirp masses as used in the `banksim` analysis.

In Fig. 3-4 we report the median and the 90% credible interval for the chirp mass (top panel), total mass (middle panel) and mass ratio (bottom panel) for the sources with $a = 0.05$ and $q = 1$, after subtracting the true values of the parameters.

We see that for the chirp mass estimate typical statistical uncertainties are of the order of few $\times\ 10^{-4}\ M_\odot$. Uncertainties are smaller for smaller chirp masses, as one would expect given that lighter objects have a longer inspiral, thus enabling a better measurement of their phasing. These uncertainties are of the same order of magnitude as the offsets from the template bank, Fig. 3-1. In fact, for some of the sources the chirp mass point-estimate from the template bank lies outside of the 90% credible interval from PE. This happens because the chirp mass is the parameter measured with the smallest statistical uncertainty for low-mass CBC signals, such as BNSs. While this implies that the biases on the chirp mass can be statistically significant, it must be stressed that one is still talking of offsets which are at most of the order of $10^{-3}\ M_\odot$. This bares no practical consequences on the strategies for EM follow-up proposed by [598], which are based on assigning BNS sources to regions in the chirp mass space with characteristic widths of $\sim 0.08\ M_\odot$ or larger. It is clear that biases smaller than $\sim 10^{-3}\ M_\odot$ will not move sources from one region to the other.

Next, we look at the mass ratio q, Fig. 3-4 bottom panel. We find that typical uncertainties are of the order of ~ 0.4 which is comparable to the size of the offsets we have found for equal-mass sources in the previous section. Finally, we look at the total mass, for which the statistical uncertainties are of the order of $\sim 0.1\ M_\odot$ or larger. Comparing this with Fig. 3-1 we see that the offsets on total mass are usually smaller than the statistical uncertainty from PE.

We find similar results for the systems with $[q = 0.8, \mathcal{M} = (1, 1.15, 1.35)\ M_\odot,$ $a = 0.05]$ with or without tidal terms, and for the systems with $[\mathcal{M} = 1.15\ M_\odot,$ $q = 0.8, a = 0.1, 0.2]$ for which tidal terms were not added.

3.3 Discussion

It has been recently argued that an estimation of the chirp mass for BNS mergers detected in gravitational-wave data could help prioritize their EM follow-up, if released in low-latency, together with a sky map and a distance estimation [598].

Two possible concerns with releasing chirp mass estimates produced in low-latency is that these are point-estimates calculated from template banks with finite resolution and that the waveforms used for the bank have missing physics; namely they do not account for neutron star spins larger than 0.05, spin precession, and tidal effects. Theoretically, this all can lead to biases in the estimation of the source parameters: that is, the template in the bank that best matches the source signal might have parameters which are different from those of the source.

In this chapter we have built a set of simulated BNSs with chirp masses in the range $[1 - 1.35]\ M_\odot$, mass ratios in the range $[0.8 - 1]$, spin magnitudes in the range $[0.05 - 0.2]$ and spin tilt angles in the range $[0° - 90°]$ and we have used the same template bank used by the `pycbc` search algorithm during the LIGO-Virgo second observing run (O2) to quantify eventual biases. All of the simulated events had a network SNR of 35 and an orbital inclination angle of $35°$. The simulations were performed using `IMRPhenomPv2_NRTidal` waveforms.

The O2 template bank consisted of $14,975$ `TaylorF2` waveforms in the BNS mass range, with spins exactly aligned with the orbital angular momentum and with magnitude $a \leq 0.05$, and no tidal deformability.

We found that the bias on the chirp mass, i.e. the difference between the point-estimate from the template bank and the true value, is always smaller than a few $\times$ $10^{-3}\ M_\odot$ (and usually much smaller than that).

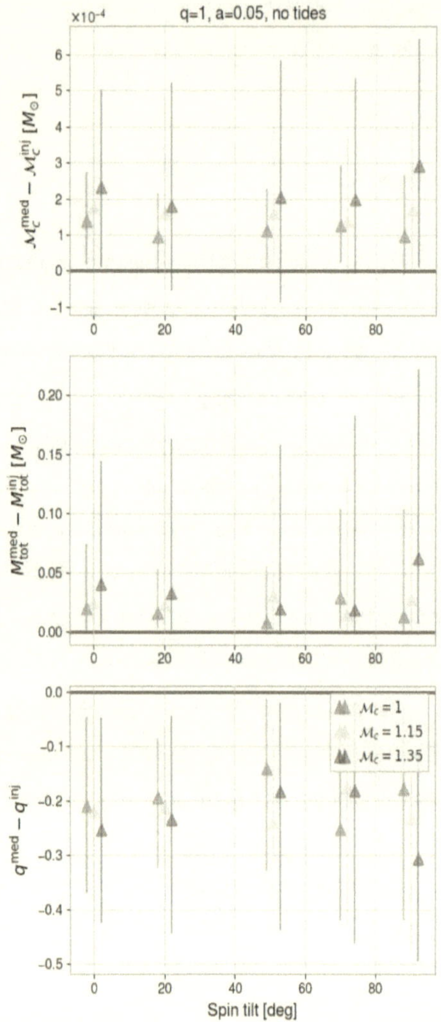

Figure 3-4: Difference between the median of the posterior from parameter estimation and the true value, as a function of the spin tilt angle. The simulated BNS have no tides, $q = 1$, and dimensionless spin $a = 0.05$. The error bars show the statistical 90% credible intervals, while the marker reports the median. The values of the true chirp mass is given in the legend.

We have also shown that the eventual presence of tidal effects in the true signal and the exact waveform model used do not modify these conclusions. The total mass of the BNS sources we simulated is also measured with a discrete accuracy by the template bank, with biases smaller than 6% in all cases. Conversely, the mass ratio and the effective inspiral spin, χ_{eff}, can be more affected by missing physics in the template bank and can suffer from large biases.

We have compared the biases obtained from the template bank with the statistical uncertainties from a stochastic sampler [828] which uses the same `IMRPhenomPv2_NRTidal` waveforms used to simulate the BNSs. We find that for the total mass, mass ratio, and chirp mass, the biases from the template bank are comparable to the size of the 90% credible intervals from PE. However, for the chirp mass the biases are small enough that they would not have any practical implication on the EM observing strategies outlined in [598]. The same is probably true for the total mass [615], for which we get biases smaller than 6%.

We conclude by mentioning a few caveats with our approach and results. We have assumed that the template which returns the largest overlap (i.e. recovers the larger fraction of SNR) will be selected by the search algorithm. In reality things are more complicated, and search algorithms go through a number of extra steps. For example, they require that the *same* template is selected in all detectors in the network [56, 891, 235, 610, 658, 776]. Specific noise realizations at the individual detectors might thus result in selecting a different template than what would be chosen in a single-detector analysis. While this is true, we notice that it is an effect that is not related to missing physics in the template bank, but rather to the properties of the noise, and that it would affect the PE step in a similar way, since this latter maximizes a joint likelihood across the detector network [906]. Thus, if the specific noise realization is a problem for the search, it will also be a problem for the PE step, and an eventual bias will not be reduced by waiting.

Indeed, the chirp mass offsets we have found in this study are consistent with those of [145], where the `GstLAL` search algorithm was run on an astrophysical population of neutron star binaries (though covering a narrower parameter space than what we considered in this study). Those authors showed that their conclusions are virtually unchanged if one uses simulated Gaussian noise or real interferometric data.

A notable exception is if short instrumental noise artifacts exist in the stretch of data that contains the signal, which can be at least partially removed in higher latency, as was the case for GW170817 [686]. In that case, depending on the shape of the artifact and how it overlaps with the inspiral signal, it might be the case that the search could give a biased result, while the PE step could give an unbiased one. Since in this chapter we specifically focus on potential biases coming from missing physics, we did not expressly account for this possibility. However, this should become less and less of a concern in the future as low-latency methods to remove noise transients are being developed.

We have checked that the results we presented do not depend on the choice of the inclination angle for the simulated signals. Rerunning all template bank simulations with an inclination angle of $85°$ yields virtually indistinguishable results, with the same template providing the best match for all simulated signals as for the runs with an inclination angle of $35°$. Because the results do not change for this choice of a nearly edge-on inclination where the effects of spin precession on the waveform are maximal [90, 923], we do not expect variations at intermediate inclination angles.

It should also be stressed that the search algorithms measure the detector-frame mass parameters, which are related to the source-frame masses (the astrophysically interesting values) by a factor of $(1 + z)$, with z the redshift of the source. Even for GW170817, whose distance was only ~ 40 Mpc, the difference between source-frame and detector-frame chirp mass was much larger than any of the biases we found in this chapter [17]. (This is not true for the total mass estimation.) This is a difference that is easy to remove, as the LVC releases a marginalized distance posterior in low-latency that, once converted to a redshift posterior assuming a cosmology, can be used to transform detector-frame masses to source-frame masses. A related question is whether the statistical uncertainty on the estimation of the source-frame chirp mass, which includes contributions from both the detector-frame chirp mass and from the distance estimation, is so large that it makes it impossible to select a single bin of the "multi-messenger matrix" from [598]. We find that the statistical uncertainty on the source-frame chirp mass is on the order of 10^{-3} $M_\odot$, consistent with the offsets on the detector-frame chirp mass found in the low-latency search.

The results of this study suggest that the finite size of the template bank and the missing physics in the waveform models that are used do not introduce biases in the estimation of the chirp mass larger than 10^{-3} $M_\odot$. These are small enough that they cannot negatively impact the EM observing strategies outlined in [598]. It might be argued that even the total mass could be used to inform the EM follow-up strategies, as it doesn't suffer from biases larger than 6%.

Chapter 4

The effect of spin mismodeling on gravitational-wave measurements of the binary neutron star mass distribution

The content of this chapter was previously published in the Monthly Notices of the Royal Astronomical Society as Ref [156] in Feb. 2022. ASB conceived the project, performed the analysis, and wrote the manuscript.

Abstract

The binary neutron star (BNS) mass distribution measured with gravitational-wave observations has the potential to reveal information about the dense matter equation of state, supernova physics, the expansion rate of the universe, and tests of General Relativity. As most current gravitational-wave analyses measuring the BNS mass distribution do not simultaneously fit the spin distribution, the implied population-level spin distribution is the same as the spin prior applied when analyzing individual sources. In this work, we demonstrate that introducing a mismatch between the implied and true BNS spin distributions can lead to biases in the inferred mass distribution. This is due to the strong correlations between the measurements of the mass ratio and spin components aligned with the orbital angular momentum for individual sources. We find that applying a low-spin prior which excludes the true spin magnitudes of some sources in the population leads to significantly overestimating the maximum neutron star mass and underestimating the minimum neutron star mass at the population level with as few as six BNS detections. The safest choice of spin prior that does not lead to biases in the inferred mass distribution is one which allows for high spin magnitudes and tilts misaligned with the orbital angular momentum.

4.1 Introduction

The growing catalog of compact binary mergers detected in gravitational waves [44] has provided a novel means to probe the properties of black holes and neutron stars [52, 937, 963, 768, 173]. The neutron star mass distribution has the potential to independently yield information on the dense matter equation of state (EoS) via the maximum mass, M_{TOV}, beyond which the internal pressure of the neutron star can no longer support it against gravitational collapse to a black hole [622, 535, 220, 546]. The value of the maximum mass depends on the yet-unknown EoS and the neutron star spin [539], although astrophysical processes may prevent the formation of neutron stars with mass up to M_{TOV} in some scenarios. The mass distribution can also be used to constrain the supernova physics leading to neutron star formation [696, 912], the astrophysical stochastic gravitational-wave background [975, 20], the rate of expansion of the Universe [230, 348, 861, 860] and alternative theories of gravity beyond General Relativity [347].

Initial measurements of the mass distribution of Galactic neutron stars detected electromagnetically suggested that their component masses follow a narrow Gaussian distribution [870], particularly for those found in binary systems with another neutron star [677, 517]. More recent measurements reveal that the overall distribution of neutron star masses including those found in binary systems with other types of companions is broader [517, 675] and better described by a double Gaussian [77, 88, 859, 801]. Including the first binary neutron star system observed in gravitational waves, GW170817 [17], Ref. [337] also find weak evidence for bimodality in the mass distribution of the double neutron star population alone.

Gravitational-wave observations of compact binary mergers involving at least one neutron star offer a complementary means to probe the neutron star mass distribution at extragalactic distances. While only two binary neutron star (BNS) systems have been detected in gravitational waves, these observations already suggest that there may be a distinction between the Galactic population accessible as pulsars and the gravitational-wave population [685, 393, 781, 52]. GW190425 represents the most massive BNS system ever detected, with a total mass five standard deviations away from the mean of the Galactic population [32]. When analyzing the full population of neutron stars detected in gravitational waves—including neutron star-black hole mergers [47]—Ref. [536] find that the masses are consistent with being uniformly distributed, although with significantly more support for high neutron star masses compared to the Galactic population. Thus, accurate and precise measurements of the neutron star mass distribution from gravitational-wave observations have the potential to elucidate the formation channels of these systems.

Such measurements utilize the framework of hierarchical Bayesian inference (e.g. [580, 874]), where the properties of the population as a whole are determined while taking into account the uncertainty on the parameters of individual sources and the selection effect introduced by the varying sensitivity of the detector to sources with different properties [566, 585, 919]. This requires both unbiased parameter estimates for individual sources and physically realistic models for the population and detector sensitivity. One potential source of systematic error is the correlation between measurements of the intrinsic parameters describing binary neutron star systems, including the masses, spins, and tidal deformabilities of the components. Refs. [948, 419] recently demonstrated that unphysical assumptions about the equation of state (or tidal deformability) when measuring the mass distribution independently can lead to biases with as few as 37 events, emphasizing the importance of fitting these distributions simultaneously.

The mass ratio and spin components aligned with the orbital angular momentum are particularly correlated for individual sources [264, 444, 144, 332, 655]. This is due to the fact that for a given chirp mass, $\mathcal{M} \equiv (m_1 m_2)^{3/5}/(m_1 + m_2)^{1/5}$, binaries with larger spins aligned with the orbital angular momentum will merge more slowly [200], while binaries with more unequal mass ratios will merge more quickly, introducing a degenerate effect on the waveform. Gravitational-wave analyses of individual BNS sources typically assume two different priors for the spin distributions, a "low-spin" prior and a "high-spin" prior [17, 32]. The "low-spin" prior restricts the maximum dimensionless spin magnitude—defined as $|\chi| = c|\mathbf{S}|/(Gm^2)$, where $\mathbf{S}$ is the spin vector of the neutron star with mass m—to $|\chi| \leq 0.05$, informed by the spins of Galactic double neutron stars that will merge within a Hubble time [567, 563, 976]. The "high-spin" prior extends to $|\chi| \lesssim 0.89$, as allowed by available waveform models without making any assumptions about the consistency of the system with the observed Galactic population. It is worth noting that pulsars have been observed with spins as high as $|\chi| \lesssim 0.4$, even in binary systems [458].

In this work, we demonstrate that mismodeling the spin distribution of BNSs can lead to a bias in the recovered mass distribution. We find that for a population of sources with moderate aligned spins extending out to $|\chi| \leq 0.4$, using individual-event mass estimates obtained with the low-spin prior without simultaneously fitting the spin distribution hierarchically leads to significant bias in both the inferred mass ratio distribution and maximum mass. Conversely, in the case of a low aligned-spin population with $|\chi| \leq 0.05$, the inferred mass ratio distribution under the high-spin prior is also biased, although this effect can be ameliorated by allowing for the spins to be misaligned with the orbital angular momentum. In Section 4.2 we describe our methodology and simulated BNS populations. In Section 4.3 we present the results of our hierarchical inference and conclude with a discussion of the implications of our findings in Section 4.4. We also include a demonstration of the importance of obtaining unbiased inferences for individual sources via the fallibility of the commonly-used "probability-probability plot" as a test of sampler performance in Appendix 4.5.1.

4.2 Methods

In addition to the component masses, spins, and tidal deformabilities, quasi-circular BNS mergers are characterized by seven extrinsic parameters including the distance, sky location, time of coalescence, and inclination angle between the orbital angular momentum and observer line-of-sight. We simulate two populations of BNS systems with distinct spin distributions. Both have spins aligned with the orbital angular momentum drawn from the implied distribution on χ_z assuming that the magnitudes are distributed uniformly on $[0, \chi_{\max}]$ and the directions are isotropic. The first population allows for medium spins consistent with the maximum observed neutron star spin [458], $\chi_{\max} = 0.4$, while the second population restricts to $\chi_{\max} = 0.05$, following the observed spins of Galactic double neutron stars [567, 563]. The mass distribution is chosen based on [337]; the mass ratio is drawn from a narrow truncated Gaussian with mean $\mu = 1$, width $\sigma = 0.1$, and lower limit $q_{\min} = 0.4$. The total mass distribution is a power-law with index $\alpha = -2.5$ between $M_{\mathrm{tot,\,min}} = 2.3 \; M_\odot$ and $M_{\mathrm{tot,\,min}} = 4.3 \; M_\odot$ with low-mass smoothing (see Eq. 7-8 of [851]) over $\delta M_{\mathrm{tot}} = 0.4 \; M_\odot$. We choose to parameterize the mass distribution in terms of total mass and mass ratio since the latter is particularly sensitive to correlations with the spin parameters [721, 655]. For the extrinsic parameters, we assume the sources are distributed uniformly in comoving volume between luminosity distances of $d_L = 10$ Mpc and $d_L = 300$ Mpc and isotropically on the sky. Standard distributions are chosen for the remaining binary parameters (see e.g., [762]).

For both of the populations described above, we generate 100 events that are detectable with a LIGO Hanford-Livingston detector network [3] operating at the sensitivity achieved during the third observing run [22, 35]. We consider an event to be detectable if it is observed with a network optimal signal-to-noise ratio (SNR) $\rho_{\mathrm{opt}}^{\mathrm{net}} \geq 9$. For each event, we perform Bayesian parameter estimation to obtain samples from the posterior probability distributions for the binary parameters, $\boldsymbol{\theta}$, describing an individual BNS system, i:

$$p(\boldsymbol{\theta_i}|d_i) \propto \mathcal{L}(d_i|\boldsymbol{\theta_i})\pi_{\mathrm{PE}}(\boldsymbol{\theta_i}), \tag{4.1}$$

where $\mathcal{L}(d_i|\boldsymbol{\theta_i})$ is the likelihood of observing data d_i given the binary parameters $\boldsymbol{\theta_i}$ [908, 761]:

$$\mathcal{L}(d_i|\boldsymbol{\theta_i}) \propto \exp\left(-\sum_k \frac{2|d_k - h_k(\boldsymbol{\theta_i})|^2}{TS_k}\right). \tag{4.2}$$

Here $h(\boldsymbol{\theta_i})$ represents the gravitational waveform for the BNS signal with parameters $\boldsymbol{\theta_i}$, T is the duration of the analyzed data segment, S_k is the noise power spectral density (PSD) characterizing the sensitivity of the detector, and k indicates the frequency dependence of the data, waveform, and PSD. The prior in Eq. 8.3 is denoted by $\pi_{\mathrm{PE}}(\boldsymbol{\theta_i})$, with the "PE" subscript indicating that this is the prior assumed during the initial parameter estimation step for each individual event.

We use the reduced order quadrature (ROQ) implementation [820] of the IMRPhenomPv2 waveform model [445, 509, 482] in the likelihood above in order to curtail the computational cost of the individual-event parameter estimation. As such, we assume that the neutron stars we simulate are point masses with no tidal deformability. While previous works have demonstrated the importance of simultaneously inferring the mass and tidal deformability distributions at the risk of introducing biases when they are fit independently [948, 419], we limit the scope of this work to mass-spin correlations and leave a full exploration of the potential correlations between the three intrinsic parameters of each neutron star to future studies. However, we comment on the effects of correlations with tides in Section 4.4. We use the PYMULTINEST [344, 345, 346, 182] and DYNESTY [828] nested samplers as implemented in the BILBY [96, 762] parameter estimation package to obtain samples from the posterior distributions for each event.

We choose priors $\pi_{\mathrm{PE}}(\boldsymbol{\theta_i})$ that are uniform in chirp mass with a width of 0.2 $M_\odot$ centered on the true value of the chirp mass for each event and uniform in mass ratio over $[0.125, 1]$. This choice of prior and mass parameterization is more convenient for sampling. The priors on the extrinsic parameters are the same as those from which the populations were drawn. We adopt a number of different choices for the spin priors, as described below and summarized in Table 4.1.

Table 4.1: Summary of the different combinations of true spin distribution and prior applied during the initial parameter estimation step for each of the scenarios we explore in this work. The p_{draw} column describes the distribution assumed for the sensitivity injections used to incorporate selection effects as detailed in Appendix 4.5.2.

$\pi_{\mathrm{pop}}(\chi_1, \chi_2)$	$\pi_{\mathrm{PE}}(\chi_1, \chi_2)$	$p_{\mathrm{draw}}(\chi_1, \chi_2)$
Aligned, $\chi_{\max} = 0.4$	Aligned, $\chi_{\max} = 0.8$	Aligned, $\chi_{\max} = 0.8$
Aligned, $\chi_{\max} = 0.4$	Aligned, $\chi_{\max} = 0.4$	Aligned, $\chi_{\max} = 0.4$
Aligned, $\chi_{\max} = 0.4$	Aligned, $\chi_{\max} = 0.05$	Aligned, $\chi_{\max} = 0.05$
Aligned, $\chi_{\max} = 0.05$	Aligned, $\chi_{\max} = 0.8$	Aligned, $\chi_{\max} = 0.8$
Aligned, $\chi_{\max} = 0.05$	Precessing, $\chi_{\max} = 0.8$	Aligned, $\chi_{\max} = 0.8$

Once we have obtained posterior samples for each of the individual BNS events, we can combine them to measure the underlying mass distribution of our simulated population. In this case, we are no longer interested in the individual binary parameters, $\boldsymbol{\theta}_i$, but rather in a set of hyperparameters $\boldsymbol{\Lambda}$ that describe the population distribution, $\pi_{\mathrm{pop}}(\boldsymbol{\theta}|\boldsymbol{\Lambda})$, which is also referred to as the hyper-prior. For our choice of population distribution, $\boldsymbol{\Lambda} = \{\alpha, M_{\mathrm{tot,min}}, M_{\mathrm{tot,max}}, \delta M_{\mathrm{tot}}, \mu, \sigma\}$. The likelihood of observing a set of individual events $\{d\}$ given the hyperparameters $\boldsymbol{\Lambda}$ is obtained by marginalizing over the binary parameters for each event and multiplying the resultant marginal likelihoods:

$$\mathcal{L}(\{d\}|\boldsymbol{\Lambda}) = \prod_i \int \mathcal{L}(d_i|\boldsymbol{\theta}_i)\pi_{\mathrm{pop}}(\boldsymbol{\theta}_i|\boldsymbol{\Lambda})d\boldsymbol{\theta}_i. \tag{4.3}$$

This joint likelihood can be constructed from the individual-event posterior samples obtained in the first parameter estimation step via the "recycling" method (e.g. [874]),

$$\mathcal{L}(\{d\}|\boldsymbol{\Lambda}) \propto \prod_i \sum_j \frac{\pi_{\mathrm{pop}}(\boldsymbol{\theta}_{i,j}|\boldsymbol{\Lambda})}{\pi_{\mathrm{PE}}(\boldsymbol{\theta}_{i,j})}, \tag{4.4}$$

so that the joint likelihood is the ratio of the hyper-prior and the original PE prior, where the subscript j denotes a sum over the individual posterior samples for each event.

The likelihood above assumes that the individual events included in the observed population are an unbiased sample of the true population found in nature. However,

Table 4.2: Hyperparameters describing the mass distribution and the maximum and minimum values allowed in the prior applied during hierarchical inference. The priors on all parameters are uniform.

Symbol	Parameter	Minimum	Maximum
α	total mass power-law index	0	4
$M_{\mathrm{tot,\,min}}$	minimum total mass	$2\ M_\odot$	$3\ M_\odot$
$M_{\mathrm{tot,\,max}}$	maximum total mass	$3.2\ M_\odot$	$5\ M_\odot$
δM_{tot}	smoothing parameter	$0\ M_\odot$	$1\ M_\odot$
μ	mass ratio mean	0.4	1
σ	mass ratio standard deviation	0.01	0.5

we know that this is not the case for observed gravitational-wave events, since the detectors are more sensitive to high-mass sources [353]. This selection effect needs to be accounted for in the likelihood in order to obtain unbiased estimates of the hyperparameters describing the astrophysical, rather than the observed, distribution [566, 585, 874, 919],

$$\mathcal{L}(\{d\}|\Lambda) = \prod_i \frac{\int \mathcal{L}(d_i|\boldsymbol{\theta_i})\pi_{\mathrm{pop}}(\Lambda|\boldsymbol{\theta_i})d\boldsymbol{\theta_i}}{\alpha(\Lambda)}, \tag{4.5}$$

$$\alpha(\Lambda) = \int d\boldsymbol{\theta_i} p_{\mathrm{det}}(\boldsymbol{\theta_i})\pi_{\mathrm{pop}}(\boldsymbol{\theta_i}|\Lambda). \tag{4.6}$$

The function $p_{\mathrm{det}}(\boldsymbol{\theta_i})$ gives the probability that an individual event with parameters $\boldsymbol{\theta_i}$ will be detected. We evaluate $\alpha(\Lambda)$ using a Monte Carlo integral over a set of simulated signals drawn from the distribution $p_{\mathrm{draw}}(\boldsymbol{\theta})$ following the approach described in Ref. [334]. More details on the simulated population used for determining the detection probability can be found in Appendix 4.5.2. We evaluate the corrected likelihood in Eq. 4.5 to obtain samples from the posterior distributions of the hyperparameters using the NESTLE sampler [111] and the GWPOPULATION package [849]. The priors on the hyperparameters are all uniform over the ranges given in Table 4.2.

So far we have not mentioned the spins in the hierarchical inference step and have restricted the hyperparameters to include only those governing the mass distribution. For any parameters that are not explicitly included in the ratio in Eq. 7.2, it is implicitly assumed that the PE prior and the hyper-prior are the same (see Section 5.4

of [919]). Even if we are not interested in measuring the hyperparameters of the spin distribution, if there is a mismatch between the underlying population distribution and the prior applied during the initial sampling, this can introduce biases in the hyperparameters that are being measured if there are significant correlations between those binary parameters, as is the case for the mass ratio and spins for binary neutron star systems, for example. To probe how such mismatches between the assumed and true population distributions for the spins can introduce biases in the mass distribution, we deliberately apply spin priors during the PE step that differ from those used to generate the population. This is consistent with what is currently done in population analyses, as will be described in detail in Section 4.4.

For the medium-spin population, we apply both high and low-spin priors with $\chi_{\max} = 0.8, 0.05$, respectively, as limited by the range of validity of the ROQ. We also perform parameter estimation and hierarchical inference on the mass distribution with a PE prior that matches the true population distribution with $\chi_{\max} = 0.4$ as a sanity check to ensure that the resultant biases we see are indeed due to a spin prior mismatch. An example corner plot showing the posteriors on the mass ratio and component spins obtained under each of these priors for one individual event is shown in Fig. 4-1. While the true values of all three parameters are included within the prior range for all three choices of spin prior, the q posterior peaks at lower values when analyzed with the low-spin prior compared to the other two due to the correlation between mass ratio and spin. For the low-spin population, we apply the high-spin prior first assuming aligned spins and then relax this assumption to allow for precessing spins. The combinations of true population and PE prior are summarized in Table 4.1.

4.3   Mass-spin correlations

The hyperparameter posteriors for $M_{\rm tot,max}$, μ, and σ obtained when analyzing the medium-spin population with true $\chi_{\max} = 0.4$ with both the low and high-spin priors are shown in the corner plot in Fig. 4-2. The blue contours show the results obtained

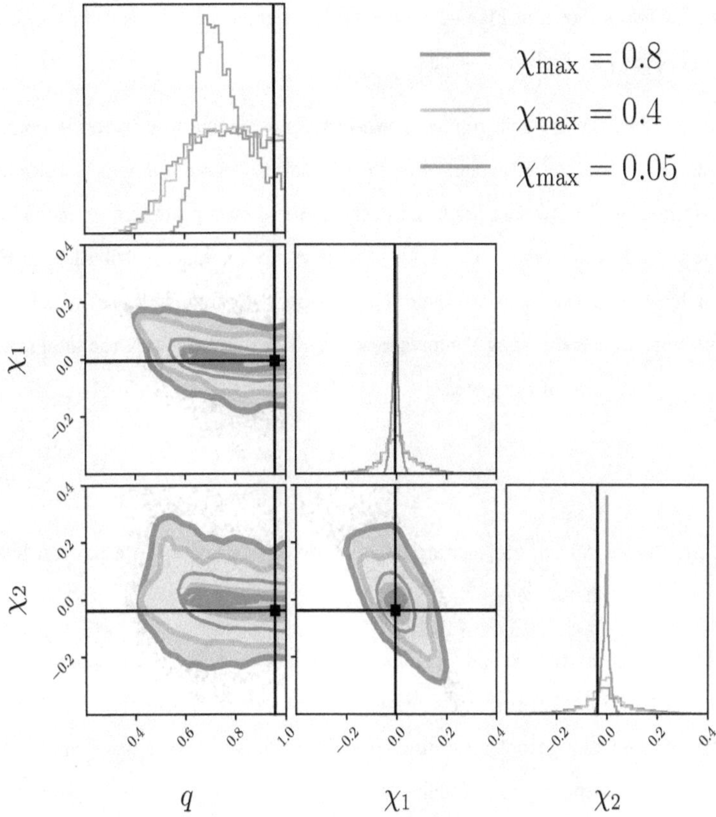

Figure 4-1: Corner plot showing the one-dimensional posteriors and 50 and 90% contours for the mass ratio and component spin magnitudes for one event recovered with aligned-spin priors for three different choices of χ_{max}. The true values are indicated with the black lines and are $q = 0.96, \chi_1 = -0.005, \chi_2 = -0.037$, within the prior range for all three choices of prior.

with the high-spin prior. The true hyperparameter values are all recovered within the 90% credible region of the posterior, demonstrating that analyzing a population with $\chi_{\max} = 0.4$ with a prior out to $\chi_{\max} = 0.8$ does not lead to biases in the inferred mass distribution. This is consistent with the fact that the analyses with the medium and high-spin priors shown in Fig. 4-1 do not lead to significant differences in the mass ratio posteriors for individual events.

Conversely, the low-spin results shown in red are significantly biased in both σ and $M_{\mathrm{tot,max}}$. This can be explained by the correlations shown in Fig. 4-1, which are exacerbated by the low-spin prior and push the mass ratio posteriors for individual events towards lower values. This leads to a preference for wider distributions in the hierarchical inference step, since there is more support for extreme mass ratios in the population. Underestimating the mass ratio results in overestimating the total mass, since the chirp mass of the system,

$$\mathcal{M} = \left(\frac{q}{(1+q)^2}\right)^{3/5} M_{\mathrm{tot}}, \tag{4.7}$$

is still well-constrained, propagating into the overestimation of the maximum total mass at the population level.

The inferred mass ratio and total mass distributions under the low-spin prior are shown in Fig. 4-3. These are represented by the posterior population distribution (PPD), which is the astrophysical distribution of the binary parameters $\boldsymbol{\theta}$ implied by the inferred hyperparameters $\boldsymbol{\Lambda}$ [49]:

$$p(\boldsymbol{\theta}|\{d\}) = \int d\boldsymbol{\Lambda}\, p(\boldsymbol{\Lambda}|\{d\}) \pi_{\mathrm{pop}}(\boldsymbol{\theta}|\boldsymbol{\Lambda}). \tag{4.8}$$

The recovery of a wider mass ratio distribution and higher maximum total mass results in overestimating the maximum mass of the primary neutron star and underestimating the minimum mass of the secondary, as can be seen in the bottom panel of Fig. 4-3. These biases can have profound implications for both single and binary neutron star formation mechanisms and their equation of state. Few equations of state are

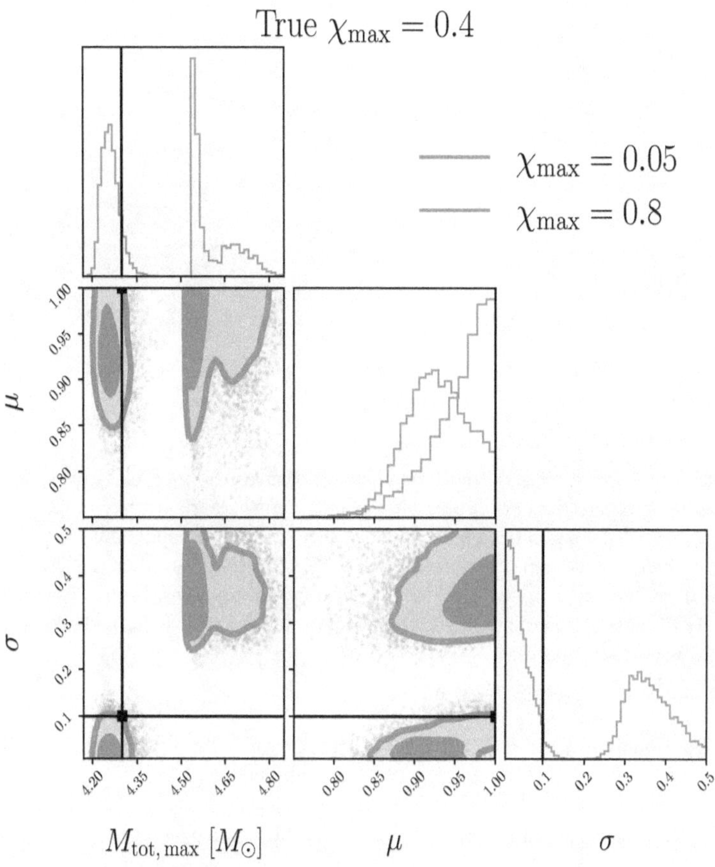

Figure 4-2: Hyperparameter posteriors on the maximum total mass and mean and width of the mass ratio distribution for two different choices of initial sampling prior applied to the medium-spin population. While the sources were drawn from an aligned-spin distribution with $\chi_{\mathrm{max}} = 0.4$, the red and blue contours show the hierarchical inference results when they are recovered with priors with $\chi_{\mathrm{max}} = 0.05, 0.8$, respectively. The black lines denote the true values of the hyperparameters. The value for $\mu = 1$ lies at the edge of the prior, and hence the black line is not visible for this parameter. The high-spin prior results are conistent with the true population, while the low-spin prior results favor a much wider mass ratio distribution and a higher maximum total mass. The true values are excluded at $> 3\sigma$ confidence.

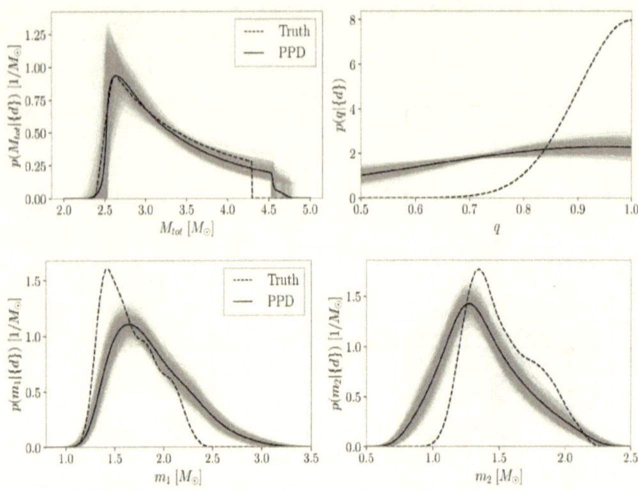

Figure 4-3: Inferred mass posterior population distributions (solid black lines) when a spin prior mismatch is introduced for a population where the true spin follows the aligned-spin distribution out to $\chi_{\mathrm{max}} = 0.4$, but the assumed population only allows $\chi_{\mathrm{max}} = 0.05$. The dashed black lines show the true distributions, while the light blue lines show individual draws from the hyperparameter posterior. Top: Total mass (left) and mass ratio (right) distributions. Bottom: Primary (left) and secondary (right) mass distributions.

able to support neutron stars with $m \geq 2.5\ M_\odot$, where 7% of the probability lies for the recovered PPD on m_1. Additionally, it is difficult to form neutron stars with $m \lesssim 1\ M_\odot$ under current stellar evolution models [912], a region of parameter space which contains 10% of the probability for the inferred PPD on m_2.

The evolution of the bias in the hyperparameters as a function of the number of individual events included in the analysis is shown in blue in Fig. 4-4. With six events, denoted by the first vertical grey line, the true value of σ is already excluded from the 3σ credible region. The same occurs with 13 events for the maximum total mass parameter. We also show the evolution of the bias in the maximum and minimum components masses, which are represented by the 99th percentile of the primary mass distribution and the first percentile of the secondary mass distribution, respectively, since we do not directly parameterize the population in terms of the component masses. We find similar constraints on m_{max} and m_{min} to those presented in Ref. [220], but these parameters become significantly biased with a similar number of events to the σ and $M_{\mathrm{tot,max}}$ hyperparameters. The bias in $M_{\mathrm{tot,max}}$ is driven by four events in particular, which correspond to the distinguishable upward jumps in the top panel of Fig. 4-4. These events all have true values of $|\chi_1| > 0.05$, which is outside the range allowed by the low-spin prior. This leads to a significant bias in the mass ratio posterior towards low values, driving the total mass upwards to keep the chirp mass constant. We emphasize that the posteriors for these individual events do not exhibit railing against the prior edges in either the spin or mass parameters and are well-converged. As such, they would not immediately be identified as problematic if they corresponded to real events. These results demonstrate that choosing a population model for the spins (even implicitly) that does not include all the sources in the population can significantly bias the BNS mass distribution. The width of the 90% credible interval (CI) for the same parameters in the case of no spin-prior mismatch is shown in the black dotted lines in Fig. 4-4 for comparison. The true values of all the hyperparameters lie within the 90% credible interval, demonstrating that there is no bias when the implied and true spin distributions match.

While we have found that analyzing the population of sources with $\chi_{\mathrm{max}} = 0.4$ with a prior going up to $\chi_{\mathrm{max}} = 0.8$ does not introduce a bias on the mass distribution, we now seek to investigate if the same conclusion holds for the low-spin population with $\chi_{\mathrm{max}} = 0.05$. The hyperparameter posteriors for the low-spin population analyzed with high-spin priors assuming both aligned and precessing spins are shown in the corner plot in Fig. 4-5. The true values of the hyperparameters are not always contained within the 90% credible region for the high aligned-spin prior shown in red, indicating a hint of a bias when 100 individual events are included in the population. The much larger difference in the prior volume between the low- and high-spin priors in this case leads to a bias even when all the observed events have spins within the allowed prior region. Allowing for the tilts to be misaligned to the orbital angular momentum introduces additional degrees of freedom that break the strong degeneracy between q and χ_z, alleviating this bias. The results obtained under the high

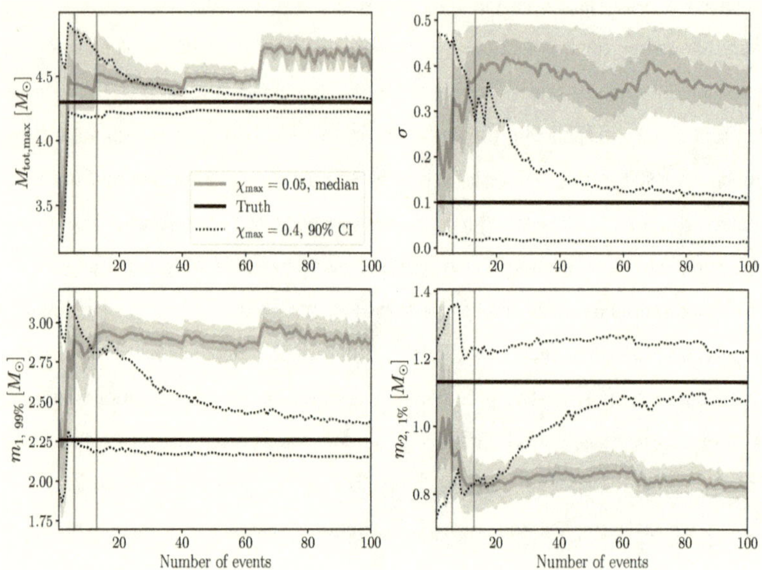

Figure 4-4: The top two panels show the evolution of the hyperparameters $M_{\rm tot,max}$ and σ as a function of the number of events included in the hierarchical analysis. The bottom two panels show the evolution of the maximum and minimum inferred component masses, represented by the 99th percentile of the m_1 distribution and the first percentile of the m_2 distribution, respectively. The solid black line shows the true value for each hyperparameter, while the solid blue line shows the median obtained when applying the low-spin prior to the medium-spin population. The blue shading gives the 50% and 90% credible intervals. The dotted black lines denote the 90% credible region when there is no spin prior mismatch applied to the medium-spin population. The vertical grey lines show the sixth and thirteenth events, where the true values of σ and $M_{\rm tot,max}$ are excluded at $> 3\sigma$, respectively.

precessing-spin prior shown in blue in Fig. 4-5 include the true hyperparameter values within the 90% credible region. Thus, we conclude that using the high, precessing-spin prior is the safest choice for BNS systems if the mass and spin distributions are not modeled simultaneously during hierarchical inference.

4.4 Conclusions

In this work, we have demonstrated that introducing a mismatch between the true, underlying spin distribution for BNS systems observed in gravitational waves and that assumed when characterizing individual systems can lead to a bias in the inferred mass distribution at the population level. If the mass and spin distributions are not fit simultaneously, the implied population model for the spin distribution is the same as the prior used when conducting parameter estimation for individual sources. To investigate the effects of such a mismatch, we simulated two distinct populations of BNS sources, one with medium aligned spins out to $\chi_{\mathrm{max}} = 0.4$ and the other with low aligned spins out to $\chi_{\mathrm{max}} = 0.05$.

The mass distribution inferred for the medium-spin population was significantly biased when the individual sources were analyzed with a low aligned-spin prior but unbiased when analyzed with a high aligned-spin prior. The bias is due to the degeneracy between the mass ratio and aligned spin components for BNS systems, which pushes the mass ratio posteriors for individual sources out towards more extreme values and also drives the total mass towards higher values. This translates into an overestimation of the maximum neutron star mass and and underestimation of the minimum mass, with adverse implications for both the inference of the nuclear equation of state and supernova mechanisms. The most massive neutron stars with posterior support in Fig. 4-3 are only supported by the stiffest equations of state. The illusion of these high-mass neutron stars in the population would falsely populate the putative lower mass gap between the heaviest neutron stars and lightest black holes. Additionally, the false presence of a significant subpopulation of sub-solar mass compact objects would affect the inferred contribution of primordial sub-solar mass black

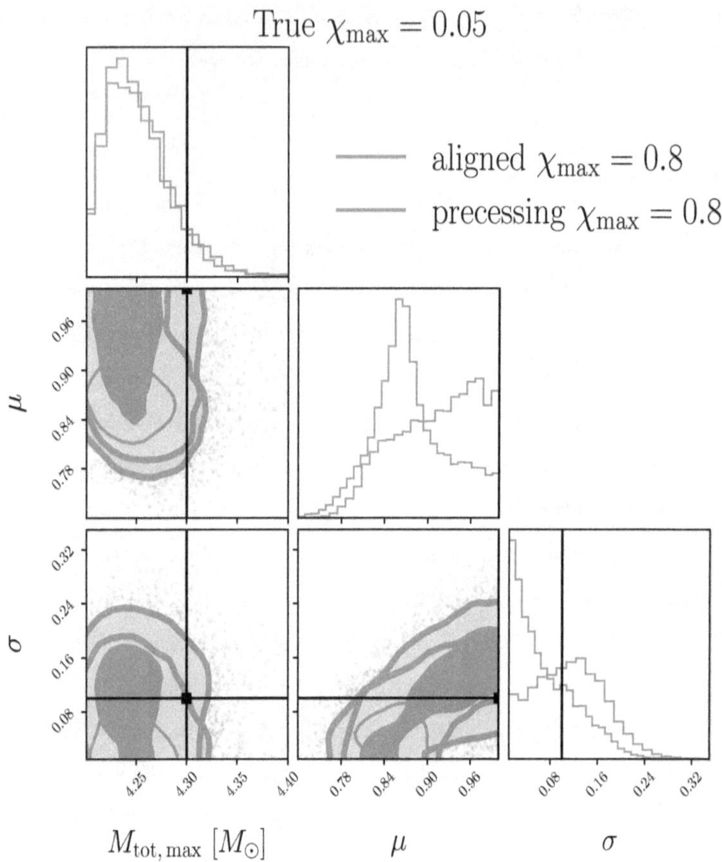

Figure 4-5: Hyperparameter posteriors on the maximum total mass and mean and width of the mass ratio distribution for two different choices of initial sampling prior applied to the low-spin population. The sources were drawn from an aligned-spin distribution with $\chi_{\mathrm{max}} = 0.05$, but both priors assume $\chi_{\mathrm{max}} = 0.8$. However, the red contours show the hierarchical inference results when an aligned-spin prior is applied, while the blue allows for spin precession. While the high aligned-spin results are only marginally consistent with the true hyperparameter values, allowing for precession breaks the strong degeneracy between q and χ_z and ameliorates the bias.

holes to the dark matter density [23, 28, 662, 51], as neutron stars with such low masses are not expected to form theoretically, and ground-based gravitational-wave detectors are not sensitive to the gravitational radiation from less massive compact objects such as white dwarfs. While in principle tidal effects could be used to distinguish sub-solar mass black holes from neutron stars, in practice gravitational-wave constraints on the tidal deformability are often weak (e.g. [32, 47]).

The mass dicitestribution inferred for the low-spin population demonstrated a hint of bias with the high aligned-spin prior, but this was alleviated when the individual sources were analyzed with a prior allowing for misaligned spin tilts. The extra degrees of freedom introduced by the precessing-spin model break the strong degeneracy between q and χ_z. These biases demonstrate the importance of fitting the mass and spin distributions simultaneously, to avoid implicitly mismodeling the spin distribution. However, if the mass distribution must be analyzed independently, we conclude that using a high, precessing-spin prior for the individual sources is the safest choice.

We note that the unbiased results we obtain in this demonstration are only robust if the choice of parameterized mass model used during the hierarchical inference step is physically realistic. If the assumed shape of the mass distribution does not match the underlying population, further biases can be introduced. However, this sort of mismatch is unlikely to affect the inferred maximum and minimum neutron star masses as significantly as the spin prior mismatch, so long as the parameterized population model covers the full range of allowed neutron star masses. This potential problem can be further ameliorated by fitting the mass distribution with several different hierarchical models, including for example the bimodal models favored by current observations and simulations (e.g. [88, 77, 859, 912, 337]), and comparing the statistical evidence obtained between them to determine which provides the best fit. Alternatively, a more flexible model that does not impose a specific functional form on the mass distribution could be used (e.g. [585, 938, 875, 554, 778, 751]).

Correlations between the tidal parameters–which we have set to zero in our analysis– and the masses and spins can also introduce systematic errors in the inferred mass distribution if not accounted for. For the posteriors of individual events, more extreme values of mass ratio allow for smaller values of the tidal parameter which enters the gravitational waveform at leading order, $\tilde{\Lambda}$ [926, 27]. Since the high-spin prior typically provides more support for more unequal mass ratios, changing the spin prior can also affect the inferred tidal parameters. Comparing our results with those of [419], which demonstrates the effect of mismodeling the tidal parameters on the inferred mass distribution, we conclude that enforcing a low-spin prior when there are larger spins in the population introduces a much more significant bias in the inferred mass distribution. However, both types of mismatches lead to increased support for higher neutron star masses.

In addition to the mismatches between the true and implied population models for the spins, masses, and tides discussed above, inadequate sampler performance when conducting parameter estimation for individual events can also manifest as a bias in the inferred population properties when multiple events are combined, as demonstrated in Appendix 4.5.1. However, this sort of bias can be diagnosed and addressed by performing hierarchical inference on a simulated population. Another potential source of systematic error is the accuracy limitations of the waveform models used to infer the properties of individual events. For the SNRs expected with the current generation of detectors, this effect should be small compared to the bias introduced by mismodeling the spin distribution [26, 305, 295, 611, 32].

The upcoming fourth observing run of the LIGO and Virgo detectors is expected to add tens of new binary neutron star sources to the catalog of compact binaries detected in gravitational waves [22]. Based on our results, a bias in the mass distribution could be imposed on the observed population with as few as four additional BNS detections if a low-spin prior is applied to a population with higher spins. We note that while current population studies do not model the BNS mass distribution independently due to the paucity of detections, a variety of models consider BNS sources as part of the compact object population as a whole [584, 352, 330, 52] or as part of the population of neutron star-containing systems [536, 52, 553]. Most of these models and analyses assume the BNS spin distribution is uniform in magnitude on the interval [0, 1) with isotropic tilts, corresponding to the safe choice identified in Section 4.3. Both the MULTI SOURCE model presented in Ref. [52] and the analysis of Ref. [553] fit the mass and spin distributions of the BNS subpopulation simultaneously. The MULTI SOURCE model assumes the spin magnitudes follow a Beta distribution with $\chi_{\mathrm{max}} = 0.05$ and the tilts are isotropically distributed. The analysis of Ref. [553] takes a similar approach, assuming the neutron star spin magnitudes follow a truncated Gaussian out to $\chi_{\mathrm{max}} = 0.05$ and fitting the tilts as a mixture model between aligned and isotropic distributions. While assuming small spins will not lead to a bias for the two BNS events currently detected, this assumption should be relaxed to avoid introducing a bias in the mass distribution with the first few events detected during the next observing run[1].

4.5   Supplementary material

4.5.1   The insufficiency of P-P plots as a diagnostic test

In order to obtain unbiased posteriors for the hyperparameters describing a population of events using the likelihood in Eq. 4.5, the individual-event posteriors must also be unbiased. A common diagnostic tool for evaluating the performance of a stochastic sampler is a "probability-probability plot", or a P-P plot [239, 852]. This provides a graphical way to verify that the true parameter values are recovered within a certain credible interval for the expected fraction of events in a population. If the likelihood correctly describes the distribution of data, the true parameter values should be recovered within the 5% credible interval 5% of the time, the 95% credible interval 95% of the time, etc. For the Whittle likelihood in Eq. 9.1 used when performing parameter estimation on individual gravitational-wave sources, this is satisfied if the data are Gaussian about the assumed noise power spectral density. In the case of unbiased individual-event posterior samples for a particular parameter, the P-P plot should be approximately diagonal, as it shows the fraction of events for which the

---

[1]A public repository with example code to reproduce the results in this manuscript is available on github.

true value of a given parameter falls within the given credible interval as a function of that credible interval.

An example P-P plot obtained using the PyMultiNest sampler for a population of 100 simulated BNS sources is shown in Fig. 4-6. The extrinsic parameters are drawn from the same prior distributions described in the main text, and the spins follow the low aligned-spin prior. The mass ratio is also drawn from the same population prior explored in the main text, namely a narrow truncated Gaussian with $\mu = 1$, $\sigma = 0.1$. The chirp masses are drawn from a uniform prior between 1.52 and 1.70 $M_\odot$. For the P-P plot to be unbiased, the distributions from which the events are drawn must match the priors applied during sampling, meaning that there is no prior mismatch and no cut based on the SNR of the individual events, as has been the case in the rest of this work. The legend shows the probability for the fraction of events within each credible interval to be drawn from a uniform distribution for individual parameters, as expected from Gaussianity. For the mass ratio, this value is $p = 0.742$, passing the P-P test (where the threshold is $p > 1/11 = 0.09$). The probability that the individual-parameter probabilities are drawn from a uniform distribution is 0.254, consistent with random chance for 11 parameters and indicative of unbiased sampling across the 11 parameters drawn from and recovered with this particular set of priors.

However, when the same 100 individual events are analyzed hierarchically to recover the true values of μ and σ, the results are biased at the 3σ level, as shown in Fig. 4-7. This demonstrates a case where the sampling algorithm passes the P-P test for a particular population but still yields biased hierarchical inference results, highlighting the insufficiency of P-P plots as a diagnostic tool for individual-event parameter estimation. In this case, the recovered hyperparameters favor a narrowly distributed population peaking away from $\mu = 1$, indicating that the sampler is unable to thoroughly explore the edge of the prior space where most of the probability lies for the nearly equal-mass events included in the population. This is due to the adapted simultaneous ellipsoidal nested sampling method used by PyMulti-Nest [645, 805, 344, 345], which bounds the iso-likelihood contours around clusters

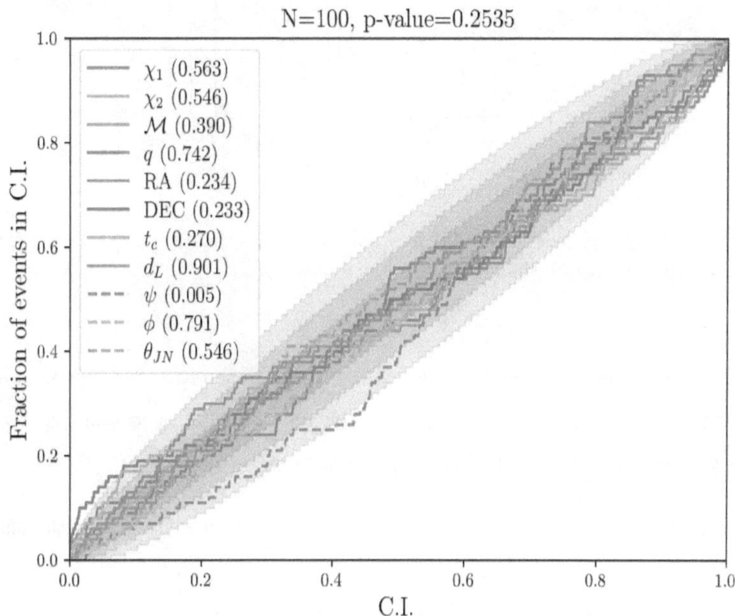

Figure 4-6: P-P plot showing the fraction of events for which the true value is recovered within a certain credible interval as a function of that credible interval for a population of 100 simulated BNS sources, sampled with the PyMultiNest package. The lines for individual parameters stay within the 3σ credible region, shaded in light gray, and the probability values quoted in the legend are consistent with passing the P-P test. The other grey shaded areas show the 1 and 2σ credible regions.

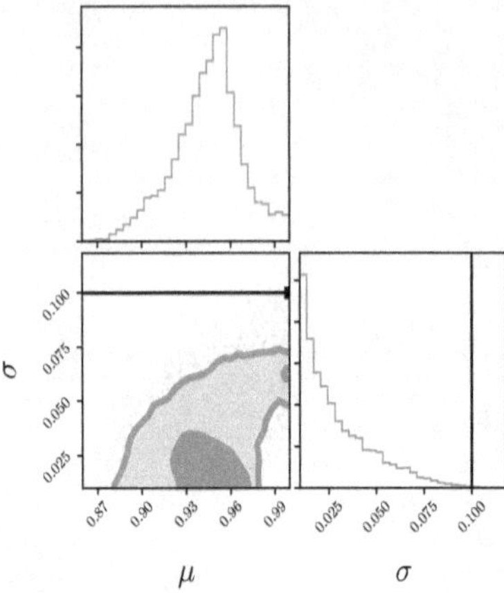

Figure 4-7: Corner plot for the inferred mass ratio hyperparameters using the 100 events used to generate the P-P plot in Fig. 4-6.

of live points with N-dimensional ellipsoids. Because the probability for mass ratio rails against the edge of the prior and the algorithm is inefficient at sampling near edges, the peak of the distribution at equal mass is undersampled.

PYMULTINEST internally works with a uniform prior for samples from the unit hyper-cube which must then be scaled to the physical parameter space such that the scaled samples are drawn from the desired physical prior distribution. In order to improve the convergence near equal masses, we propose to use a two-stage mapping that shifts the peak of the probability at $q = 1$ away from the edge of the prior in the frame of the sampler. Typically samples from the unit cube are rescaled onto the appropriate prior distribution for a given parameter via the inverse of the prior's cumulative distribution function, such that for a sample from the unit cube, x, $q(x) = \mathrm{CDF}^{-1}(x)$ (although other methods have been proposed, e.g., [76]). Here, we propose to add an intermediate step,

$$u = 2\min(x, 1 - x), \tag{4.9}$$

$$q = \mathrm{CDF}^{-1}(u), \tag{4.10}$$

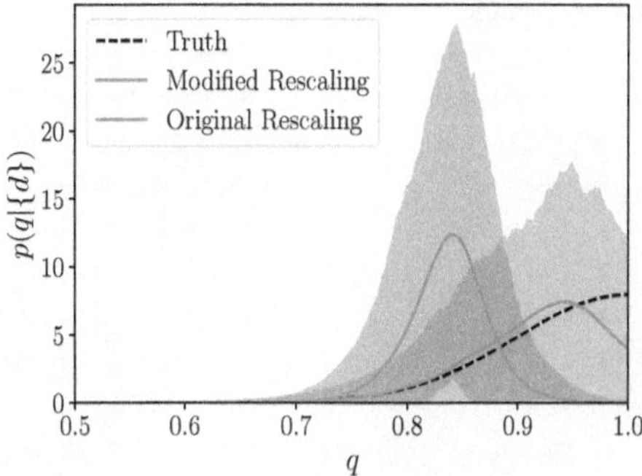

Figure 4-8: Inferred mass ratio PPDs when applying the original inverse CDF rescaling method (red) and the modified rescaling method described in Eq. 4.10 (blue) to the medium-spin population described in the main text analyzed with no spin prior mismatch. The dotted line shows the true distribution, and the shading shows the 90% credible region. Without the modified rescaling, the mass ratio distribution is biased towards lower q values and more narrowly peaked.

where u still takes on values within the unit interval, but instead of $q(x = 1) = 1$, $q(x = 0.5) = 1$. This transformation maps equal mass–where the peak of the probability lies–to the center of the sampled space rather than the edge, which is more difficult to sample. In Fig. 4-8, we show the mass ratio PPDs for the medium-spin population described in the main text with no spin prior mismatch obtained with and without this modified rescaling method. Similarly to the corner plot in Fig. 4-7, without the modified rescaling, the PPD shown in red peaks at lower mass ratios and is more narrowly distributed. The true hyperparameters are excluded from the recovered posteriors at $> 3\sigma$ credibility. Once the modified rescaling is implemented for the individual-event parameter estimation, the hierarchical inference becomes unbiased as shown in blue. This stealth bias introduced by the stochastic sampling algorithm that is not caught with a P-P test demonstrates the importance of verifying hierarchical inference analyses with synthetic populations where the true hyperparameter values are known and controllable before conducting the analysis on real data.

4.5.2 Selection Effects

In order to evaluate the selection function in Eq. 4.6, we calculate the detection probability, $p_{\mathrm{det}}(\boldsymbol{\theta_i})$, using an injection campaign. $\alpha(\boldsymbol{\Lambda})$ gives the fraction of signals that will be detected drawn from a population model with hyperparameters $\boldsymbol{\Lambda}$. We generate 194953 total simulated signals, calculating the network optimal SNR, $\rho_{\mathrm{opt}}^{\mathrm{net}}$ for each. Of these, 40000 of are above the threshold for detection, $\rho_{\mathrm{opt}}^{\mathrm{net}} \geq 9$. The distribution from which the injections are drawn, $p_{\mathrm{draw}}(\boldsymbol{\theta})$, is uniform in total mass over the range $[2, 5]$ $M_\odot$, and uses the same prior distributions described in the main text for all the extrinsic parameters. The mass ratio distribution is

$$p_{\mathrm{draw}}(q) = 0.5\left(1.67 + \frac{1}{\sigma}\frac{\mathcal{N}(q|\mu,\sigma)}{\Phi\left(\frac{q_{\max}-\mu}{\sigma}\right) - \Phi\left(\frac{q_{\min}-\mu}{\sigma}\right)}\right), \tag{4.11}$$

$$q_{\min} < q < q_{\max}$$

which is the normalized superposition of a uniform distribution and truncated Gaussian distribution with $\mu = 1, \sigma = 0.1$ between $q_{\min} = 0.4$ and $q_{\max} = 1$. The truncated Gaussian is added to enhance the number of injections with nearly equal mass ratios, since this is the part of the parameter space that should have the most support given the true distribution we used for the simulated populations described in Sec. 4.2. The spins are drawn following the aligned-spin prior with $\chi_{\max} = 0.99$. The injections are reweighted so that $\chi_{\max}$ matches the corresponding value used during the first parameter estimation step for each of the spin-prior mismatches we consider above. The different combinations of true population distribution, PE prior, and injection distribution are summarized in Table 4.1.

We can then estimate the detection probability as follows:

$$\alpha(\boldsymbol{\Lambda}) = \int d\boldsymbol{\theta} p_{\text{det}}(\boldsymbol{\theta}) \pi_{\text{pop}}(\boldsymbol{\theta}|\boldsymbol{\Lambda}), \tag{4.12}$$

$$\alpha_{\text{draw}} = \int d\boldsymbol{\theta} p_{\text{det}}(\boldsymbol{\theta}) p_{\text{draw}}(\boldsymbol{\theta}) \approx \frac{N_{\text{found}}}{N_{\text{draw}}} \tag{4.13}$$

$$p_{\text{found}}(\boldsymbol{\theta}) = \frac{p_{\text{draw}}(\boldsymbol{\theta}) p_{\text{det}}(\boldsymbol{\theta})}{\alpha_{\text{draw}}}, \tag{4.14}$$

$$\alpha(\boldsymbol{\Lambda}) = \alpha_{\text{draw}} \int d\boldsymbol{\theta} \frac{p_{\text{found}}(\boldsymbol{\theta})}{p_{\text{draw}}(\boldsymbol{\theta})} \pi_{\text{pop}}(\boldsymbol{\theta}|\boldsymbol{\Lambda}), \tag{4.15}$$

$$\alpha(\boldsymbol{\Lambda}) \approx \frac{1}{N_{\text{found}}} \sum_{j} \frac{\pi_{\text{pop}}(\boldsymbol{\theta}_j|\boldsymbol{\Lambda})}{p_{\text{draw}}(\boldsymbol{\theta}_j)}. \tag{4.16}$$

We account for the uncertainty in the Monte Carlo integral in Eq. 4.16 following the method in Ref. [334] and reject parts of the hyperparameter space during sampling that do not have enough injections to meet the accuracy requirements therein. We note that our population model, $\pi_{\text{pop}}(\boldsymbol{\theta}_j|\boldsymbol{\Lambda})$, in Eq. 4.16 includes only the total mass distribution, and not the mass ratio distribution, as the latter has a negligible effect on the detectability of the source and would require a much higher number of injections to meet the accuracy requirements for our choice of narrow truncated Gaussian population distribution.

Chapter 5

Constraining short gamma-ray burst jet properties with gravitational waves and gamma rays

The content of this chapter was previously published in the Astrophysical Journal as Ref. [157] in April 2020. ASB performed the analysis, wrote the manuscript, and contributed to project development.

Abstract

Gamma-ray burst (GRB) prompt emission is highly beamed, and understanding the jet geometry and beaming configuration can provide information on the poorly understood central engine and circum-burst environment. Prior to the advent of gravitational-wave astronomy, astronomers relied on observations of jet breaks in the multi-wavelength afterglow to determine the GRB opening angle, since the observer's viewing angle relative to the system cannot be determined from the electromagnetic data alone. Gravitational-wave observations, however, provide an independent measurement of the viewing angle. We describe a Bayesian method for determining the geometry of short GRBs using coincident electromagnetic and gravitational-wave observations. We demonstrate how an ensemble of multi-messenger detections can be used to measure the distributions of the jet energy, opening angle, Lorentz factor, and angular profile of short GRBs; we find that for a population of 100 such observations, we can constrain the mean of the opening angle distribution to within $10°$ regardless of the angular emission profile. Conversely, the constraint on the energy distribution depends on the shape of the profile, which can be distinguished.

5.1 Introduction

Understanding the emission profile and jet geometry of gamma-ray bursts (GRBs) has wide-ranging implications for the energetics, rates, and luminosity function of these relativistic explosions, all of which ultimately provide insight into the nature of the central engine. The GRB population is bimodal in duration and hardness, with long-soft and short-hard bursts defined by a transition at ~ 2 s [526]. Short GRB (sGRB) afterglows are uniformly fainter than their long GRB counterparts [e.g., 399, 362] (see [647, 143] for reviews) and were first observed in 2005 [468, 369, 400, 914, 116]. The lack of an associated supernova [e.g., 369, 520, 825, 280] together with the localization of some short GRBs to early-type galaxies [360, 361, 717] provided early evidence in support of the binary neutron star or neutron star-black hole merger progenitor model [319, 649]. The recent coincident detection [15] of gravitational-wave event GW170817 from a binary neutron star merger [17] and the short, hard burst GRB 170817A [417, 793] has confirmed the compact binary progenitor model for at least some sGRBs.

In this chapter, we describe a Bayesian method to combine gravitational-wave and electromagnetic observations of sGRBs from binary neutron star coalescences to infer the total energy and Lorentz factor of the jet as well as the opening angle and power law index of the jet emission profile. Because gravitational-wave data provides an independent measurement of the inclination angle between the jet axis and the observer's line of sight[1], the opening angle can be inferred directly from the prompt emission, eliminating the dependence on afterglow observations imposed by the traditional jet break calculation.

Our analysis builds on previous Bayesian methods for combining GW and EM data, which have been used to provide improved estimates of the neutron star parameters like the mass ratio and tidal deformability for GW170817 [248, 734, 206, 250, 732].

---

[1]The gravitational-wave signal provides an independent measurement of the angles between the binary angular momentum and the line of sight. Based on the results of fully general relativistic magnetohydrodynamical simulations, the jet is believed to be emitted along the spin axis of the remnant black hole due to the presence of a strong poloidal magnetic field, so the viewing angle of the GRB is expected to coincide with the inclination angle measured with gravitational waves [746, 747, 411].

Refs. [327, 328] have previously shown that combining gravitational-wave and GRB data can also be used to determine the GRB luminosity function and host galaxy and offer improved inference of parameters already constrained by the gravitational-wave data alone like the inclination angle and distance to the source system. While previous studies have offered constraints on the jet opening angle using estimates of the coincident gravitational-wave/GRB detection rate for top-hat jets [226, 237, 934] and by fitting the GRB luminosity assuming a structured jet geometry in conjunction with estimates of the binary neutron star merger rate [631], we seek to measure the GRB energy, Lorentz factor, opening angle, and power law index directly by parameterizing the measured fluence of the GRB prompt emission in terms of these four parameters and additional parameters inferred from the GW data. We analyze a simulated population of coincident GW and GRB detections to determine what type of constraints can be derived on the distributions of these parameters by combining an ensemble of multiple coincident events.

The rest of this chapter is organized as follows. We first discuss the jet break method for estimating the jet opening angle and its applications to GRB 170817A in Section 5.2, and then describe the top-hat and universal structured jet energy models, as well as the prescription for calculating the observed GRB fluence for a given jet geometry and inclination angle in Section 5.3. In Section 5.4 we outline the Bayesian parameter estimation method that we use to combine the GW and EM measurements for individual events and the hierarchical model used to determine the population hyper-parameters. We present results for simulated top-hat and structured power-law jet populations in Section 5.5, and conclude in Section 5.6 with a discussion of the implications of this study.

5.2 Existing Observational Constraints

The observational signature of collimated jets in GRBs is an achromatic jet break in the afterglow light curve that occurs at t_j after the prompt emission, when the bulk Lorentz factor of the outflow has decreased to $\Gamma \approx 1/\theta_j$, where θ_j is the half-opening angle of the jet [748, 749, 788]. The break is caused by a combination of two effects. The first is an edge effect that occurs when the entire emitting surface of the jet becomes visible. Due to relativistic beaming, the emission appears to come from a small fraction of the visible area, so the "missing" component relative to the expectation from a spherical outflow manifests itself as a steepening in the light curve. Simultaneously, when the jet edge comes into causal contact with the jet center as the Lorentz factor decreases, the jet begins to spread laterally, and the energy per solid angle decreases with time and radius, again resulting in a steepening of the light curve (see [429] for a review of GRB jets). The jet break is observable from the X-ray to the radio bands, and the opening angle can be calculated via [788]:

$$\theta_j \approx 9.5^\circ \left(\frac{t_{j,\mathrm{d}}}{1+z} \right)^{3/8} \left(\frac{n_0}{E_{\mathrm{K,iso},52}} \right)^{1/8}, \tag{5.1}$$

where $t_{j,\mathrm{d}}$ is the jet break time measured in days, n_0 is the density of the circumburst medium in cm^{-3}, and $E_{\mathrm{K,iso},52}$ is the isotropic equivalent kinetic energy of the ejecta in units of 10^{52} erg.

Because the afterglow emission of most sGRBs decays much faster and at a uniform rate compared to long GRBs, jet breaks have only been reported for five sGRBs (GRBs 051221A [194, 825], 090426A [656], 111020A [363], 130603B [364], and 140903A [885]). Furthermore, the observation of a jet break provides no information on the structure of the jet. The two leading jet structure models that both predict a jet break in the afterglow light curve are the uniform, or *top-hat jet*, where the energy per solid angle $\mathcal{E}$ and the Lorentz factor Γ are constant over the entire emitting surface [748, 749, 788, 430, 683], and the *universal structured jet*, where $\mathcal{E}$ and Γ decay as a power law with the angle from the jet axis, θ^{-k} [968, 766]. While the jet break can be explained in terms of the intrinsic opening angle of the jet in the top-hat model, the universal structured jet model explains the jet break in terms of the viewing angle of the observer, implying that the opening angle of the jet is much wider than in the top-hat case. Other profiles, like a Gaussian structured jet or a radially stratified jet, can also reproduce the jet break behavior. All of these models make simplifying assumptions about the true angular emission profile, which would be obtained from hydrodynamical simulations in the ideal case where such simulations could reliably produce estimates of the GRB jet evolution.

The jet geometry of GRB 170817A [417, 15] has been studied extensively. Its low luminosity together with the lack of *early* X-ray [883] and radio afterglow [883] disfavors both of the simple top-hat and power-law universal structured jet models [502], and is instead better explained by "cocoon" emission; as the jet drills through the merger ejecta surrounding the central engine, it inflates a mildly relativistic cocoon. In addition to the internal shocks that arise in the jet, the interaction of the jet and the merger ejecta forms another set of forward and reverse shocks. The reverse shock heats the jet material and creates an inner cocoon surrounding the jet. The forward shock propagating into the merger ejecta forms the outer cocoon, which is only mildly relativistic, with Lorentz factors of a few [425, 544]. The forward shock continues to propagate through the ejecta as long as the medium is optically thick enough to sustain its width, at which point the radiation inside the shock layer breaks out, producing the observed γ-rays as the residual photons diffuse out of the cocoon ([426, 648]; see also the shock breakout model of [139]).

If the initial jet has a very short duration, low energy, or wide opening angle, it may be "choked" by the cocoon. In this scenario, the jet does not manage to escape from the merger ejecta, and all of the initial energy of the jet is deposited into the cocoon. The observed γ-rays come entirely from the cocoon fireball [703]. If the initial jet launched by the central engine does manage to escape, it will still inflate a cocoon, so the observed γ-ray emission will consist of an ultra-relativistic, narrow core in addition to mildly relativistic cocoon "wings" [426]. In this sense, the cocoon model provides physical motivation for the universal structured jet model.

In the case of GRB 170817A, the jet is under-luminous by several orders of magnitude compared to E_{iso} measurements for the rest of the short GRB population [15]. Light curve modeling revealed that it is impossible to reproduce the observed emission using a top-hat jet model viewed off axis for physically realistic values of the circum-burst density [502]. Instead, emission from a wide-angled cocoon that fades on the order of a few hours can explain both the prompt emission and the lack of early observations in the X-ray and radio bands that would be expected from the afterglow emission of a standard top-hat jet [443, 883, 633, 426]. It was impossible to determine if the gamma-ray emission from the cocoon was accompanied by a successful jet from the observed prompt emission alone. Follow-up radio observations using very long-baseline interferometry determined that the jet exhibited superluminal motion, indicating that a collimated jet with opening angle $\theta_j < 5$ deg successfully broke out of the cocoon [632]. This picture was confirmed by further analyses covering the entire afterglow spectrum [599, 534, 407, 977].

5.3 GRB Energy Models

5.3.1 Isotropic Equivalent Energy

The isotropic equivalent energy of the GRB in the source frame, E^{iso}, is calculated by assuming the gamma-ray fluence, F^γ, measured by the observer is the same in all directions: $E^{\text{iso}} = 4\pi F^\gamma d_L^2 (1 + z)^{-1}$, where d_L is the luminosity distance and z is the redshift of the source. The measured fluence and thus the isotropic equivalent energy depend on the observer's inclination angle, the total energy of the jet, and the emission profile. Consider a jet with fixed, uniform Lorentz factor Γ, that emits total energy $E_0^\gamma/2$ in the source frame[2]. Each jet element is moving radially in the direction $\hat{n}$ at angle $\theta = \arccos{(\hat{n} \cdot \hat{z})}$ from the jet axis $\hat{z}$ and emits isotropically. The energy per solid angle emitted in the rest frame of each element is then:

$$\mathcal{E}_R(\theta) \equiv \frac{dE}{d\Omega_R} = \frac{E_0^\gamma/2}{4\pi\Gamma} f_R(\theta) \tag{5.2}$$

---

[2] The source frame and the observer frame are the same except for the redshift correction.

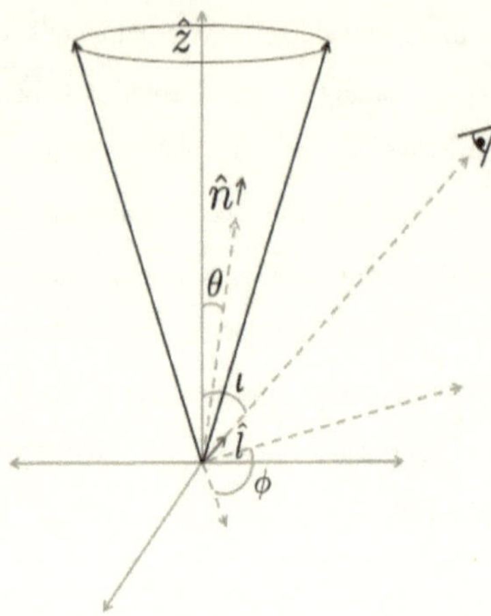

Figure 5-1: Coordinate system for the jet seen by an observer in the $\hat{l}$ direction. The angles θ and ι are defined such that $\hat{l} \cdot \hat{z} = \cos\iota$ and $\hat{n} \cdot \hat{z} = \cos\theta$ and ϕ is the angle between the projections of $\hat{l}$ and $\hat{n}$ in the $x - y$ plane.

The profile function $f_R(\theta)$ is azimuthally symmetric around the jet axis, and is normalized so that

$$\oint f_R(\theta)d\Omega_R = 2\pi \int_0^{\pi/2} f_R(\theta)\sin\theta d\theta = 1. \tag{5.3}$$

The profile function is defined in the rest frame and determines the brightness of each jet element as a function of its angular distance from the jet axis. The emission from each element, while isotropic in the rest frame, will appear highly beamed into a cone of angular width $\sim 1/\Gamma$ in the source frame due to relativistic beaming. The source-frame energy per solid angle can be calculated by applying the relativistic Doppler shift to Eq. 5.2 [318, 431, 782]:

$$\mathcal{E}_S(\theta) = \oint \mathcal{E}_R(\theta)\Gamma^{-3}[1 - \beta(\hat{l} \cdot \hat{n})]^{-3}d\Omega_R \tag{5.4}$$

where $\beta = v/c$ is the speed of the merger ejecta, and the Doppler factor is applied once for the energy and once for each angular dimension. The integral sums the

Doppler-boosted contribution to the total energy from each jet element at angle θ relative to the jet axis. The inclination angle ι of the observer is encoded in the dot product, $\hat{l} \cdot \hat{n} = \cos\theta\cos\iota + \sin\theta\sin\iota\cos\phi$, between the unit vector in the direction of the jet element, $\hat{n}$, and the unit vector pointing towards the observer, $\hat{l}$, as illustrated in Fig. 8-1. The isotropic equivalent energy is calculated by assuming that the energy per unit solid angle measured at some inclination angle, ι is the same in all directions:

$$E^{\mathrm{iso}}(\iota) = 4\pi\mathcal{E}_S(\theta)$$

$$= \frac{E_0^\gamma}{2\Gamma^4} \int_0^{2\pi} \int_0^{\pi/2} \frac{f_R(\theta)\sin\theta\, d\theta\, d\phi}{[1 - \beta(\cos\theta\cos\iota + \sin\theta\sin\iota\cos\phi)]^3}, \tag{5.5}$$

where we have substituted the definition of $\mathcal{E}_R(\theta)$ given in Eq. 5.2. If the Lorentz factor is also allowed to depend on the angle from the jet axis, the fluence at a particular inclination angle then becomes

$$F^\gamma(\iota) = \frac{E_0^\gamma(1+z)}{8\pi d_L^2} \times \tag{5.6}$$

$$\int_0^{2\pi} \int_0^{\pi/2} \frac{f_R(\theta)\sin\theta\, d\theta\, d\phi}{\Gamma^4(\theta)[1 - \beta(\theta)(\cos\theta\cos\iota + \sin\theta\sin\iota\cos\phi)]^3}.$$

5.3.2 Uniform Jet Model

Under the uniform jet or top-hat model, the energy per solid angle in the rest frame is expected to be constant within some jet opening angle θ_j [430, 683]:

$$\frac{dE}{d\Omega_{\dot{R}}} = \mathcal{E}_R(\theta) = \begin{cases} \mathcal{E}_0, & \theta \leq \theta_j \\ 0, & \theta > \theta_j \end{cases}. \tag{5.7}$$

The total gamma-ray energy emitted in the rest frame can then be calculated by integrating over all solid angles and multiplying by a factor of 2 to account for both jets:

$$E_0^\gamma = 2 \int_0^{2\pi} \int_0^{\theta_j} \mathcal{E}_0 \sin\theta' d\theta' \, d\phi$$

$$= 4\pi \mathcal{E}_0 (1 - \cos\theta_j)$$

$$= 4\pi \frac{dE}{d\Omega_R} f_b, \tag{5.8}$$

where we have defined the beaming factor $f_b \equiv (1 - \cos\theta_j)$. For the uniform jet model, we recover the typical relationship between the total energy and the isotropic equivalent energy [374]:

$$E^{\mathrm{iso}}(\theta_j) \approx E_0^\gamma / f_b. \tag{5.9}$$

5.3.3 Universal Structured Jet Model

The universal structured jet model assumes that all GRBs have a quasi-universal beaming configuration and that the variability in jet break time is due to the inclination angle rather than the intrinsic opening angle of the jet itself. Both the energy per solid angle and the Lorentz factor fall off as power laws as a function of the angle from the jet axis [968, 766]:

$$\frac{dE}{d\Omega_R} = \mathcal{E}_R(\theta) = \begin{cases} \mathcal{E}_0, & \theta \le \theta_c \\ \mathcal{E}_0 (\theta/\theta_c)^{-k}, & \theta_c < \theta \le \theta_j \\ 0, & \theta > \theta_j, \end{cases} \tag{5.10}$$

$$\Gamma(\theta) = \begin{cases} \Gamma_0, & \theta \le \theta_c \\ \Gamma_0 (\theta/\theta_c)^{-k}, & \theta_c < \theta \le \theta_j \\ 0, & \theta > \theta_j, \end{cases} \tag{5.11}$$

where θ_c is introduced to avoid the divergence at $\theta = 0$, and the power law index k is taken to be the same for both the energy and Lorentz factor for simplicity. Geometric constraints impose the limit $\theta_j \le \pi/2$, and θ_c is chosen to be much smaller than any of

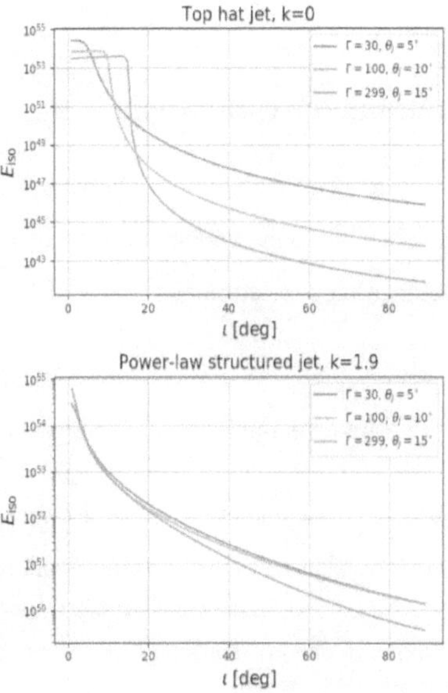

Figure 5-2: Isotropic equivalent energy as a function of inclination angle for the top-hat (top) and universal structured jet model with k=1.9 (bottom) for a range of opening angles and Lorentz factors for a jet with total energy $E_0^\gamma = 10^{52}$ erg.

the other angles of interest, with a lower limit of $\theta_c > 1/\Gamma_{\max}$. In theory, the universal structured jet model should restrict k to $1.5 \lesssim k \leq 2$ in order to recover the properties of the uniform jet model and to guarantee a standard energy reservoir [968, 766], but recent attempts to fit k from data have given much larger values (e.g., $k \sim 8$, [698]). For a given Γ_0 and θ_j, k is constrained so that $\Gamma(\theta_j) \geq 1$:

$$k_{\max} = \frac{\log \Gamma_0}{\log(\theta_j/\theta_c)}. \tag{5.12}$$

The isotropic equivalent energy as a function of the inclination angle is shown in Fig 5-2. The energy drops off quickly once the inclination angle exceeds the opening angle of the jet, and higher Lorentz factors beam the emission more efficiently, making detection more difficult for off-axis observers. The shape of the profile is nearly indistinguishable at high inclination angles.

5.4   Methods

5.4.1   Bayesian parameter estimation

We aim to measure the posterior probability distributions for the opening angle, θ_j, power law index, k, total source-frame gamma-ray energy E_0^γ, and Lorentz factor Γ_0 through the joint observation of electromagnetic and GW data. We define three relevant parameter sets:

- $\mathbf{x} = \{E_0^\gamma, \Gamma_0, \theta_j, k\}$ – Parameters unique to the electromagnetic data

- $\boldsymbol{\eta} = \{\iota, d_L, z\}$ – Parameters common to both the electromagnetic and gravitational-wave data: orbital inclination, luminosity distance and redshift [3]

- $\boldsymbol{\lambda}$ – Parameters unique to the gravitational-wave data such as the masses and spins of the neutron stars

We then define two likelihood functions in terms of these parameters:

$$\mathcal{L}^{\mathrm{GW}}(h|\boldsymbol{\eta}, \boldsymbol{\lambda}) \propto \exp\left(-\sum_k \frac{2\left(h(f_k) - \widehat{h}(f_k|\boldsymbol{\eta}, \boldsymbol{\lambda})\right)^2}{TS_h(f_k)^2}\right) \tag{5.13}$$

$$\mathcal{L}^{\mathrm{EM}}(F^\gamma|\mathbf{x}, \boldsymbol{\eta}) \propto \exp\left(-\frac{\left(F^\gamma - \widehat{F^\gamma}(\mathbf{x}, \boldsymbol{\eta})\right)^2}{2\sigma_{F^\gamma}^2}\right). \tag{5.14}$$

The first likelihood, $\mathcal{L}^{\mathrm{GW}}(h|\boldsymbol{\eta}, \boldsymbol{\lambda})$, is the gravitational-wave likelihood function for the strain data $h(f_k)$ in each frequency bin f_k for an analysis segment of duration T given the waveform model $\widehat{h}(f_k|\boldsymbol{\eta}, \boldsymbol{\lambda})$ [906]. The second likelihood, $\mathcal{L}^{\mathrm{EM}}(F^\gamma|\mathbf{x}, \boldsymbol{\eta})$, is the electromagnetic likelihood function for the fluence data, which depends on the purely electromagnetic parameters and some of the binary parameters given by the subset $\boldsymbol{\eta}$. We assume that the fluence is measured with an uncertainty of $\sigma_{F^\gamma} = 0.3 \times 10^{-7}\,[\mathrm{erg/cm^2}]$, which is the average reported fluence uncertainty for

---

[3] The gravitational-wave data only provides the luminosity distance, from which the redshift can be obtained if the cosmology is known.

GRB 170817A [417, 793]. We assume the gravitational-wave noise is Gaussian, where $S_h(f_k)$ is the noise power spectral density at Advanced LIGO design sensitivity [3].

Combining the two likelihoods, we obtain a posterior for the EM parameters $\mathbf{x}$:

$$p(\mathbf{x}|h, F^\gamma) = \frac{\pi(\mathbf{x})}{\mathcal{Z}_\mathbf{x}} \int \left[d\boldsymbol{\eta}\, \mathcal{L}^{\mathrm{EM}}(F^\gamma|\mathbf{x}, \boldsymbol{\eta})\pi(\boldsymbol{\eta}) \right. \tag{5.15}$$

$$\left. \times \left(\int \mathcal{L}^{\mathrm{GW}}(h|\boldsymbol{\eta}, \boldsymbol{\lambda})\pi(\boldsymbol{\lambda})d\boldsymbol{\lambda} \right) \right]$$

$$\equiv \frac{1}{\mathcal{Z}_\mathbf{x}}\mathcal{L}^{\mathrm{GW+EM}}(h, F^\gamma|\mathbf{x})\pi(\mathbf{x}) \tag{5.16}$$

where $\pi(\mathbf{x})$, $\pi(\boldsymbol{\eta})$, and $\pi(\boldsymbol{\lambda})$ are the priors for each set of parameters defined above. In the first step we marginalize separately over the gravitational-wave nuisance parameters $\boldsymbol{\lambda}$ [4] and the common parameters $\boldsymbol{\eta}$, and in the second step we define the joint GW+EM likelihood:

$$\mathcal{L}^{\mathrm{GW+EM}}(h, F^\gamma|\mathbf{x}) = \int \mathcal{L}^{\mathrm{GW}}(h|\boldsymbol{\eta})\mathcal{L}^{\mathrm{EM}}(F^\gamma|\mathbf{x}, \boldsymbol{\eta})\pi(\boldsymbol{\eta})d\boldsymbol{\eta}. \tag{5.17}$$

$\mathcal{Z}_\mathbf{x}$ is the Bayesian evidence obtained by marginalizing the joint likelihood over the GRB parameters $\mathbf{x}$:

$$\mathcal{Z}_\mathbf{x} = \int \mathcal{L}^{\mathrm{GW+EM}}(h, F^\gamma|\mathbf{x})\pi(\mathbf{x})d\mathbf{x}. \tag{5.18}$$

5.4.2 Simulated coincident event population

We simulate 100 binary neutron star gravitational-wave events. The masses are drawn uniformly in chirp mass:

$$\mathcal{M} = \frac{(m_1 m_2)^{3/5}}{(m_1 + m_2)^{1/5}}, \tag{5.19}$$

---

[4]We stress that we use the same symbol for the marginalized likelihood, just removing the marginalized parameters from the list of parameters that the likelihood depends on. So for example the marginalization over the gravitational-wave parameters $\boldsymbol{\lambda}$ follows from $\mathcal{L}^{\mathrm{GW}}(h|\boldsymbol{\eta}) \equiv \int \mathcal{L}^{\mathrm{GW}}(h|\boldsymbol{\eta}, \boldsymbol{\lambda})\pi(\boldsymbol{\lambda})d\boldsymbol{\lambda}$.

between 0.888 and 1.63 $M_\odot$ and in mass ratio:

$$q = m_2/m_1, \ m_1 \geq m_2, \tag{5.20}$$

between 0.7 and 1. These ranges are chosen to be consistent with the domain of validity of the reduced order quadrature model [820] for the IMRPhenomPv2 [445] waveform, which we employ to keep the computational cost under control.

The events are distributed uniformly in comoving volume between 10 and 80 Mpc and added into a Hanford-Livingston detector network using simulated design sensitivity Gaussian noise. While this network will be sensitive to BNS mergers at larger distances out to $\sim$ 200 Mpc [22], we choose to limit the maximum event distance for this analysis so that all the simulated events are detectable. We comment on the consequences of this setup in Sec. 5.6. At these small distances, the redshift is calculated from the luminosity distance posterior as $z = d_L/D_H$, where $D_H = 9.26 \times 10^{27}/h_0$ cm is the Hubble distance, and we take $h_0 = 0.68$. The inclination angle and sky position are distributed isotropically. The priors used when running the sampler, $\pi(\boldsymbol{\lambda})$ and $\pi(\boldsymbol{\eta})$, are identical to those from which the event parameters are drawn. The neutron stars are assumed to be non-spinning point masses with no tidal deformability, which does not have a significant effect on the inference of the common GW+EM parameters, $\boldsymbol{\eta}$.

Each of the 100 simulated gravitational-wave sources is randomly associated with a GRB event. We simulate two GRB populations, each with 100 events – one with only top-hat jets and one with power-law jets with $k = 1.9$. These fiducial models are chosen to demonstrate our method, and we leave consideration of other structures like the Gaussian jet to future work. In both cases, the energy is drawn from a truncated log-normal distribution in $\log_{10} E_0$ between 10^{47} and 10^{54} erg with a mean of 10^{50} erg and a width of one dex. For the top-hat population, the opening angle θ_j is drawn from a truncated Gaussian between $2°$ and $50°$ with a mean of $25°$ and a width of $5°$, and the Lorentz factor is also drawn from a truncated Gaussian with $\mu_\Gamma = 100$, $\sigma_\Gamma = 50$, $2 \leq \Gamma \leq 299$. For the power-law population, the distributions of

Parameter	μ	σ	min	max
$\log E_0$	50	1	47	54
Γ	100	50	2	299
θ_j	25°	5°	2°	50°
μ_k	0	0	0	0

Table 5.1: Parameters describing the distributions used for simulating the population of top-hat jets.

Parameter	μ	σ	min	max
$\log E_0$	50	1	47	54
Γ	270	20	2	299
θ_j	7°	4°	2°	50°
μ_k	1.9	0	0	0

Table 5.2: Parameters describing the distributions used for simulating the population of power-law jets.

Γ and θ_j are chosen to preserve the constraint imposed by Eq. 5.12 for $k = 1.9$. Both the Lorentz factor and opening angle are drawn from truncated Gaussians with the same boundaries described above and $\mu_\Gamma = 270$, $\sigma_\Gamma = 20$, $\mu_{\theta_j} = 7°$, $\sigma_{\theta_j} = 4°$. The distributions used to simulate the top-hat and power-law populations are summarized in Tables 5.1 and 5.2, respectively. In all cases the parameter boundaries are chosen to be consistent with theory and the observed population of short GRBs.

The fluence for each joint GW+EM event is calculated by evaluating the expression in Eq. 5.6 at the simulated event parameters. Since the integral is costly to evaluate analytically, we use a lookup table to calculate it efficiently, see Appendix 5.7.1.

To simulate the GRB detector noise, the "measured" value of the fluence for each GW+EM event is drawn from a Gaussian distribution centered on the "true" fluence value calculated as described above for each event's simulated parameters with the width given by $\sigma_{F\gamma}$. The distribution of "true" fluences is shown in Fig. 5-3. This means that some events with sub-threshold "true" fluence values will end up with negative values for the "measured" fluence, which corresponds to a dearth of counts after background subtraction. While the gamma-ray photons arriving at the GRB detector are actually Poisson-distributed, the Gaussian approximation we make in simulating the detector noise and in the likelihood in Eq. 5.14 is valid in the limit of large numbers of counts. Because we run our analysis only on detectable BNS

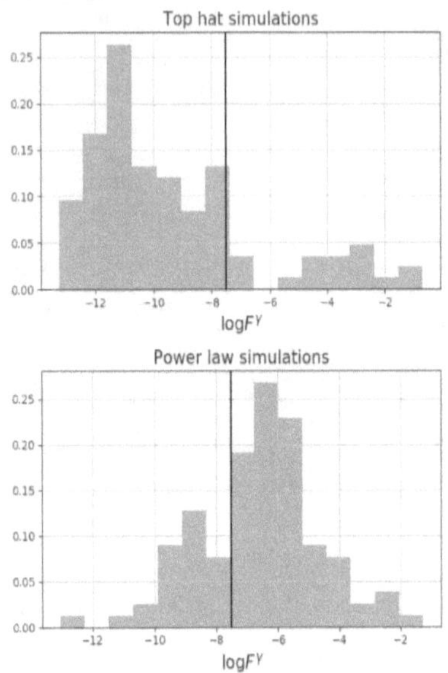

Figure 5-3: Distributions of $\log F^\gamma$ for the top-hat (left) and power-law (right) simulated populations. The vertical line indicates the value of σ_{F^γ}, which serves as a threshold for the detection of the sGRB emission. There are 21 detectable top-hat events and 73 detectable power-law events out of the 100 events in each population.

gravitational-wave events but include non-detections of associated short GRBs when the "measured" fluence is sub-threshold, this corresponds to a GW-triggered search including upper limits on fluence obtained by GRB satellites. This does not include GW events for which the sky region is outside the GRB satellite's field of view or cases where the GRB satellites are not in observing mode at the time of the GW trigger.

For each event, we obtain posteriors for the GW parameters (η, λ) and the EM-only parameters $\mathbf{x}$ using the bilby parameter estimation package [96] and the dynesty nested sampler [828]. Because we marginalize over the uncertainty in the GW parameters following the prescription detailed in Appendix 5.7.2, the posteriors for the EM-only parameters include the effects of the correlation between distance and inclination in the gravitational-wave posteriors and the uncertainty in these parameters.

We use uniform priors, $\pi(\mathbf{x})$, for all parameters in the ranges covered by the simulated event distributions (see Tables 5.1 and 5.2) except for k, which has a conditional prior that is uniform between 0 and k_{max} as defined in Eq. 5.12 for each prior sample in Γ and θ_j.

5.4.3   Hierarchical modeling

While each GRB will have a different value of the energy, Lorentz factor, and opening angle, we can use the population of events to measure the properties of the underlying distributions of which the individual parameters are a representative sample. This is usually referred to as hierarchical modeling. We assume that the underlying distribution for the parameters $\mathbf{x}$ can be characterized by a set of hyper-parameters $\boldsymbol{\Lambda} = \{\mu_{E_0^\gamma}, \sigma_{E_0^\gamma}, \mu_{\Gamma_0}, \sigma_{\Gamma_0}, \mu_{\theta_j}, \sigma_{\theta_j}, \mu_k, \sigma_k\}$, i.e. we assume individual GRB sources have parameters drawn from truncated Gaussian distributions with unknown means and standard deviations. This underlying distribution is called the hyper-prior, $\pi(\mathbf{x}|\boldsymbol{\Lambda})$. We stress that while we have fixed the true value of k to be the same for all simulated GRB events ($k = 0$ for the top-hat model and $k = 1.9$ for the power-law model) we still measure the hyper-parameters associated with k in order to determine whether a universal angular emission profile can be inferred from the data. For individual events, the joint likelihood in Eq. 5.17 depends on the hyper-parameters only implicitly through the distributions of the individual-event parameters $\mathbf{x}$. The likelihood for the hyper-parameters is thus obtained by marginalizing the joint likelihood for the jth event over the EM-only parameters:

$$\mathcal{L}(h_j, F_j^\gamma | \boldsymbol{\Lambda}) = \int d\mathbf{x}\, \mathcal{L}^{\mathrm{GW+EM}}(h_j, F_j^\gamma | \mathbf{x}, \boldsymbol{\Lambda}) \pi(\mathbf{x}|\boldsymbol{\Lambda}). \tag{5.21}$$

Hierarchical modeling must typically take selection biases into account due to the fact that only detected events, which have different properties than the population as a whole, are included in the analysis sample, thus affecting the inference of the hyper-parameters [566, 7, 357, 946, 638, 585]. However, we do not introduce a cut based on the detectability of either the gravitational-wave or electromagnetic data

Parameter	Shape	min	max
$\mu_{\log E_0}$	Uniform	47	54
$\sigma_{\log E_0}$	Uniform	0.1	5
μ_Γ	Uniform	2	299
σ_Γ	Uniform	5	200
μ_{θ_j}	Uniform	2°	50°
σ_{θ_j}	Uniform	1°	15°
μ_k	Uniform	0	8
σ_k	Log-Uniform	10^{-4}	1

Table 5.3: Priors on the hyper-parameters used in the hierarchical modeling step, $\pi(\boldsymbol{\Lambda})$.

so that we don't need to account for selection biases in our analysis. Even if a cut were introduced on the gravitational-wave data, this would not affect the inference of the hyper-parameters $\boldsymbol{\Lambda}$ describing the electromagnetic parameters $\mathbf{x}$, as explained in detail in Appendix 5.7.2. Therefore, the posterior on the hyper-parameters for an ensemble of N events is obtained by multiplying the individual event likelihoods without any modifications to account for the probability of detection:

$$p(\boldsymbol{\Lambda}|\{h\},\{F^\gamma\}) = \frac{\pi(\boldsymbol{\Lambda})}{\mathcal{Z}_{\boldsymbol{\Lambda}}} \prod_j^N \mathcal{L}(h_j, F_j^\gamma|\boldsymbol{\Lambda}), \qquad (5.22)$$

where $\pi(\boldsymbol{\Lambda})$ is the prior on the hyper-parameters given in Table 5.3, $\mathcal{L}(h_j, F_j^\gamma|\boldsymbol{\Lambda})$ is the joint EM-GW likelihood marginalized over the hyper-prior for the j-th event from Eq. 5.21, and $\mathcal{Z}_{\boldsymbol{\Lambda}}$ is the hyper-evidence (see Appendix 5.7.2). We produce samples from this distribution using the `bilby` implementation of the `pymultinest` [344, 345, 346, 182] and `cpnest` samplers [909].

5.5 Results

5.5.1 Individual event analysis

The morphology of the individual event posteriors for the EM-only parameters $\mathbf{x}$ varies depending on the SNR of the GRB signal. The corner plot for an uninformative (sub-threshold) power-law event is shown in Fig. 5-4. The posterior for k, which is highly peaked around $k = 0$, essentially returns the prior for this uninformative event. Additionally, there is more support for higher values of k for narrower opening angles and higher Lorentz factors, since the k_{max} condition is more easily satisfied in that part of the parameter space. The energy posterior favors lower values since a lower energy results in a lower fluence, and the posterior for the Lorentz factor slightly favors higher values because this causes the fluence to drop off more steeply for inclination angles outside the jet edge (see Fig. 5-2). The posteriors for uninformative top-hat events show similar trends.

For informative events, the best-constrained parameter is E_0^γ, since it decouples from the integral expression encoding the dependence of the fluence on the other three parameters in Eq. 5.6. The corner plot for a relatively informative power-law event is shown in Fig. 5-5. Because we are trying to constrain four parameters with only one piece of data (the fluence), there are degenerate regions of parameter space that can produce the same fluence value. A wider opening angle but a steeper drop-off of the fluence due to a higher Lorentz factor could yield the same fluence value as a narrower opening angle with a more gradual drop-off for a particular inclination. A higher value of k also causes the fluence to drop off more quickly, which could be compensated for by increasing the energy of the event. Because of these parameter-space degeneracies, we only observe very weak deviations from the prior in the posteriors for Γ and θ_j. The posteriors for informative top-hat events look very similar to the power-law posteriors, since the prior for k very strongly disfavors values of $k \gtrsim 2$, so we do not show an example corner plot here.

5.5.2 Population analysis

While individual event posteriors are not very informative for the parameters encoding the jet geometry even for events with a high GRB SNR, we can use a population of events to place constraints on the hyper-parameters.

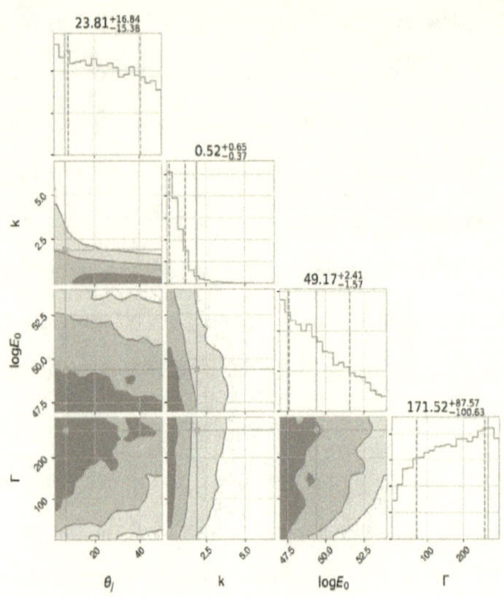

Figure 5-4: Corner plot for an uninformative power-law event with true fluence $F^\gamma = 2.88 \times 10^{-9}$ erg/cm^2. The orange lines represent the true parameter values, while the blue dashed lines are the 1σ uncertainties. These are also indicated above each marginalized posterior along with the median value for each parameter.

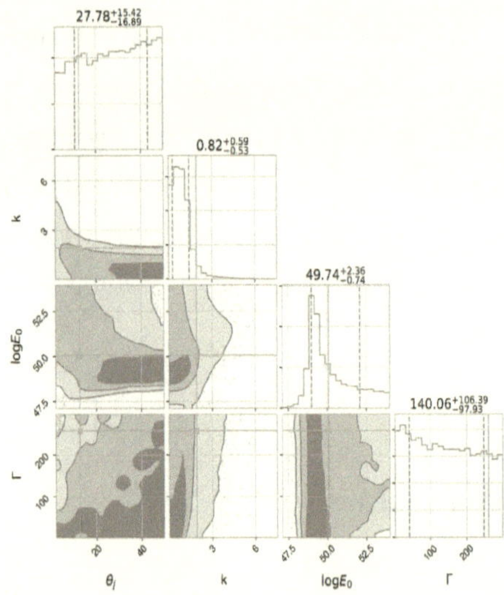

Figure 5-5: Corner plot for an informative power-law event with true fluence $F^\gamma = 1.46 \times 10^{-4}$ erg/cm^2.

Simulated top-hat population

Fig. 5-6 shows the 1, 2, and 3σ confidence intervals for the 8 hyper-parameters in our model for the top-hat population as a function of the number of BNS events accompanied by a GRB fluence measurement or upper limit included in the analysis. The energy hyper-parameters are well constrained to within ~ 2 dex in $\log E_0$ at the 1σ level with relatively few events, which is consistent with the energy being the most informative parameter in the individual event analysis presented earlier. The true values of both $\mu_{\log E_0}$ and $\sigma_{\log E_0}$ are contained within the 2σ confidence interval for all 100 events in the population. The hyper-parameter μ_{θ_j} is also well constrained, with the 1σ region spanning about $10°$, even though relatively little information can be gained about the opening angle of individual GRBs from the first step of PE.

The posterior for the σ_{θ_j} parameter is less informative, due in part to the fact that the prior range is narrower. The μ_Γ posterior is slightly offset towards higher values of Γ because of the shape of the Γ posterior for the uninformative individual events, which dominate the population. As described in the previous section, higher Lorentz factors lead to lower fluence values for observers outside the jet edge. The true value of μ_Γ is contained within the 3σ confidence interval, however. We note that the posterior for μ_Γ is also strongly dependent on the particular realization of the hyper-prior for the 100 true values that were chosen for our simulated population. We repeated the analysis with 100 different draws from the hyper-prior, and the offset in this parameter disappeared. This adds support to the idea that this posterior could converge to the true value with a larger population of events. The posterior for σ_Γ is more informative because deviations from the uninformative posterior in Fig. 5-4 indicate the spread of the true Γ values. The true values of both μ_k and σ_k are included in the 1σ confidence intervals for these parameters.

We also calculate the posterior predictive distributions (PPDs) for the parameters $\mathbf{x}$ from their hyper-parameters,

$$p_\Lambda(\mathbf{x}|\{d\}, \{F^\gamma\}) = \int p(\Lambda|\{d\}, \{F^\gamma\})\pi(\mathbf{x}|\Lambda)d\Lambda, \tag{5.23}$$

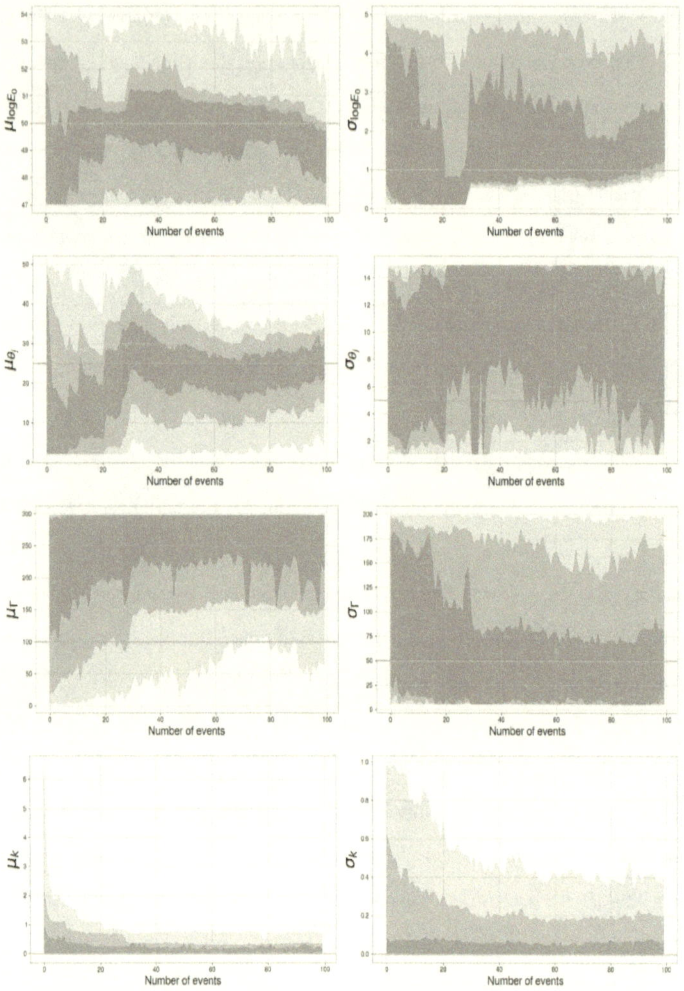

Figure 5-6: 1, 2, and 3σ intervals for all hyper-parameters for the top-hat population, with the true value shown in orange.

which can be written for a discrete set of m hyper-parameter posterior samples as:

$$p_{\boldsymbol{\Lambda}}(\mathbf{x}|\{d\}, \{F^\gamma\}) = \frac{1}{m}\sum_i^m \pi(\mathbf{x}|\boldsymbol{\Lambda}_i). \tag{5.24}$$

The posterior predictive distribution represents the updated prior on $\mathbf{x}$ after incorporating the information gained from the data via the posteriors on the hyper-parameters $\boldsymbol{\Lambda}$ [25]. The PPDs for the $\mathbf{x}$ parameters are shown in Fig. 5-7 using the hyper-parameter posteriors inferred from all 100 events in our simulated population along with the 50% and 90% confidence regions and the true distributions used for simulating the events. As expected from the hyper-parameter posteriors presented in Fig. 5-6, the distribution for θ_j is the best recovered, and the distribution for Γ peaks above the true value. The distribution for k peaks away from 0 but is consistent with the true value within error, and the width of the PPD can be attributed to the sampling error encompassed in non-zero values of σ_k. The PPD for $\log E_0$ is slightly wider with a lower peak than the true value, consistent with the hyper-parameter posteriors for $\mu_{\log E_0}$ and $\sigma_{\log E_0}$.

Simulated power-law population

Fig. 5-8 shows the 1, 2, and 3σ regions for all 8 hyper-parameters for the power-law population. The mean of the energy distribution is wider than for the top-hat simulations, while the width is constrained at a similar level. The 1σ region for the μ_{θ_j} posterior is constrained to $< 10°$ and includes the true value, and the σ_{θ_j} posterior is again less informative. σ_Γ is constrained to within ~ 70 at the 1σ level, which is slightly narrower than in the top-hat case even though the μ_Γ posterior spans nearly the entire prior range. Because the true Γ distribution is narrower for the power-law population, there is less deviation in the shape of the individual-event Γ posteriors, which are not very informative to begin with. This leads to more uncertainty in the peak of the distribution but a better measurement of the spread. The μ_k posterior does not peak at the true value because higher values of k are strongly disfavored by the prior in the individual-event PE, but the true value is included in the 3σ

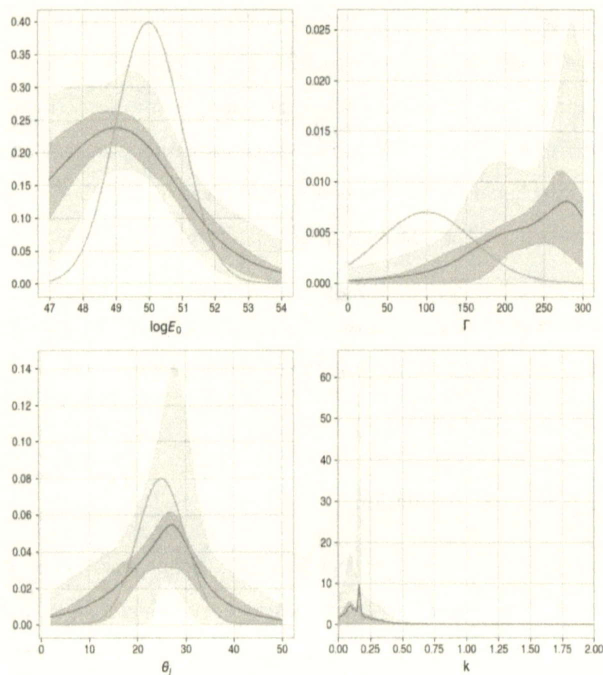

Figure 5-7: Posterior predictive distributions for the EM-only parameters for the top-hat population (dark blue lines) along with the 50% and 90% confidence intervals and the true distributions (orange lines).

region. When compared to the μ_k posterior for the top-hat population, this is clear evidence for a structured jet, even though the exact value of the power-law index is not recovered accurately. The σ_k posterior is again consistent with 0, as expected for a delta function distribution.

The PPDs for the power-law population are shown in Fig. 5-7. The distribution for θ_j is again the best recovered. Even though σ_Γ and $\sigma_{\log E_0}$ are well constrained, the PPDs for Γ and $\log E_0$ are very broad because of the large uncertainty in μ_Γ and $\mu_{\log E_0}$. The distribution for k does not peak at the true value, but the top-hat model with $k = 0$ is excluded at 90% confidence for this population, again clearly indicating the presence of jet structure for these simulations.

5.6　Discussion

In this chapter, we have developed a new method for determining the energy, Lorentz factor, opening angle, and power-law index for individual short GRBs as well as the distributions of these parameters for a given population of detected binary neutron star gravitational-wave events with an associated gamma-ray burst observation or fluence upper limit. Our method is completely independent of afterglow observations and uses Bayesian inference to combine the information provided by gravitational-wave parameter estimation on the inclination angle and distance to the source with the fluence measured by GRB satellites. We have simulated two populations of short GRBs– one with top-hat jet geometry, and another with a power-law structured jet geometry with $k = 1.9$. For individual events, little information is obtained for the jet geometry parameters θ_j, Γ, and k because of the degeneracy of the parameter space, but the $\log E_0$ of the jet can be constrained with an uncertainty of ~ 2.5 dex. The hyper-parameters that describe the population as a whole are better constrained by combining the information from all 100 events in our simulation. For both jet structures, the peak of the opening angle distribution can be measured to within $10°$. The peak of the energy distribution is also relatively well-reconstructed with an uncertainty of ~ 2 dex for the top-hat population. The Γ distribution is the most

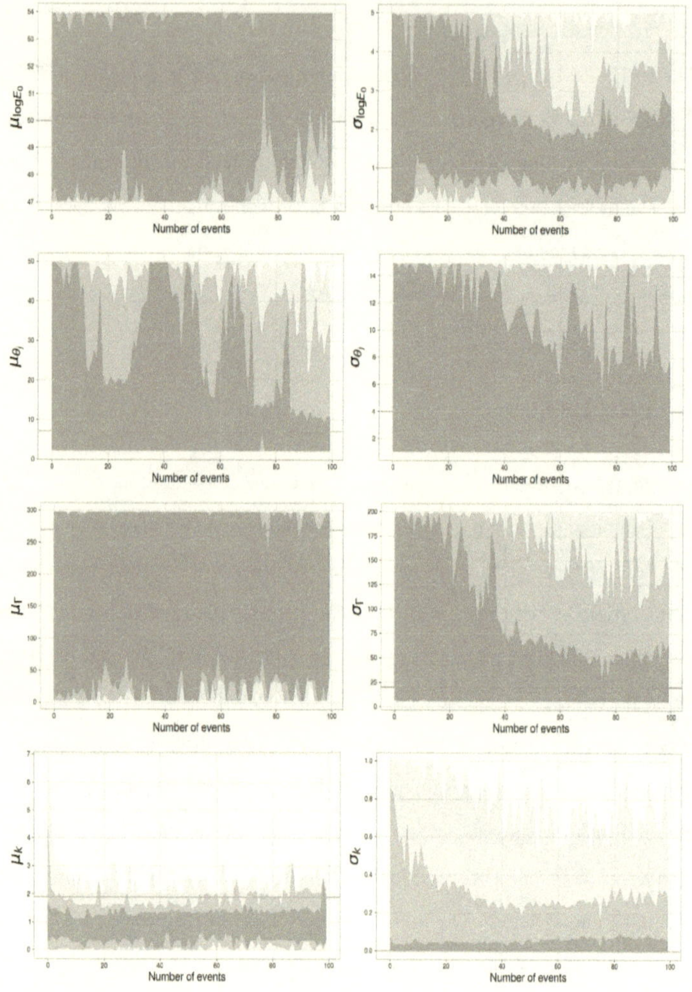

Figure 5-8: 1, 2, and 3σ intervals for all hyper-parameters for the power-law population, with the true value shown in orange.

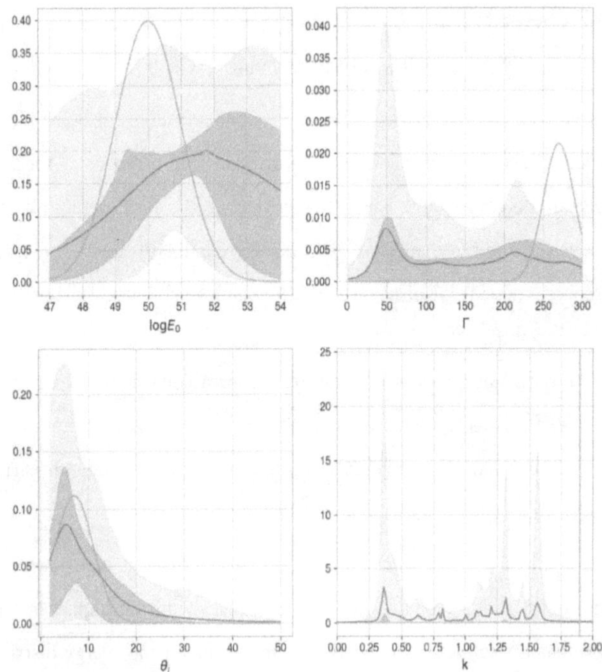

Figure 5-9: Posterior predictive distributions for the EM-only parameters for the power-law population set (dark blue lines) along with the 50% and 90% confidence intervals and the true distributions (orange lines).

difficult to constrain, since informative posteriors on this parameter in the individual event analysis depend on the observer looking right at the jet edge. Because the top-hat model is a subset of the power-law model as we have defined it with $k = 0$, the hyper-parameter posterior for μ_k allows for us to distinguish between the two jet structures. The μ_k distribution does not accurately recover the true value for the power-law jet simulation because values of $k \gtrsim 1$ are strongly disfavored by the prior in the individual-event PE, but it does provide clear evidence for the universal structured jet model, ruling out a power law index of $k = 0$ at 90% confidence.

The method we have developed offers the unique advantage of being independent of observations of the multi-wavelength GRB afterglow that traditional measurements of GRB jet parameters rely on. Even though the current alert system allows for the rapid follow-up of gravitational-wave BNS merger candidates by X-ray, optical, and radio telescopes, the sky localization is often prohibitively large for telescopes with small fields of view [5, 22]. Our method will provide increased statistics since the GRB satellite data can be searched a-posteriori for a coincident detection or upper limit. This also holds as the sensitivity of gravitational-wave detectors improves and the number of BNS candidate events increases, since electromagnetic partners will have to prioritize which events to follow up.

Measurements of the energy, opening angle, Lorentz factor, and power-law index distributions of the short GRB population will have a significant impact on the theory of how these jets are launched and evolve. Constraints on the energy distribution to the level that we have demonstrated will enable distinguishing if the range of isotropic equivalent energies and luminosities of observed sGRBs is due to differences in the intrinsic brightness of the jet or if it is an artifact of observing the emission at different inclination angles. Measurements of the opening angle distribution could be used to determine the efficiency of jet collimation and could reveal two distinct populations of jets. Successful jets that manage to drill through, and eventually break out of, the merger ejecta will have narrower opening angles and smaller power-law indexes. On the other hand, "choked" jets that cannot escape the merger ejecta will still inflate cocoons, but the resulting gamma-ray emission will appear to have a wider opening angle and a much lower Lorentz factor, which is an effect that we have shown can be measured. Together with the Lorentz factor distribution and power-law index, the opening angle distribution can also provide information on the density of circum-burst environment, since more interaction between the jet material and merger ejecta leads to slower and wider jets.

We conclude by considering some caveats to our analysis. The first is that neither of the two populations of GRBs that we have simulated matches the observed fluence distribution [147]. This is in part due to the fact that the observed distribution is convolved with the instrument selection function, which we have not considered here since we run our analysis even on undetectable GRBs for which only upper limits on the fluence would be available. For this analysis, we also ignore the cosmological "k-correction" that should be applied to account for the limited bandwidth of the GRB satellite and the effect of redshift on the observed photon frequency [169]. However, for the small distances we consider here, the effects of redshift are negligible and we argue that the method developed in Section 5.3 in terms of the bolometric fluence holds in this regime. Selection effects aside, the discrepancy in the fluence distributions indicates that the distributions we have chosen for the EM-only parameters do not correspond to the actual astrophysical distributions. Since the goal of our method is to measure these distributions, we had to make some assumptions for the purposes of our simulation. We chose to use Gaussian distributions since they are straightforward to parameterize, but our method could be extended to other distributions with different hyper-parameters.

We also only consider two angular profiles, the top-hat and power-law universal structured jet, both of which are simplified models that do not encompass the full details of the evolution of the jet emission. We emphasize that this analysis is a proof of principle and that our method could be extended to use models that parameterize more astrophysically-motivated scenarios like Gaussian structured jets [949] or the generic "boosted fireball" [944, 945] as these more realistic models are starting to become available. While currently our method requires the jet structure model to be chosen before running the analysis, it could be applied to multiple jet structures to conduct Bayesian model selection between them at the individual-event or population level. We leave this to future studies.

By only including BNS progenitors out to 80 Mpc in our simulations, we have avoided the problem of gravitational-wave selection effects since all sources will be detectable out to this distance once the detectors reach design sensitivity (and we argue that even if a detection threshold were applied on the gravitational-wave data, it would not affect the inference of our chosen electromagnetic hyper-parameters in Appendix 5.7.2). The SNR of the GW signal has very little impact on the shape of the individual event posteriors shown in Section 5.5.1, which are instead dominated by the SNR of the GRB signal. We therefore expect that the results obtained here would also hold for a population of 100 detectable BNS sources even out to farther distances. Assuming even the most optimistic BNS merger rate, it would take much longer than the proposed lifetime of second-generation gravitational-wave detectors to reach 100 detections of BNSs within 80 Mpc. However, we could reach 100 total BNS detections out to farther distances by the end of advanced LIGO's fifth observing run (O5) with a five-detector network operating at a BNS detection range of at least ~ 200 Mpc for a realistic merger rate [26, 22]. A fluence measurement or upper limit will not be available for every GRB associated with a detectable BNS, since the current GRB satellite network consisting of *Fermi*-GBM, *Swift*-BAT, *INTEGRAL*, and the *Interplanetary Network* (IPN) has an all-sky duty cycle of $\sim 65\%$ [61, 473]. Reaching the 100 gravitational-wave events with a GRB detection or fluence upper limit we have simulated here could be accomplished by the end of O5 if the detectors are running at the upgraded A+ sensitivity with a BNS detection range of 330 Mpc [4] for a more optimistic merger rate of ~ 2000 Gpc^{-3} yr^{-1} assuming one year of observation [26], and is definitely achievable at LIGO Voyager sensitivity–a proposed upgrade to the existing advanced LIGO facilities in the late 2020s–which has a projected BNS range out to 1100 Mpc [4, 59]. Because we use both joint detections and non-detections with fluence upper limits in our analysis to constrain the population hyper-parameters, the rates we quote here are more optimistic than the joint GW-GRB detection rates in other work [473]. The addition of next-generation GRB instruments like *THESEUS* [80, 840], *BurstCube* [730], and *HERMES* [389] coincident with the gravitational-wave detector upgrades to Voyager or even A+ sensitivity will greatly increase the all-sky duty cycle of the GRB satellite network and the joint detection rate. We note that based on the results in Figs. 5-6 and 5-8, the opening angle, power-law index, and energy distributions are well-constrained even with 40 coincident events, which is achievable with second-generation detectors for realistic merger rates.

5.7 Supplementary material

5.7.1 Evaluation of the fluence integral

The integral for the fluence in Eq. 5.6 is costly to evaluate, and must be computed a prohibitively large number of times while sampling. To contain the cost, we calculated it numerically using a Riemann sum with 1000 bins in θ between 0 and θ_j and 1000 bins in ϕ between 0 and 2π. The core angle θ_c is chosen to be 1° for power law jets with $k > 0$. Because even the numerical integration would be prohibitively time-consuming when evaluating the fluence for each prior sample, a lookup table is constructed by evaluating the integral on a 4-dimensional grid in Lorentz factor, opening angle, power-law index, and inclination angle. The grid spacing for each parameter is detailed in Table 5.4. Points on the grid that violate the condition set in Eq. 5.12 and thus represent unphysical parts of parameter space are left blank. When the fluence is evaluated for each prior sample, the value of the integral is then obtained via a nearest-neighbor interpolation using the lookup table. We have verified that discretizing the parameter space in this way does not impact the results.

5.7.2 Details of the Bayesian analysis implementation

Individual-event analysis

While Eq. 5.15 is valid for constructing the posterior for continuous parameters, we obtain a series of discrete posterior samples for each of gravitational-wave parameters,

Parameter	min	max	Δ
Γ	2	299	3
θ_j	$2°$	$50°$	$1°$
k	0	8	0.1
ι	$0°$	$90°$	$1°$

Table 5.4: Minimum and maximum values and grid spacing for the 4-dimensional grid constructed for interpolating the value of the integral in the fluence expression, Eq. 5.6.

so the value of the marginalized GW likelihood $\mathcal{L}^{\text{GW}}(h|\boldsymbol{\eta})$ is not known directly but can be extracted if the prior and the gravitational-wave evidence,

$$\mathcal{Z}_{\text{GW}} = \int \mathcal{L}^{\text{GW}}(h|\boldsymbol{\eta}, \boldsymbol{\lambda})\pi(\boldsymbol{\eta})\pi(\boldsymbol{\lambda}) \, d\boldsymbol{\eta} \, d\boldsymbol{\lambda}, \tag{5.25}$$

and priors, $\pi(\boldsymbol{\eta})$ and $\pi(\boldsymbol{\lambda})$, are known. $\mathcal{Z}_{\text{GW}}$ is calculated by the sampler in the gravitational-wave parameter estimation step described above, so the likelihood marginalized over the parameters unique to the gravitational-wave analysis, $\boldsymbol{\lambda}$, can be rewritten as:

$$\mathcal{L}^{\text{GW}}(h|\boldsymbol{\eta}) = \frac{\mathcal{Z}_{\text{GW}}p(\boldsymbol{\eta}|h)}{\pi(\boldsymbol{\eta})}. \tag{5.26}$$

The parameter vector $\boldsymbol{\eta}$ is not continuous, but rather a list of k n-tuples, where $n = 4$ for this analysis. The probability of each $\boldsymbol{\eta}_i$ is $p(\boldsymbol{\eta}_i) = 1/k$. If we substitute the likelihood from Eq. 5.26 into the joint likelihood function defined in Eq. 5.17, we obtain

$$\mathcal{L}^{\text{EM+GW}}(h, F^\gamma|\mathbf{x}) = \mathcal{Z}_{\text{GW}} \int p(\boldsymbol{\eta}|h)\mathcal{L}^{\text{EM}}(F^\gamma|\mathbf{x}, \boldsymbol{\eta}) \, d\boldsymbol{\eta}, \tag{5.27}$$

which is just the expectation value of the EM likelihood

$$\mathcal{L}^{\text{EM+GW}}(h, F^\gamma|\mathbf{x}) = \mathcal{Z}_{\text{GW}}\langle\mathcal{L}^{\text{EM}}(F^\gamma|\mathbf{x}, \boldsymbol{\eta})\rangle, \tag{5.28}$$

since $p(\boldsymbol{\eta}|h)$ is a normalized probability distribution. For a discrete set of posterior samples, this expression becomes

$$\mathcal{L}^{\mathrm{EM+GW}}(h, F^\gamma|\mathbf{x}) = \frac{\mathcal{Z}_{\mathrm{GW}}}{k} \sum_{i=1}^{k} \mathcal{L}^{\mathrm{EM}}(F^\gamma|\mathbf{x}, \boldsymbol{\eta}_i), \tag{5.29}$$

so the posteriors for the EM-only parameters $\mathbf{x}$ are obtained by "recycling" the posteriors on the common parameters obtained from the gravitational-wave parameter estimation step [25, 874]:

$$p(\mathbf{x}|h, F^\gamma) = \frac{\pi(\mathbf{x})}{\mathcal{Z}_\mathbf{x}} \frac{\mathcal{Z}_{\mathrm{GW}}}{k} \sum_{i=1}^{k} \mathcal{L}^{\mathrm{EM}}(F^\gamma|\mathbf{x}, \boldsymbol{\eta}_i). \tag{5.30}$$

Hierarchical Modeling

The likelihood for the hyper-parameters defined in Eq. 5.21 can be recast in terms of the posterior on the EM-only parameters $\mathbf{x}$ that we have already obtained in the previous sampling step:

$$\mathcal{L}(h, F^\gamma|\boldsymbol{\Lambda}) = \int d\mathbf{x}\, \mathcal{L}^{\mathrm{GW+EM}}(h, F^\gamma|\mathbf{x}, \boldsymbol{\Lambda})\pi(\mathbf{x}|\boldsymbol{\Lambda}) \tag{5.31}$$

$$= \int d\mathbf{x} \frac{p(\mathbf{x}|h, F^\gamma)\mathcal{Z}_\mathbf{x}}{\pi_0(\mathbf{x})}\pi(\mathbf{x}|\boldsymbol{\Lambda}), \tag{5.32}$$

where $\mathcal{Z}_\mathbf{x}$ is the EM evidence defined in Eq. 5.18 and $\pi_0(\mathbf{x})$ is the prior used in sampling the EM-only parameters first presented in Eq. 5.15. The likelihood for the hyper-parameters is the expectation value of the ratio of the hyper-prior to the original prior because the posterior $p(\mathbf{x}|h, F^\gamma)$ is a normalized probability distribution function,

$$\mathcal{L}(h, F^\gamma|\boldsymbol{\Lambda}) = \mathcal{Z}_\mathbf{x} \left\langle \frac{\pi(\mathbf{x}|\boldsymbol{\Lambda})}{\pi_0(\mathbf{x})} \right\rangle, \tag{5.33}$$

which can be written as a sum over the posterior samples obtained for the EM-only parameters $\mathbf{x}_i$ in the second sampling step described above for an individual event:

$$\mathcal{L}(h, F^\gamma | \Lambda) = \frac{\mathcal{Z}_\mathbf{x}}{n} \sum_i \frac{\pi(\mathbf{x}_i | \Lambda)}{\pi_0(\mathbf{x}_i)}. \tag{5.34}$$

The hyper-parameter posterior for a population of N events defined in Eq. 5.22 can then be written in terms of the sum over samples as:

$$p(\Lambda | \{h\}, \{F^\gamma\}) = \frac{\pi(\Lambda)}{\mathcal{Z}_\Lambda} \prod_j^N \mathcal{L}(h_j, F_j^\gamma | \Lambda) \tag{5.35}$$

$$= \frac{\pi(\Lambda)}{\mathcal{Z}_\Lambda} \prod_j^N \frac{\mathcal{Z}_{\mathbf{x}_j}}{n_j} \sum_i^{n_j} \frac{\mathcal{N}(\mathbf{x}_{ij}, \Lambda)}{\pi_0(\mathbf{x}_{ij})}, \tag{5.36}$$

where we have substituted the truncated multivariate Gaussian $\mathcal{N}(\mathbf{x}_{ij}, \Lambda)$ for the hyper-prior $\pi(\mathbf{x}_i | \Lambda)$, and the hyper evidence, $\mathcal{Z}_\Lambda$ is given by marginalizing the likelihood in Eq. 5.21:

$$\mathcal{Z}_\Lambda = \int d\Lambda \; \pi(\Lambda) \prod_j^N \mathcal{L}(h_j, F_j^\gamma | \Lambda). \tag{5.37}$$

Selection Effects

Below we demonstrate that as long as a detection threshold based on the GRB parameters is never introduced, selection effects do not enter the method we've developed. Even if a detection threshold is imposed based on the gravitational wave parameters, selection biases do not affect our analysis because the population properties we seek to characterize are independent of the GW parameters. We follow the arguments presented in Appendix E1 of [874]. If we impose an arbitrary detection threshold on the gravitational-wave matched filter SNR, $\rho_\mathrm{mf} > \rho_\mathrm{min}$, where

$$\rho_\mathrm{mf} = \frac{\langle h, \hat{h}(\boldsymbol{\eta}, \boldsymbol{\lambda}) \rangle}{\sqrt{\langle \hat{h}(\boldsymbol{\eta}, \boldsymbol{\lambda}), \hat{h}(\boldsymbol{\eta}, \boldsymbol{\lambda}) \rangle}} \tag{5.38}$$

for the inner product defined as

$$\langle a, b\rangle = \frac{4}{T}\sum_k \Re\left(\frac{a^*(f_k)b^*(f_k)}{S_h(f_k)}\right), \tag{5.39}$$

then the likelihood in Eq. 5.26 needs to be modified so that it remains properly normalized with respect to the data, h:

$$\mathcal{L}^{\mathrm{GW}}(h|\boldsymbol{\eta}, \boldsymbol{\lambda}, \det) = \begin{cases} \frac{1}{p_{\det}(\boldsymbol{\eta}, \boldsymbol{\lambda})}\mathcal{L}^{\mathrm{GW}}(h|\boldsymbol{\eta}, \boldsymbol{\lambda}) & \rho_{\mathrm{mf}} \geq \rho_{\min} \\ 0 & \rho_{\mathrm{mf}} < \rho_{\min} \end{cases}. \tag{5.40}$$

The detection probability, $p_{\det}(\boldsymbol{\eta}, \boldsymbol{\lambda})$ is defined as:

$$p_{\det}(\boldsymbol{\eta}, \boldsymbol{\lambda}) = \int_{\rho_{\mathrm{mf}}>\rho_{\min}} dh\ \mathcal{L}^{\mathrm{GW}}(h|\boldsymbol{\eta}, \boldsymbol{\lambda}). \tag{5.41}$$

The prior on the GW parameters needs to be similarly modified to account for the preference of detecting sources in certain parts of the sky with certain masses, etc. which can be quantified *a priori* using simulations:

$$\pi(\boldsymbol{\eta}, \boldsymbol{\lambda}|\det) = \frac{\pi(\boldsymbol{\eta})\pi(\boldsymbol{\lambda})p_{\det}(\boldsymbol{\eta}, \boldsymbol{\lambda})}{\int \pi(\boldsymbol{\eta})\pi(\boldsymbol{\lambda})p_{\det}(\boldsymbol{\eta}, \boldsymbol{\lambda})d\eta d\lambda}. \tag{5.42}$$

The denominator is a constant normalization factor, which is the same for all coincident events in our population as long as the same GW prior is used. For a detected event, the joint likelihood in Eq. 5.17 is now

$$\mathcal{L}^{\mathrm{GW+EM}}(h, F^\gamma|\mathbf{x}, \det) = \int \frac{\mathcal{L}^{\mathrm{GW}}(h|\boldsymbol{\eta}, \boldsymbol{\lambda})}{p_{\det}(\boldsymbol{\eta}, \boldsymbol{\lambda})}\mathcal{L}^{\mathrm{EM}}(F^\gamma|\mathbf{x}, \boldsymbol{\eta})\frac{\pi(\boldsymbol{\eta})\pi(\boldsymbol{\lambda})p_{\det}(\boldsymbol{\eta}, \boldsymbol{\lambda})}{\int \pi(\boldsymbol{\eta})\pi(\boldsymbol{\lambda})p_{\det}(\boldsymbol{\eta}, \boldsymbol{\lambda})d\eta d\lambda}d\eta d\lambda \tag{5.43}$$

$$= \frac{1}{C_{\det}}\int \mathcal{L}^{\mathrm{GW}}(h|\boldsymbol{\eta}, \boldsymbol{\lambda})\mathcal{L}^{\mathrm{EM}}(F^\gamma|\mathbf{x}, \boldsymbol{\eta})\pi(\boldsymbol{\eta})\pi(\boldsymbol{\lambda})d\eta\ d\lambda \tag{5.44}$$

$$= \frac{1}{C_{\det}}\mathcal{L}^{\mathrm{GW+EM}}(h, F^\gamma|\mathbf{x}) \tag{5.45}$$

where we have defined

$$C_{\text{det}} = \int \pi(\boldsymbol{\eta})\pi(\boldsymbol{\lambda})p_{\text{det}}(\boldsymbol{\eta}, \boldsymbol{\lambda})d\boldsymbol{\eta}d\boldsymbol{\lambda}. \tag{5.46}$$

Because C_{det} is a constant for all events in the population, this extra normalization factor can be practically ignored and the method described in Section 5.4 is unaffected by the detection threshold introduced on the gravitational-wave signal. We stress that this is only true if the population parameters are independent of the GW parameters, i.e.:

$$\pi(\boldsymbol{\eta}, \boldsymbol{\lambda}, \mathbf{x}|\boldsymbol{\Lambda}) = \pi(\mathbf{x}|\boldsymbol{\Lambda})\pi(\boldsymbol{\eta})\pi(\boldsymbol{\lambda}), \tag{5.47}$$

and if there is no cut applied on the GRB fluence in order for the event to be included in our analysis. We have enforced this in our analysis by randomly assigning GRB parameters drawn from the hyper-prior $\pi(\mathbf{x}|\boldsymbol{\Lambda})$ to each gravitational-wave source and by running on GRBs with arbitrarily low fluences.

Chapter 6

An Infrared Search for Kilonovae with the WINTER Telescope: Binary Neutron Star Mergers

The content of this chapter was previously published in the Astrophysical Journal as Ref. [382] in Feb. 2022. ASB generated the input data, performed the compact binary parameter estimation and simulated tiling of the observing field, and wrote most of Section 6.2.

Abstract

The Wide-Field Infrared Transient Explorer (WINTER) is a new 1 deg^2 seeing-limited time-domain survey instrument designed for dedicated near-infrared follow-up of kilonovae from binary neutron star (BNS) and neutron star-black hole mergers. WINTER will observe in the near-infrared Y, J, and short-H bands (0.9-1.7 microns, to $J_{AB} = 21$ magnitudes) on a dedicated 1-meter telescope at Palomar Observatory. To date, most prompt kilonova follow-up has been in optical wavelengths; however, near-infrared emission fades more slowly and depends less on geometry and viewing angle than optical emission. We present an end-to-end simulation of a follow-up campaign during the fourth observing run (O4) of the LIGO, Virgo, and KAGRA interferometers, including simulating 625 BNS mergers, their detection in gravitational waves, low-latency and full parameter estimation skymaps, and a suite of kilonova lightcurves from two different model grids. We predict up to five new kilonovae independently discovered by WINTER during O4, given a realistic BNS merger rate. Using a larger grid of kilonova parameters, we find that kilonova emission is ≈ 2 times longer-lived and red kilonovae are detected $\approx$1.5 times further in the infrared than in the optical. For 90% localization areas smaller than 150 (450) deg^2, WINTER will be sensitive to more than 10% of the kilonova model grid out to 350 (200) Mpc. We develop a generalized toolkit to create an optimal BNS follow-up strategy with any electromagnetic telescope and present WINTER's observing strategy with this framework. This toolkit, all simulated gravitational-wave events, and skymaps are made available for use by the community.

6.1 Introduction

During the second Advanced LIGO-Virgo observing run (O2; [3, 53]), the binary neutron star (BNS) merger GW170817 led to the first coincident detection of gravitational waves (GWs) with electromagnetic (EM) radiation spanning X-ray to radio frequencies [17, 18]. GW170817 marked the first direct evidence of a kilonova—a thermal transient powered by rapid neutron capture (r-process) nucleosynthesis in the neutron-rich ejecta of BNS mergers [17, 114, 253, 113, 324, 417, 437, 442, 443, 496, 501, 552, 600, 619, 616, 752, 767, 853, 883]. Kilonovae offer a unique laboratory to study the production of heavy elements via the r-process, probe the neutron star equation of state [597, 250, 251, 175], and provide a new class of standard sirens for resolving the Hubble tension [470, 14, 225, 249, 294]. The combination of GW alerts and electromagnetic follow-up provides a new tool to expand the small dataset of known kilonovae from BNS (although see [856, 884, 365]) and neutron star-black hole (NSBH) mergers [47].

Despite the successful observation of GW170817 in O2, no new, confirmed electromagnetic counterparts to BNS or NSBH mergers were discovered during the third Advanced LIGO-Virgo observing run (O3; [41, 47, 42, 246, 499, 971]). Many ultraviolet/visible/infrared (UVOIR) teams undertook follow-up observations of the O3 LIGO alerts, including the ASAS-SN [522], ATLAS [880], BOOTES [476], DDOTI [129], DES-GW [824], ENGRAVE [549], GOTO [421], GRANDMA [85], GROWTH [499], MASTER-Net [556], SAGUARO [693], Swift UVOT [664], and VINROUGE [54] teams. In addition to searches for UVOIR kilonovae, the Fermi and Swift [681] spacecraft conducted gamma-ray and X-ray follow-up, respectively, and the Australian

Square Kilometre Array Pathfinder searched in radio wavelengths [298]. During O3, spurious GW alerts and alert retractions complicated electromagnetic follow-up strategies [247]. Additionally, models predict many of the BNS mergers detected in GWs in O3 produced kilonovae that were too faint to be detected by the UVOIR facilities conducting follow-up at the time [499, 971].

Optical-wavelength searches dominated the UVOIR O3 GW follow-up landscape, with dozens of optical telescopes across the globe and in orbit triggered during O3 for GW follow-up. Theoretical models and observational constraints from O3 nonde-tections predict that the blue, optical emission from a kilonova is angle-dependent, fades rapidly (<1 week), and may not be present in all BNS or NSBH mergers [616, 496, 498, 114, 499]. In contrast, the near-infrared emission is expected to be isotropic, long-lived (>1 week), and ubiquitous in models regardless of mass ratio, viewing an-gle, or remnant lifetime [498]. Models predict the detection rates of kilonovae in the near-infrared could be up to $\sim 8 - 10$ times higher than in optical wave bands [970].

However, due to the high cost-per-pixel of detectors and bright sky backgrounds, the dynamic infrared sky remains largely underexplored compared to optical wave-lengths. Existing time-domain, infrared surveys are either restricted to small areas on sky (e.g., the VISTA Variables in the Via Lactea survey covering 520 deg^2 [212] or the UKIRT Deep Extragalactic Survey covering 35 deg^2 [543]) or are relatively shallow (e.g., Palomar Gattini IR, J $\sim$ 16 mag; [282]). The lack of deep, all-sky reference images limits the number of GW events that can be followed up in the infrared.

The Wide-field Infrared Transient Explorer (WINTER) is a new instrument that will perform the first near-infrared, all-sky survey to $J_{AB} = 21$ magnitudes and is specially built for GW follow-up [380, 568]. WINTER will operate on a dedicated 1-meter telescope at Palomar Observatory that was commissioned in June 2021 with an optical camera on one port, with the infrared WINTER instrument to be added on the second port in late 2021. WINTER's wide, 1 deg^2 field of view can quickly tile the median expected fourth observing run (O4) 33 deg^2 BNS localization contour [22] with rapid-response robotic observing. With three near-infrared filters in the Y, J, and a shortened-H bands (centered at 1.0, 1.2, and 1.6 µm, respectively), WINTER is designed to discover kilonovae and observe them for two weeks or more [380].

As an alternative to the traditional, but expensive, mercury-cadmium-telluride (HgCdTe) detectors that dominate the near-infrared landscape, WINTER employs cheaper indium-gallium-arsenide (InGaAs) detectors new to astronomical instrumentation. A prototype instrument confirmed InGaAs detectors achieve background-limited near-infrared photometry without cryogenic cooling [813], which is required for HgCdTe detectors. WINTER will head a new class of InGaAs-based near-infrared astronomical instruments coming online in the next decade, including the DREAMS telescope in the Southern Hemisphere [881]. The upcoming HgCdTe-based PRime-focus Infrared Microlensing Experiment telescope will also join WINTER and DREAMS in an upcoming effort to deepen our understanding of the near-infrared time-domain sky.

In this chapter, we present an end-to-end simulation of WINTER's performance in O4 and make a case for infrared follow-up of BNS GW signals. Existing simulations such as [22] also present predictions for BNS merger rates detected by the global GW network in O4, but do not model any electromagnetic counterparts. [699] repeats the analysis in [22] with more realistic parameters and studies the resulting BNS lightcurves using only optical-wavelength kilonova models. Our study additionally takes into account uncertainty introduced by the use of realistic GW matched-filter pipelines, compares any potential differences between the use of low- and medium-latency skymaps, and is primarily focused on the infrared. We include only BNS mergers in this work and leave the study of NSBH mergers, which are expected to be especially promising to follow up in the infrared [82, 340, 972], for future work.

In Section 6.2 we describe the simulation, including modeling a population of BNS mergers, their respective skymaps, and WINTER follow-up observations. We present the results in Section 6.3. This leads to a discussion of the merits of studying kilonovae in the infrared and the planning of WINTER observations in Section 6.4, and we conclude in Section 6.5.

6.2 Methods

We simulate follow-up of BNS gravitational-wave signals with WINTER using the
following procedure:

1. We generate a simulated BNS population and model a set of electromagnetic
   lightcurves from two different model grids [498, 184] for each merger event.

2. To model the gravitational-wave response to each event, we simulate Advanced
   LIGO [3], and Virgo [53] detector noise and network configurations expected for
   O4. We also include the Japanese interferometer KAGRA [826, 99, 67], which
   joined the global network of ground-based gravitational-wave detectors at the
   end of the third observing run.

3. We perform a low-latency search for gravitational-wave events in our simulated data using the PyCBC Live pipeline [658, 269]. Low-latency skymaps
   are generated using the BAYESTAR software [814] to obtain the localization uncertainty contour for each event recovered by PyCBC Live. Full
   source characterization is then performed using the Bilby parameter estimation
   pipeline [96, 762] to obtain medium-latency skymaps. Our simulated PyCBC
   Live results, BAYESTAR skymaps, and Bilby posteriors are publicly available on Zenodo at [381].

4. We simulate WINTER follow-up observations searching the resultant BAYESTAR
   and Bilby skymaps. If WINTER successfully observes the true location of the
   event, we then verify that the event is bright enough to qualify as a statistically
   significant detection.

6.2.1 Simulated BNS population

The fourth observing run of the Advanced LIGO gravitational-wave interferometers [3] is expected to have a duration of one year and include a four-detector network
consisting of the two Advanced LIGO detectors [3], Advanced Virgo [53], and KA-
GRA [826, 99, 67, 22]. One major goal of this study is to determine how many BNS

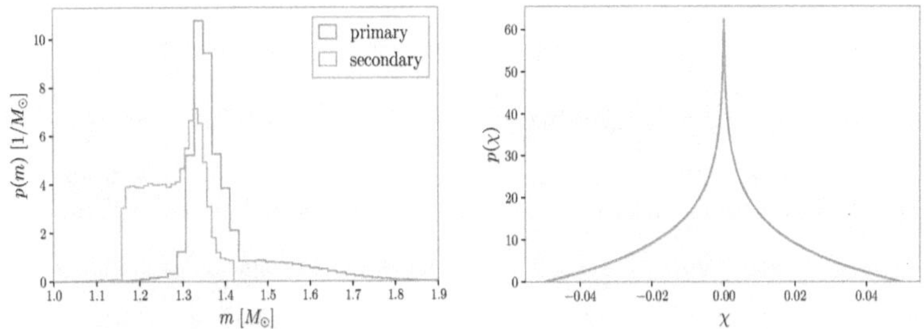

Figure 6-1: *Left*: Probability density function for the distribution of primary and secondary source-frame BNS masses from [337]. *Right*: Probability density function for the distribution of dimensionless spins.

systems would be detected in gravitational waves and then successfully followed up with WINTER during O4. To this end, we simulate a realistic population of BNS systems that could be detected over the course of one year of observing with this detector network operating at its expected sensitivity during this observing run [35]. A (quasi-circular) BNS system is fully described by 17 binary parameters, $\boldsymbol{\theta}$, including the component masses, six-dimensional spins, and tidal deformabilities of the individual neutron stars, and the extrinsic parameters—the distance, inclination angle, right ascension, declination, polarization angle, and time and phase at coalescence.

We simulate systems following the mass distribution providing the best fit to the population of Galactic double neutron stars and the components of GW170817 as determined in [337], which is parameterized in terms of the "slow" and "recycled" binary neutron star components. In the isolated channel for BNS formation [e.g., 859], the recycled neutron star forms first and gets spun up to a period of $\sim 10 - 100$ ms via accretion from its companion [731, 75, 459]. Conversely, the second-born neutron star does not experience this period of accretion, and spins down to a "slow" period of $\mathcal{O}(1)$ s on a timescale of ~ 1 Myr. The Galactic double pulsar PSR J0737-3039A/B serves as the canonical example for this type of system [193, 570].

We model the mass distribution of the recycled neutron star, m_r, as a Gaussian mixture model, with mixing fraction $\alpha = 0.68$ determining the fraction of systems in the first Gaussian,

$$p(m_s) = \alpha \mathcal{N}(m_r; \mu_1, \sigma_1) + (1 - \alpha)\mathcal{N}(m_r; \mu_2, \sigma_2). \tag{6.1}$$

The two Gaussians have means and widths given by $\mu_1 = 1.34\ M_\odot, \sigma_1 = 0.02\ M_\odot, \mu_2 = 1.47\ M_\odot$, and $\sigma_2 = 0.15\ M_\odot$. We draw the masses of the slow neutron star, m_s, in each binary from a uniform distribution between $[1.16\ M_\odot, 1.42\ M_\odot]$ [337]. The primary and secondary masses are assigned via $m_1 = \max(m_r, m_s), m_2 = \min(m_r, m_s)$, the distributions for which are shown in the top panel of Figure 6-1. The tidal deformabilities of the neutron stars, Λ_i, are calculated from the masses assuming the AP4 equation of state [65, 66].

We assume the dimensionless spins of the neutron stars χ are aligned with the orbital angular momentum and drawn from the implied distribution on χ_z assuming that the magnitudes are distributed uniformly on $[0, 0.05]$ and the directions are isotropic. This spin prior choice is motivated by the maximum spin of observed Galactic double neutron star systems [567]. The resultant distribution is shown in the bottom panel of Figure 6-1. The inclination angle, θ_{JN}, between the total angular momentum and the line of sight of the observer is drawn uniformly in $\cos\theta_{\mathrm{JN}}$, and the orbital phase at coalescence and polarization angle also follow uniform distributions.

The detector-frame coalescence times, t_d, of the sources are distributed uniformly over the course of one year starting on January 1st 2022. We choose the sky locations and distances (and hence the redshifts) assuming the sources are distributed isotropically on the sky and uniformly in comoving volume and source-frame time, such that $p(z, t_d) \propto \frac{dV_c}{dz}(1+z)^{-1}$ out to a luminosity distance of 400 Mpc, where V_c is the comoving volume. We consider three different merger rates consistent with those presented in [52]: pessimistic, realistic, and optimistic rates of 100, 500, and 1000 $\mathrm{Gpc}^{-3}\mathrm{yr}^{-1}$, respectively. This results in 21, 105, and 210 sources per year out to $d_L = 400$ Mpc (assuming a merger rate unevolving with redshift; see Table 8.1). We generate 625 unique BNS merger events following the distributions specified above. For each of the three rates considered, we draw 101 different realizations of the corresponding total

Table 6.1: Number of BNS mergers occurring and detected in gravitational-waves for each rate considered in this manuscript during one calendar year of O4. The first column indicates the total number of mergers, while the second column gives the median and 90% symmetric credible interval on the number of systems detected in gravitational waves obtained by averaging over the 101 different realizations of BNS merger combinations. Also included are the number of unique pairs of events that are found in gravitational waves within one day and one week of each other.

Rate	Mergers	Found	1 day	1 week
Pessimistic	21	3^{+3}_{-2}	0^{+0}_{-0}	0^{+1}_{-0}
Realistic	105	16^{+6}_{-5}	1^{+1}_{-1}	6^{+7}_{-5}
Optimistic	210	33^{+7}_{-7}	4^{+4}_{-3}	26^{+19}_{-11}

yearly number of events from these 625 systems. This allows us to marginalize over many different realizations of the "universe" described by each merger rate and more realistically account for uncertainties.

6.2.2   Detection in gravitational waves

For each simulated system, we randomly assign a detector configuration assuming each of the four interferometers are independently operating with a duty cycle of 0.7 [22]. Each BNS is added to 296 s of simulated Gaussian noise sampled at 8192 Hz colored by the expected O4 power spectral density (PSD, shown in Figure 6-2) for each of the detectors that are "observing" during that event [35] starting at a frequency of 17 Hz.

PyCBC Live

Low-latency searches for gravitational waves from compact binary mergers rely on the matched filtering technique (e.g. 263, 70), where a template bank of waveforms is used to filter the data to search for candidate events. In order to more accurately account for the uncertainties associated with the matched-filter detection of gravitational-wave events, we use the PYCBC LIVE [658, 269] pipeline—which is one of the pipelines used to search for gravitational-wave events in LIGO and Virgo data in real time—to identify candidates in our simulated population. The discreteness of the template bank is the dominant source of uncertainty on the source parameter

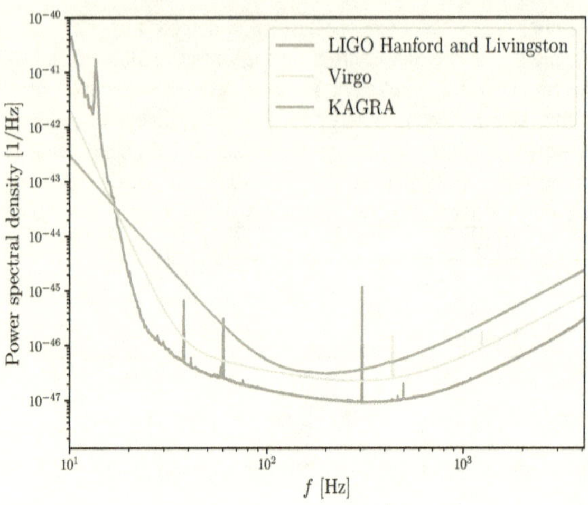

Figure 6-2: Predicted power spectral densities during O4 of the four interferometers included in this study. These curves correspond to BNS ranges of 160-190 Mpc for the LIGO detectors, 90-120 Mpc for Virgo, and 80 Mpc for KAGRA [22].

estimates obtained from matched filtering. It is constructed using a stochastic placement algorithm [103, 448, 716] so that the maximum loss in the signal-to-noise ratio ($\mathrm{SNR_{GW}}$) calculated from a starting frequency of 20 Hz between adjacent templates is 3%. The templates cover detector-frame component masses of between $1 - 3\ M_\odot$ and aligned spins out to $\chi = 0.05$.

For convenience, the search is configured to run on data from all four detectors in our O4-like network. It calculates the PSD for each detector in real time using the simulated data for each event. A trigger is generated if the criterion $\mathrm{SNR_{GW}} \geq 4.5$ is met in at least two of the four detectors. In post-processing, we then calculate the network matched-filter $\mathrm{SNR_{GW}}$ using the trigger information returned by PYCBC LIVE by taking the square root of the quadrature sum of the $\mathrm{SNR_{GW}}$ in each detector that was actually "observing" at the time of the event in question.

Because we only simulate data in short segments around the times containing a BNS merger rather than a continuous one-year data stream, the false alarm rate (FAR) cannot be meaningfully calculated by PYCBC LIVE. Rather than using

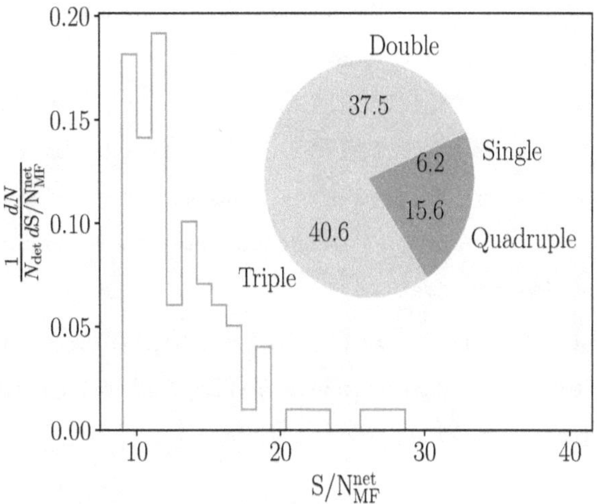

Figure 6-3: Distribution of the network matched-filter $\mathrm{SNR_{GW}}$ recovered by PyCBC Live for the 96 total found events. The inset shows the fraction of the total number of found events detected with each number of interferometers.

an FAR threshold to indicate detection, we instead set a threshold on the network matched-filter $\mathrm{SNR_{GW}}$ recovered by the search for each event of $\mathrm{SNR_{GW}} > 9$ in order for the source to count as "found." Of the 625 independent BNS mergers, 96 are found by PyCBC Live. By averaging over the 101 different realizations of BNS merger combinations for each rate as described in the previous section, we obtain the median and 90% symmetric credible interval on the number of found events shown in Table 8.1. The fraction of the total number of events detected with each number of interferometers is shown in Figure 6-3, along with the recovered $\mathrm{SNR_{GW}}$ distribution. The farthest event is found at a distance of 372.2 Mpc and the nearest at 59.0 Mpc.

BAYESTAR Map Generation

We localize each event identified by PyCBC Live (i.e., with network $\mathrm{SNR_{GW}} > 9$) using the BAYESTAR rapid localization algorithm [814]. During O3, skymaps produced using BAYESTAR were released in low-latency for each GW event that passed the public-alert threshold.

BAYESTAR takes as inputs the estimated masses of the neutron stars, the coalescence time, and the SNR time series from each detector as calculated by PyCBC Live. From this time series, BAYESTAR extracts the timing, relative phases, and amplitudes from each detector. BAYESTAR then constructs a three-dimensional sky localization posterior distribution, with a probability for each pixel on the two-dimensional sky and an additional distance component that is approximated by a Gaussian along each line of sight.

In O3 and in this work, BAYESTAR sky localizations for each event were computed in $\mathcal{O}(10\ \mathrm{s})$. Further efforts to optimize BAYESTAR have led to improvements resulting in runtimes of $\mathcal{O}(1\ \mathrm{s})$ [574].

Parameter estimation

After the initial low-latency skymap using the BAYESTAR algorithm is sent out, further source characterization is performed using full parameter estimation, whereby posterior probability distributions for the binary parameters are obtained using stochastic sampling methods. While BAYESTAR fixes the intrinsic parameter values to those corresponding to the maximum-likelihood template returned by the search pipeline, parameter estimation enables marginalization over the uncertainty in the intrinsic parameters and the extrinsic parameters that are not necessary for the skymap. Accounting for this uncertainty and allowing for correlations between parameters as done using full parameter estimation should lead to skymaps that have systematically smaller sky areas and include the true location of the source at smaller confidence intervals [815]. We present a comparison of the skymaps from the two localization algorithms in Section 6.3.1.

The likelihood of observing gravitational-wave data d given binary parameters $\boldsymbol{\theta}$ is given by [908, 761, 96]:

$$\mathcal{L}(d|\boldsymbol{\theta}) \propto \exp\left(-\sum_k \frac{2|d_k - h_k(\boldsymbol{\theta})|^2}{TS_k}\right), \tag{6.2}$$

where $h(\boldsymbol{\theta})$ represents the gravitational waveform for the BNS signal with parameters $\boldsymbol{\theta}$, T is the duration of the analyzed segment, S_k is the PSD, and k indicates the frequency dependence of the data, waveform, and PSD. The posterior probability distribution for the binary parameters characterizing each systems is then given by Bayes' Theorem:

$$p(\boldsymbol{\theta}|d) \propto \mathcal{L}(d|\boldsymbol{\theta})\pi(\boldsymbol{\theta}), \tag{6.3}$$

where $\pi(\boldsymbol{\theta})$ represents the prior probability distribution assumed for $\boldsymbol{\theta}$.

We simulate the progression of skymaps that could occur during a real observing run by performing parameter estimation using the `Bilby` software [96, 762] for all of the events that were found by PyCBC Live. We use the `PyMultiNest` nested sampler [344, 345, 346, 182] to generate samples from the posterior probability distribution for the source parameters, where the likelihood is analytically marginalized over the distance to the source and the phase at coalescence [906, 814, 874]. We make the assumption that the neutron stars are point masses with tidal deformability $\Lambda = 0$, which is unlikely to affect the inference of the sky location of the source. This assumption enables the use of the reduced order quadrature implementation [820] of the IMRPhenomPv2 waveform [445, 509, 482], significantly reducing the computational cost of the parameter estimation. The likelihood uses the true PSD used to color the Gaussian noise in each detector. In order to keep the computational cost low, we do not marginalize over uncertainty in the detector calibration.

Because the search pipelines recover the chirp mass of the source extremely well [158], we use a uniform prior on the detector-frame chirp mass with a width of 0.2 $M_\odot$ centered on the true value. The prior on the mass ratio, $q \equiv m_2/m_1$, is uniform over the interval [0.125, 1]. We restrict the analysis to aligned spins only, and the spin prior is the same as that used for drawing the simulated sources, shown in Figure 6-1. The luminosity distance prior is uniform in the source frame between 10 and 500 Mpc, and the prior on the coalescence time in the geocentric frame is uniform over a width of 0.2 s centered on the true coalescence time. We use standard priors on all other parameters [see, e.g., 762].

With these parameter estimation settings, the sampling stage takes 63^{+61}_{-32} mins across all of the recovered events. This depends most sensitively on the $\mathrm{SNR_{GW}}$ of the signal, since louder signals take longer to analyze due to the way the nested sampler explores the prior volume. We note that there are additional timing overheads for generating the weights for calculating the reduced order quadrature likelihood, reconstructing the posterior for the analytically marginalized distance parameter, and generating a skymap of the appropriate format that can be used by observers, although these additional steps generally take less time than the sampling. This timing is not necessarily indicative of what will occur in O4. There are a number of alternative techniques being explored that can be used to further accelerate parameter estimation, particularly for low-mass sources, like focused reduced order quadrature [634], relative binning [958, 350], parallelization [684, 849, 947, 819], prior restrictions [955], and machine learning based approaches [391, 434, 435, 935].

6.2.3 Electromagnetic follow-up

Lightcurve modeling

In order to capture some of the current uncertainty in kilonova modeling, we use two different prescriptions to predict the lightcurves of kilonovae from our simulated BNS mergers.

Bulla model: The first is a grid of kilonova models generated using the Monte Carlo radiative transfer code POSSIS [184]. The grids assume a two-component ejecta comprised of mass ejected on a dynamical timescale and mass ejected in the form of a wind from the debris disk post merger. The dynamical ejecta is further divided into a lanthanide-rich equatorial component and a lanthanide-free polar component. The lightcurves are calculated using four parameters: the total dynamical ejecta mass ($M_{\mathrm{ej}}^{\mathrm{dyn}}$), the total wind ejecta mass ($M_{\mathrm{ej}}^{\mathrm{wind}}$), the opening angle of the lanthanide-rich component (Φ) and the observer viewing angle ($\theta_{\mathrm{obs}} = \theta_{\mathrm{JN}}$ if $\theta_{\mathrm{JN}} < 90°$ and

$\theta_{\rm obs} = 180° - \theta_{\rm JN}$ if $\theta_{\rm JN} > 90°$). A larger value of the opening angle Φ increases the amount of lanthanide-rich material, making the kilonova redder. Further details about the BNS grids are presented in Ref. [294].

Kasen model: The second model uses the kilonova simulations presented in [498]. While these simulations only include the effects of a single, spherically symmetric ejecta component with radial density given by a broken power law, they feature a more rigorous treatment of the microphysics determining the radiation transport in the system and allow for a range of compositions. The model parameters are the total ejecta mass, which we calculate as $M_{\rm ej} = M_{\rm ej}^{\rm dyn} + M_{\rm ej}^{\rm wind}$, the expansion velocity of the ejecta, $v_{\rm ej}$, and the mass fraction of lanthanides, $X_{\rm lan}$. A higher $M_{\rm ej}$ leads to lightcurves that peak at brighter magnitudes, while higher $v_{\rm ej}$ leads to faster-fading lightcurves. More neutron-rich ejecta with a higher lanthanide fraction, $X_{\rm lan} \gtrsim 10^{-2}$, produce a redder and longer-lived kilonova than ejecta composed primarily of light r-process material, $10^{-6} \lesssim X_{\rm lan} \lesssim 10^{-2}$ [498].

To estimate the ejecta masses and velocity from the neutron star masses, we use the prescription outlined in Ref. [833]. First, we assume an AP4 equation of state [65, 66] for the merging neutron stars to calculate their radii and compactness, although recent works have developed [833] and employed [47] methods for marginalizing over the uncertainty in the equation of state. We then calculate the dynamical ejecta mass using the fitting formula from Ref. [250]:

$$\log_{10} M_{\rm ej}^{\rm dyn} = \left[a\frac{(1 - 2C_1)\, m_1}{C_1} + b\, m_2 \left(\frac{m_1}{m_2}\right)^n + \frac{d}{2} \right] + [1 \leftrightarrow 2] \qquad (6.4)$$

where m_i and C_i are the masses and the compactnesses of the two neutron stars, respectively, and $a = -0.0719$, $b = 0.2116$, $d = -2.42$ and $n = -2.905$. The ejecta velocity is similarly calculated using a fitting formula from Ref. [250]:

$$v_{\rm ej} = \left[a''\frac{m_1}{m_2}(1 + c''\, C_1) + \frac{b''}{2} \right] + [1 \leftrightarrow 2], \qquad (6.5)$$

where the coefficients are given by $a'' = -0.3090$, $b'' = 0.657$, and $c'' = -1.879$.

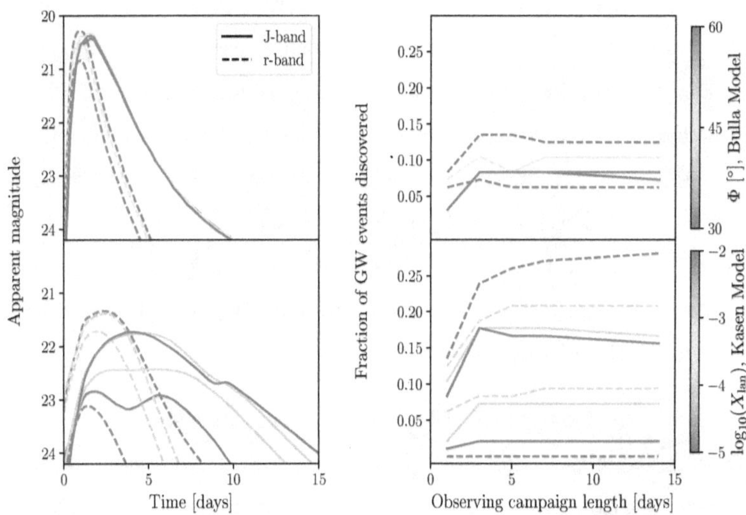

Figure 6-4: Variation due to lightcurve models. *Left:* Example lightcurves in the J (solid lines) and r (dashed lines) filters for the Bulla model (top) with different opening angles and the Kasen model (bottom) with different lanthanide fractions as shown by the top and bottom color bars, respectively. The example BNS system has masses $m_1 = 1.36\ M_\odot$, $m_2 = 1.22\ M_\odot$ and is at a distance of $d_L = 162.97$ Mpc. *Right:* The fraction of events discovered in the J band by WINTER and an equally sensitive r-band telescope, as described in Section 6.2.3, plotted against the number of nights allocated to a kilonova search, marginalized over all BNS merger event rates. Unique lightcurves are modeled for each event from the same set of Bulla (top) and Kasen (bottom) models.

We estimate the wind ejecta mass as a fraction of the disk mass $M_{\mathrm{ej}}^{\mathrm{wind}} = \zeta M_{\mathrm{ej}}^{\mathrm{disk}}$ and set $\zeta = 0.15$ [294]. We calculate the disk mass using the fitting formula

$$\log_{10} M_{\mathrm{ej}}^{\mathrm{disk}} = \max\left[-3, a'\left(1 + b' \tanh\left(\frac{c' - (m_1 + m_2)/M_{\mathrm{thresh}}}{d'}\right)\right)\right] \tag{6.6}$$

where M_{thresh} is the minimum total mass that results in a prompt collapse post merger and is calculated as in Ref. [119]; a', b', c', and d', are calculated as in Ref. [294].

We interpolate between the parameters of standard grids for both the Bulla and Kasen models using the Python package GWEMLIGHTCURVES [251, 250, 294] to predict the J-band and r-band lightcurves of the kilonovae. The left panel of Figure 6-4 shows the lightcurves for an example system. The differences in peak magnitude and decay time for the lightcurves calculated with the Kasen and Bulla models demonstrate the uncertainties currently underlying kilonova modeling. We present results for both models to conservatively account for this uncertainty in our predictions. Additionally, we note that the ejecta masses calculated from the fitting formulae are approximate and depend strongly on the equation of state, fitting parameters, and assumptions of the numerical relativistic models that they are based on. It is thus possible that the ejecta masses are not entirely representative of realistic kilonovae. We investigate the effect of varying ejecta masses in Section 6.4.1.

WINTER simulated observing

For each found gravitational-wave event, we use the corresponding skymap and simulated time of the event to create a realistic observing schedule with the GWEMOPT package [252, 244, 409]. The package takes gravitational-wave probability maps as inputs, such as the BAYESTAR and Bilby skymaps, subdivides the sky into tiles sized to the telescope field of view, and generates an optimized observing schedule. During O3, a network of telescopes including the Zwicky Transient Facility [137] created GW follow-up schedules with gwemopt (e.g., 243; 500) and we expect to use the package during nominal WINTER operations in O4.

We run a set of observing simulations allowing one, three, five, seven, and fourteen nights of dedicated telescope time searching for the kilonova from each event. For the scope of this study, we limit observations to the J-band to match WINTER's planned J-band reference images and all-sky survey and only study one gravitational-wave event at a time, even though multiple events may occur during the same night or same week (see Table 8.1). Each WINTER observation lasts 450 seconds to match the $J_{AB} = 21$ magnitude reference images, split across five dithers, with the time for dithering at approximately one second per dither represented as an overhead time in the simulation. Time to slew between each field is calculated based on the telescope and dome slew rates measured at WINTER's host telescope at Palomar Observatory. WINTER's InGaAs sensors read out continuously during each exposure, leading to no overhead time due to sensor readout [579]. Combining overhead and exposure times, for an eight hour night WINTER covers ~ 63 deg^2 to $J_{AB} = 21$ magnitudes. The WINTER data processing pipeline will subtract new science images from prebuilt reference images (constructed well before the GW alert) and detect candidate kilonovae in near real time.

We follow up each skymap with a ranked search strategy, in which we subdivide the sky into a fixed grid of telescope pointings. The center of each tile corresponds to a WINTER reference image for easy image subtraction during data processing. We prioritize the tiles based on the skymap-generated probability. Finally, we create a greedy observing schedule, where the highest probability tiles are observed first, as described in [252]. We prioritize scheduling at least two observations of each field, spread out over the length of an observing campaign to study the lightcurve evolving over time. In this simplified simulation, one observing schedule is created and executed without modification for each follow-up campaign. The simulation schedules multiple visits for each field, but fields are not preferentially revisited. During real O4 follow-up observing, WINTER will iteratively compare observations to prebuilt reference images and revisit any fields containing promising kilonova candidates (see Section 6.4.3 for details)

Next, we check the observing schedules for a successful observation of the kilonova, which can be defined in multiple ways. For the sake of clarity in this study, we define an event as *localized* if WINTER takes at least one image of the kilonova's true location on the sky. An event may not be localized if the skymap is too large for WINTER to search it efficiently or if the event is not overhead at the given time of year. Additionally, poor weather can hinder follow-up observations; however, weather simulations are outside of the scope of this study.

Even if an event is localized, the kilonova must be sufficiently bright at the time of imaging to be detected. The event qualifies as a *discovery* if it is localized, imaged at least twice, and observed to a signal-to-noise ratio $(\mathrm{SNR}_{\mathrm{EM}}) \geq 5$. We employ the lightcurve models described in Section 6.2.3 to calculate the magnitude of the event at the time of each observation and scale the magnitude based on the airmass of the observation. We follow Equation 1 as described in Section 3.1 of [380] to calculate the $\mathrm{SNR}_{\mathrm{EM}}$ for each event based on historic Palomar data and predicted instrument noises. For comparison, we repeat this exercise for WINTER observing in J band and for a fictitious optical telescope observing in the Sloan Digital Sky Survey r filter with equivalent sensitivity, field of view, exposure times, and overhead times as the WINTER J-band observations.

For the scope of this study, we use two observations to $\mathrm{SNR}_{\mathrm{EM}} \geq 5$ as a simplified proxy for true kilonova discovery, as two observations are the minimum number required to distinguish a kilonova from asteroids or other transients. In reality, to confirm a new kilonova candidate, more than two observations may be required, and promising candidates will be selected based on their color evolution, how quickly the lightcurve fades, and if the event is associated with a probable host galaxy. Discovery will then be confirmed with further photometric and spectroscopic follow-up. See Section 6.4.3 for further discussion of electromagnetic follow-up observations.

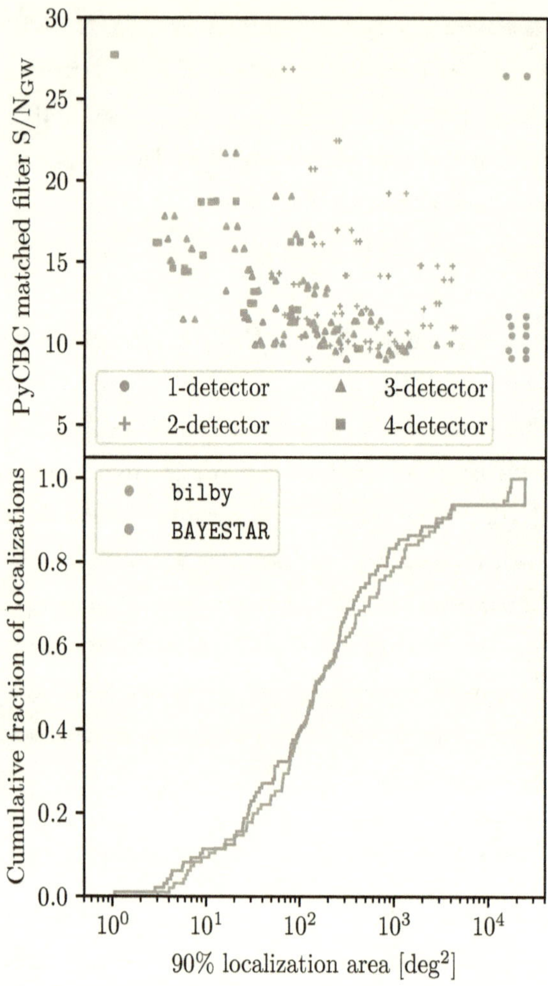

Figure 6-5: *Top*: 90% credible localization areas of Bilby and BAYESTAR skymaps for each event vs. its PyCBC LIVE matched filter SNR_{GW}. Different markers represent the number of GW detectors that identified the event. *Bottom*: Cumulative distribution of 90% credible localization areas for skymaps generated by both algorithms.

6.3 Results

6.3.1 Skymap comparisons

A total of 96 events were found by PyCBC Live out of the 625 total independent simulated mergers, as shown in Table 8.1. Each event was localized by both BAYESTAR and Bilby, as described in Section 6.2. We present a comparison of the two algorithms. This is especially relevant for electromagnetic follow-up purposes, since in O3, BAYESTAR skymaps were distributed hours or days before the more comprehensive Bilby or LALInference [906] skymap. If this delay persists in O4 and if there are significant discrepancies in the two localizations, electromagnetic observers might switch partway through the night, or might even prefer to wait for the Bilby skymap before beginning observations.

The distribution of 90% credible localization areas for all Bilby and BAYESTAR skymaps is shown in Figure 6-5. Events that are recovered by PyCBC Live with larger values of matched-filter SNR_{GW} tend to have smaller localization areas. Some events have network $SNR_{GW} > 20$ but localizations larger than 100 deg^2; these are typically one- or two-detector events where the low number of detectors results in poor constraints on timing and phase. The cumulative distributions of 90% localization areas from each algorithm are similar, with 24% of Bilby and 27% of BAYESTAR localizations falling under 50 deg^2. The BAYESTAR distribution lies slightly to the left (i.e., to smaller areas) of the Bilby distribution.

Figure 6-6 shows a comparison between the BAYESTAR and Bilby skymaps for two particular events which are generally indicative of the performance observed in the larger sample. In the left skymap, which has a 90% localization area of order $\mathcal{O}(1000)\mathrm{deg}^2$ from a two-detector event at a distance of $d_L = 141.1$ Mpc, the BAYESTAR and Bilby skymaps have some overlap, but with probability concentrated in different areas of the sky. In the right skymap, which stems from a four-detector event at a distance of $d_L = 113.8$ Mpc, the localizations are small and in very good agreement.

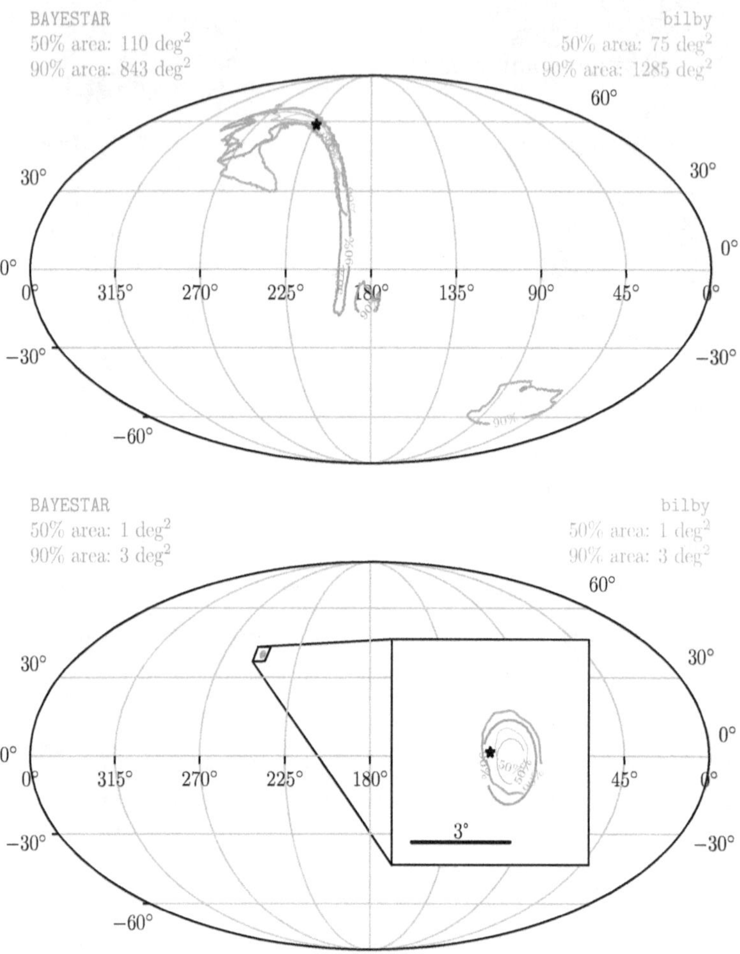

Figure 6-6: Comparison skymaps showing the 50 and 90% credible regions obtained using Bilby and BAYESTAR for two different events. The true source location is marked with the black star. The event on the left was observed with the Hanford and KAGRA interferometers at a distance of 141 Mpc, although it was not confidently detected with KAGRA since the optimal SNR in that detector is 2.6. The one on the right was detected with Hanford, Livingston, Virgo, and KAGRA at a distance of 114 Mpc. In the skymap to the right, the BAYESTAR and Bilby localizations almost completely overlap.

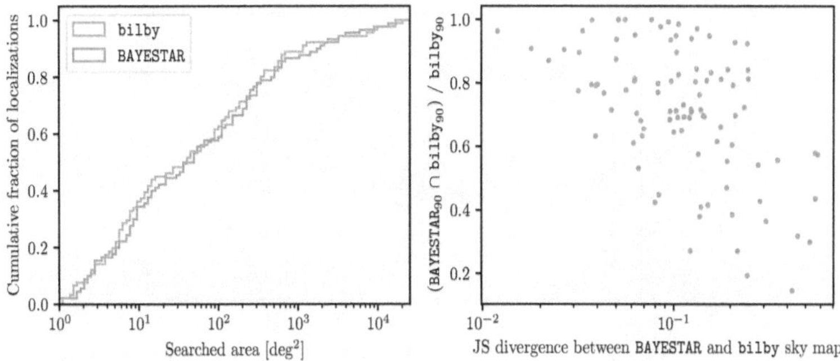

Figure 6-7: *Left*: Cumulative distributions of searched area for skymaps generated using `Bilby` and BAYESTAR. *Right*: The intersection of the BAYESTAR 90% and `Bilby` 90% credible areas, normalized by the `Bilby` 90% credible area vs. Jensen-Shannon (JS) divergences between `Bilby` and BAYESTAR skymaps for each event

A useful metric for the accuracy of localizations of simulated events is the searched area, which is the amount of sky area that is covered by integrating in order of decreasing probability from the highest probability pixel until reaching the true location of the source. The left panel of Figure 6-7 shows the distributions of searched areas for `Bilby` and BAYESTAR. The two algorithms perform very similarly.

Another convenient statistic for comparing probability distributions is the Jensen-Shannon (JS) divergence. The JS divergence measures how similar two distributions are and ranges from 0 bit for identical distributions to 1 bit for completely divergent distributions. JS divergence values greater than 0.002 are considered to be statistically significant [762]. The right panel of Figure 6-7 shows the intersection of the two 90% credible areas normalized by the `Bilby` 90% area versus the JS divergence between the two skymap posterior distributions for each event. These distributions describe the probability of the event being in each pixel of the two-dimensional skymap. For skymaps that have almost complete 90% area overlap, the JS divergence is very small and thus the maps share significant information content, and vice versa. However, there are also events for which the normalized intersections are close to unity, but have relatively large JS divergences. This is due to two types of events: events that have similar 90% localization areas, but with different probability distributions for

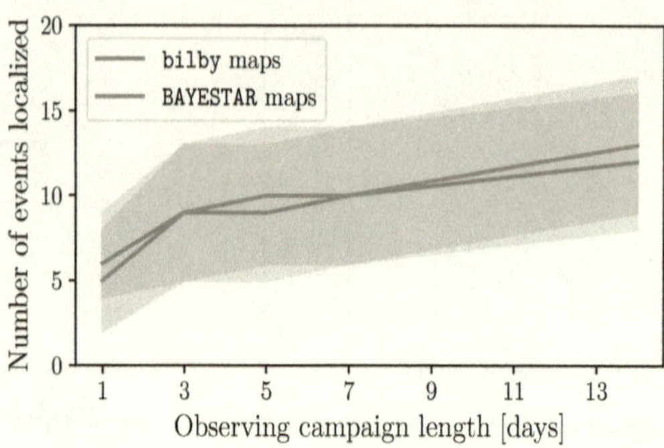

Figure 6-8: The number of events localized by WINTER for an optimistic BNS merger rate given the length of the observing campaign for following up each event with the Bilby and BAYESTAR skymaps. The solid lines represent the median number of events localized and the shaded regions show the 90% symmetric credible interval.

the pixels within that area; and events for which the intersection of the Bilby 90% area is small compared to the BAYESTAR 90% area, so the intersection divided by the Bilby area is almost unity.

Previous studies have found that localizations from full parameter estimation pipelines such as LALInference, when compared to those from BAYESTAR, have systematically smaller sky areas and include the true location of the source at smaller confidence intervals (see, e.g. Figure 3 of [815]). In this study we find that the discrepancy between the two algorithms is significantly reduced, since our BAYESTAR skymaps use data from all online detectors, while those in [815] use data only from detectors that register an $\mathrm{SNR_{GW}}$ above the detection threshold. Furthermore, since matched-filter searches recover the true chirp mass of BNS systems with extremely high accuracy, to within $\sim \mathcal{O}(10^{-4})\ M_\odot$ [158], not marginalizing over the uncertainty in the mass parameters should lead to less significant biases in the low-latency skymap compared to higher-mass sources. We expect the difference between the skymaps obtained with the two algorithms to be more significant for NSBH sources, which will be explored in future work.

Table 6.2: Number of gravitational-wave triggers leading to various categories of WINTER observations given pessimistic, realistic, and optimistic event rates. An event is accessible if it is overhead at Palomar Observatory at the given time of year, it is localized if the telescope takes an image of the kilonova's location on the sky, and the kilonova is discovered if detected at least twice to $\mathrm{SNR_{EM}} \geq 5$ in the J band. All simulations in this table use the ranked search follow-up strategy with five nights of searching the BAYESTAR skymaps.

Rate	GW triggers	EM Accessible	Localized	Discovered			
				Bulla		Kasen	
	Events	Events	Events	Φ [°]	Events	X_{lan}	Events
Pessimistic	3^{+3}_{-2}	2^{+2}_{-2}	1^{+1}_{-1}	30	0^{+1}_{-0}	10^{-2}	0^{+2}_{-0}
				45	0^{+1}_{-0}	10^{-3}	0^{+2}_{-0}
				60	0^{+1}_{-0}	10^{-4}	0^{+1}_{-0}
						10^{-5}	0^{+0}_{-0}
Realistic	16^{+6}_{-5}	11^{+5}_{-5}	5^{+3}_{-3}	30	1^{+2}_{-1}	10^{-2}	2^{+3}_{-2}
				45	1^{+2}_{-1}	10^{-3}	3^{+2}_{-2}
				60	1^{+2}_{-1}	10^{-4}	1^{+2}_{-1}
						10^{-5}	0^{+1}_{-0}
Optimistic	33^{+7}_{-7}	23^{+5}_{-7}	10^{+4}_{-4}	30	3^{+1}_{-2}	10^{-2}	6^{+3}_{-4}
				45	3^{+2}_{-2}	10^{-3}	6^{+4}_{-3}
				60	3^{+2}_{-2}	10^{-4}	2^{+2}_{-2}
						10^{-5}	1^{+1}_{-1}

6.3.2 Results from WINTER simulated observing

Figure 6-8 shows the results of searching the `Bilby` and BAYESTAR maps with WINTER, varying the length of the observing campaign from one through fourteen nights of telescope time, given an optimistic BNS merger rate. The simulation produces separate observing schedules based on the amount of time allowed to follow up the event. In agreement with Section 6.3.1, WINTER localizes the BNS merger events at similar rates for both the `Bilby` and BAYESTAR skymaps. The number of events localized with WINTER steadily increases given more nights of observing time, with ~ 2 times more events localized with a fourteen-night search strategy than a one-night search strategy.

Varying the number of nights allowed searching for the kilonova changes the ordering and prioritization of each observing schedule. For example, given a one-night observing campaign, 100% of the images of the kilonova are taken before the J-band peak of the lightcurve, regardless of model. For a five-night campaign, 80% (86%) of the localized events have their first images taken before the peak and 30% (27%) have two images taken before the peak for the Kasen (Bulla) lightcurve model. Increasing observing time allows WINTER to cover a greater portion of the skymap, but risks tiling the high-probability area of the skymap less efficiently or observing the kilonova later in its evolution when it may have already faded. Therefore, discovery of new kilonovae (defined above as observing the event at least twice to $SNR_{EM} \geq 5$) does not increase steadily with time, but levels off as the kilonova eventually fades beyond detection. In the right panel of Figure 6-4, we compare discovery of new kilonovae in the WINTER J band and for an equally sensitive r-band telescope given the number of nights each telescope is allowed to search for the kilonovae.

Given the Kasen lightcurve models, discovery in both J band and r band peaks around three nights of tiling and levels off given more nights of observing, with the exception of lanthanide-poor ($X_{lan} = 10^{-5}$) kilonova discovery in the r band continuing to increase with fourteen nights of observing (bottom right panel of Figure 6-4). In the J band with three nights of observing, WINTER discovers 8.5 times more lanthanide-rich ($X_{lan} = 10^{-2}$) events than lanthanide-poor events ($X_{lan} = 10^{-5}$). In contrast, an equally sensitive r band telescope discovers zero lanthanide-rich events and up to 28% of lanthanide-poor events.

Despite differences in the lightcurve models, J-band and r-band discovery also levels off after three nights of observing for the Bulla models (top right panel of Figure 6-4). Varying the opening angle of the lanthanide-rich component changes the resultant lightcurves less than varying the lanthanide fraction in the Kasen models. At three nights of observing, a large opening angle of the lanthanide-rich component ($\Phi = 60°$) decreases r-band discovery by 46% and does not change J-band discovery.

Table 6.2 displays the results of the end-to-end simulation studying how many kilonovae WINTER will discover during one year of follow-up observations, including the median and 90% symmetric credible interval on the number of events as described in Section 6.2.2. Some subset of the GW triggers will not be observable by a telescope at Palomar Observatory at the given trigger time, either because the event is too far south or too near the sun. For a given year of GW triggers, on average $\sim 70\%$ of the events are visible above 20° altitude for WINTER at some point during the night following the event trigger time. These events are denoted as EM accessible in the table. For a portion of those events, WINTER observes the correct patch of sky and localizes the events, with some events missed due to large localization areas or events with true locations outside of the 90% localization areas. Localized events only qualify as a discovery ($\mathrm{SNR_{EM}} \geq 5$ in at least two observations) if the event is sufficiently bright, which can vary based on the model grid used to simulate the event. WINTER discoveries range from as low as zero new kilonovae per year with a pessimistic BNS merger rate to as high as ten new kilonovae discovered per year to 90% confidence with an optimistic BNS merger rate. Given a realistic BNS merger rate, WINTER discovers up to five new kilonovae per year to 90% confidence.

6.4   Discussion

6.4.1   Advantages of infrared follow-up

In Section 6.3, we demonstrate that a 1 deg^2 J-band survey like WINTER can detect up to ten kilonovae during O4. Here, we examine the advantages of a J-band search over a similar optical search. We use a grid of realistic kilonova lightcurves calculated with the Bulla model to identify the parameter space where an infrared search outperforms an optical search. To generate the grid, we do not use the fitting formulae from Section 6.2.3, as they may not be entirely representative of the underlying physical population of kilonovae. Instead, we calculate lightcurves for a wide range of possible ejecta masses derived from numerical relativistic simulations: $M_{\mathrm{ej}}^{\mathrm{dyn}} = [0.001, 0.01]\ M_\odot$ and $M_{\mathrm{ej}}^{\mathrm{wind}} = [0.01, 0.13]\ M_\odot$ [84]. We set the dynamical ejecta opening angles in the range $\Phi = [30°, 60°]$ and calculate lightcurves for viewing angles sampled uniformly in $\cos(\theta_{\mathrm{obs}})$.

Red kilonovae are brighter at infrared wavelengths

Kilonovae with a larger opening angle of the dynamical ejecta will have more lanthanide-rich material and hence will be brighter at redder wavelengths. We quantify this in

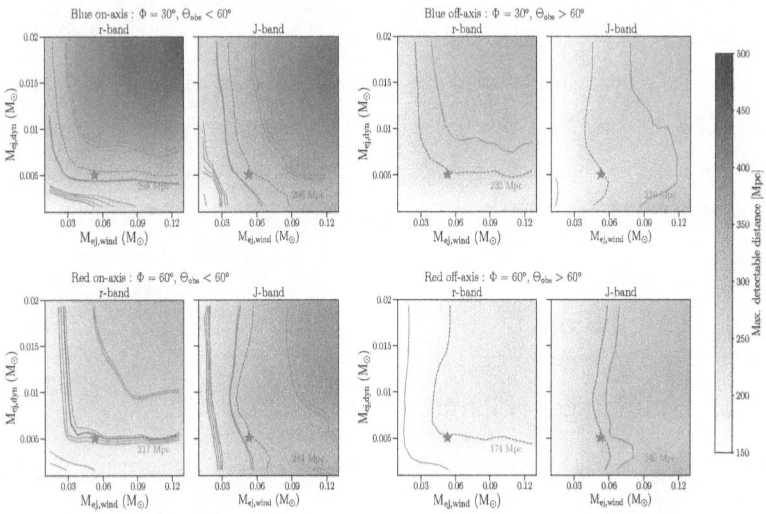

Figure 6-9: The maximum distance out to which a kilonova can be detected by an $m_{\rm lim} = 21$ survey for a given combination of dynamical and wind ejecta masses, based on the Bulla models. We plot the distances separately for J and r bands. We further distinguish the kilonova models as blue and on-axis (top left), blue and off-axis (top right), red and on-axis (bottom left) and red and off-axis (bottom right). The ejecta masses for GW170817 are plotted as a red star. The red dashed line shows the contour at the distance at which a kilonova with GW170817-like ejecta masses is detectable in each case (distance indicated in red). We also plot the contours at the distances of 15 events followed up by WINTER from one realization of our realistic-rates simulation. Twelve of these events are on-axis and three are off-axis. Two off-axis events have distances < 150 Mpc and lie off the plots. It is evident that for the same set of ejecta masses, the J band can detect kilonovae out to larger distances than r band if the kilonova is red.

213

Figure 6-9 which shows the maximum distance out to which a kilonova can be detected in the r and J bands by a telescope with a limiting depth of 21 magnitudes. We distinguish between blue ($\Phi = 30°$) and red ($\Phi = 60°$) kilonovae, and an on-axis ($\theta_{\mathrm{obs}} < 60°$) and off-axis ($\theta_{\mathrm{obs}} > 60°$) kilonova. We use the maximum-likelihood estimates of the ejecta masses of GW170817 as a benchmark ($M_{\mathrm{ej}}^{\mathrm{dyn}} \sim 0.005\ M_{\odot}$, $M_{\mathrm{ej}}^{\mathrm{wind}} \sim 0.05\ M_{\odot}$, 294) and indicate the distances out to which a kilonova with these masses can be detected. Finally, we plot contours corresponding to the distances of 15 events from one realization of our uniformly distributed, realistic-rate simulation that were detected in GWs and followed up with WINTER (see Table 8.1). Twelve of these events are on-axis and three are off-axis.

For a blue kilonova ($\Phi = 30°$), an infrared search does not offer a significant advantage over an optical search. A blue on-axis (off-axis) kilonova with GW170817-like ejecta would be detectable out to $\approx$ 300 (230) Mpc in both r and J bands. This is because the peak values of r- and J-band lightcurves for a blue kilonova are not significantly different (see top left panel of Figure 6-4). Off-axis kilonovae are generally fainter than on-axis ones, which explains the reduced sensitivity (300 Mpc vs. 220 Mpc) for off-axis kilonovae. All 12 on-axis events from our simulation are detectable in both r and J bands if they are blue. We note that the detectable region of the ejecta mass phase space (i.e., the region to the right of each contour in Figure 6-9) is slightly larger for r band than for the J band. Ten of these events have a detectable region that includes GW170817 ejecta masses in both r and J bands. All three off-axis events from our simulation are detectable in both r and J bands. Of these three, two are closer than 150 Mpc, and are detectable for the entire range of ejecta masses in both r and J bands.

However, an infrared search performs significantly better than an optical search for red kilonovae. Both on- and off-axis red kilonovae are detectable out to larger distances in the infrared than in the optical. A red on-axis (off-axis) kilonova with GW170817-like ejecta is detectable to 284 (246) Mpc in the infrared but only to 217 (174) Mpc in the optical. If the 12 on-axis kilonovae from our simulations are red, only 10 are detectable in the r band while all 12 are detectable in the J band. GW170817

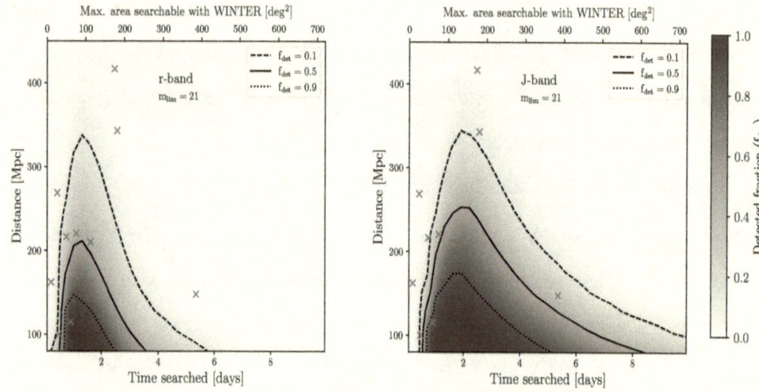

Figure 6-10: The fraction of kilonovae from our entire model grid that can be detected in the r and J bands by an $m_{\mathrm{lim}} = 21$ telescope, as a function of distance and number of days since the merger. The contours corresponding to detection fractions (f_{det}) of 0.1, 0.5, and 0.9 are plotted as dashed, solid, and dotted lines, respectively. Red crosses mark the distances and areas enclosing 90% localization probability of the events from our realistic-rate simulation (of the 16 events, only 10 are shown, as the remaining six lie outside the bounds of the axes). It is clear that kilonovae can be detected for much longer in the J band compared to r band. The right panel can also be used to select GW triggers that are worth following with WINTER. We will only follow events that have median distance estimates and localization areas such that the detection fraction is at least 10%. We note that the detection fraction drops at very early search times ($t < 0.5$ day), as the kilonova is still brightening in our models at these times. However, we will observe events with localization areas that can be tiled within 0.5 day, observing them repeatedly until the kilonova becomes bright enough to be detected.

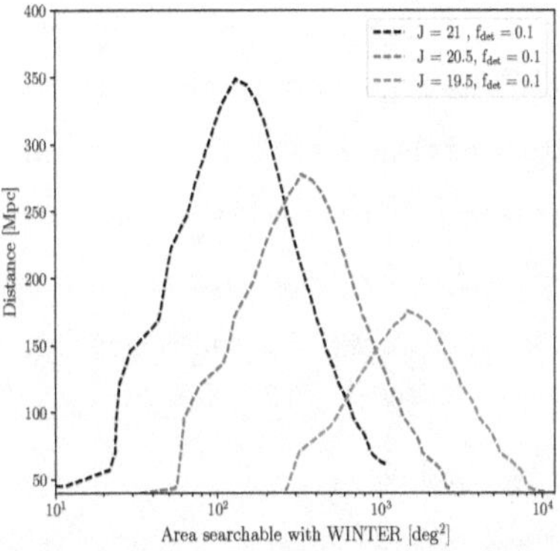

Figure 6-11: Contours marking the regions for which the kilonova detection fraction is at least 0.1 ($f_{det} > 0.1$), as a function of the localization area. We plot separate contours for searches with limiting magnitude of 21, 20.5, and 19.5 mag (black, blue, and red, respectively). With WINTER, we will follow up all events with localization areas < 450 deg^2. down to a depth of 21 mag. If the localization area is larger than 450 deg^2, we will follow only those events that have median distance estimates < 200 Mpc down to a depth of 20.5 mag. If the localization area is larger than 1000 deg^2, we will follow only those events with median distances < 150 Mpc down to a depth of 19.5 mag.

ejecta masses lie in the detectable region for only six kilonovae in the r-band but for 10 kilonovae in J band. If the three off-axis simulated events are red, only two are detectable in the r band while all three can be detected in the J band. Finally, we note that if a particular kilonova is not detected in WINTER observations, Figure 6-9 can be used to place constraints on the ejecta masses and opening angles associated with it.

All kilonovae are longer-lived in the infrared

A second advantage of infrared searches is that kilonova emission is longer-lived in the infrared than in the optical. We quantify this in Figure 6-10, which shows the fraction of kilonovae from our entire model grid that can be detected in the r and J bands by an $m_{lim} = 21$ telescope as a function of distance and number of days since the merger. We plot contours corresponding to detection fractions (f_{det}) of 0.1, 0.5, and 0.9.

Figure 6-10 clearly shows that kilonovae can be detected for much longer in the J band compared to r band. For example, at 200 Mpc more than 10% (i.e. $f_{\rm det} > 0.1$) of kilonovae are detectable in the r band for a maximum duration of three days. However, in the J band the same events can be detected for almost six days. A WINTER-like telescope with a 1 deg^2 field of view can thus search localization regions of ≈ 450 deg^2 in the J band for kilonovae at 200 Mpc, but is limited to only ≈ 200 deg^2 in the r band. At a distance of 100 (300) Mpc, these values change to ≈ 350 (120) deg^2 for r band and 750 (220) deg^2 for J band.

In Figure 6-10, we also plot the distances and areas enclosing 90% localization probabilities from the skymaps of the 15 events from one realization of our realistic BNS rate simulation. For an r-band search, seven of the 15 events have 90% areas lying in the $f_{\rm det} > 0.1$ region. In the J band, eight events lie in this region.

6.4.2   WINTER GW follow-up strategy

WINTER is a dedicated instrument for follow-up of slowly fading infrared kilonovae and can spend weeks searching for a single event. However, even though longer searches cover larger fractions of the skymaps, they risk covering the high-probability areas less efficiently or observing the kilonova once it has already faded. Based on the results from simulating WINTER observations with the Kasen and Bulla models, WINTER will dedicate up to seven nights of searching for each event. WINTER will observe compelling transients until they fade, and if there are no candidate kilonovae in the data, will stop searching after seven nights.

Furthermore, WINTER will not follow up all BNS GW alerts. Figure 6-10 provides a prescription to select which GW triggers are worth following up with WINTER during O4 based on information that is available at the time of the trigger. We will only follow events that have median distance estimates and localization areas such that the chance of detecting a kilonova is at least 10% (i.e., lying within the $f_{\rm det} > 0.1$ region of Figure 6-10). With a depth of $J_{AB} = 21$ (matching the WINTER reference images with a single exposure time $t_{\rm exp} = 450$ s), this means we can follow up events with distances of up to 350 Mpc if the time to tile the localization area is less than two days. We can follow up nearby events for much longer, with tiling times of six days for 200 Mpc. With WINTER's 1 deg^2 FOV, this corresponds to $\approx$ 150 deg^2 at 350 Mpc and 300 deg^2 at 250 Mpc.

If the localization areas are larger and the events are nearer, we will reduce our exposure times to tile the localizations faster and increase the chances of detecting a kilonova. Figure 6-11 shows the $f_{\mathrm{det}} = 0.1$ contours for searches with depths of 21 mag ($t_{\mathrm{exp}} = 450$ s), 20.5 mag ($t_{\mathrm{exp}} = 180$ s) and 19.5 mag ($t_{\mathrm{exp}} = 40$ s) as a function of the localization area that can be searched with WINTER. With WINTER, we will follow up all events with localization areas < 300 deg^2 down to a depth of 21 magnitudes, matching WINTER's J-band reference images. For events with localization areas larger than 300 deg^2 and distances < 250 Mpc, we will reduce exposure time to 180 seconds (i.e. to a depth of 20.5 magnitudes). With 180 second exposures, we can tile areas as large as 1000 deg^2 while maintaining $f_{\mathrm{det}} = 0.1$. For events with localization areas larger than 1000 deg^2, we will follow only those events that are closer than 150 Mpc to a depth of $J_{AB} = 19.5$ magnitudes. If there are multiple gravitational-wave triggers of interest on the same night (as discussed in Table 8.1), we will prioritize events that have a higher f_{det} and are easier to observe with WINTER.

Finally, we have assumed that the areas mentioned above enclose all of the BNS localization probability. If the full skymap areas are larger than the limits mentioned above, we will cover only those events where the area enclosing 50% of the localization probability can be tiled with $f_{\mathrm{det}} > 0.2$. We also note that in Figure 6-10, the detection fraction f_{det} drops at very early search times ($t < 0.5$ day), as the kilonova is still brightening in our models at these times. However, we will observe events with localization areas that can be tiled within 0.5 day, observing them repeatedly until the kilonova becomes bright enough to be detected. Similar detectability constraints were derived by ([216]; their Figure 17) using the Los Alamos National Laboratory (LANL, 936) grid of kilonova models. The constraints presented in Figure 6-10 are broadly consistent with their constraints, with minor variations attributable to the differences in the underlying Bulla and LANL models (see, for example, the differences in the analyses of Refs. [82, 864, 291]).

The methods described above can be used by other surveys to select GW triggers for follow-up during O4. We include a `python` notebook with the code to reproduce Figures 9–11 in the `Zenodo` repository at [381].

6.4.3   The realities of electromagnetic follow-up observing

The methods outlined in Section 6.2.3 represent a simplified approach to follow-up observations, where one observing strategy is decided at the outset and followed unchangingly throughout the campaign. In reality, follow-up observing is an iterative process where new decisions are made nightly or even multiple times per night. Instead of solely following one search strategy as shown in the above simulations, we will prioritize repeat observations of compelling transients found in the follow-up data from WINTER and other telescopes. In simulations, WINTER observes at the zenith to a limiting magnitude of $J_{AB} = 21$ in 450 seconds, $J_{AB} = 21.8$ in 30 minutes, and $J_{AB} = 22.7$ in 3 hours, with many repeat observations allowing for more significant constraints on kilonova detection.

Furthermore, if another telescope discovers a new kilonova candidate, infrared localizations, not just discoveries, add unique data to kilonova science. A nondetection in WINTER of an optical kilonova constrains both the dynamical and wind ejecta masses [216]. Additionally, faint detections ($SNR_{EM} < 5$) or single images of the event in the near infrared contribute to models of chemical evolution, ejecta mass, and wind speeds [114, 247, 496].

However, with no confirmed discoveries from other telescopes, there can be thousands of candidate transient events to sort through in a week of near infrared survey data. There are many tools available for classifying transients in survey data with machine-learning algorithms becoming a standard tool in the field, particularly for wide-field optical surveys, such as ZTF and the Vera C. Rubin Observatory Legacy Survey of Space and Time (LSST; [390, 441, 418, 578, 827, 888]).

Carefully planning the color and cadence of WINTER follow-up observations can assist these transient classification techniques in narrowing down the number of candidate events. For example, the characteristic ~ 1 week fading time distinguishes a kilonova lightcurve from longer-lasting supernovae or short-lived asteroids [255]. Additionally, observations in at least two filters assist in studying the reddening of the transient over time, a distinctive feature of kilonovae seen in GW170817 [92, 702, 818]. For optical telescopes, the i band provides the reddest images, and models predict kilonova discovery is maximized with the g — i filter pair for LSST [83] and a g, r, i filter cycle for ZTF [74]. In the simulation described in Section 6.2.3, we only observe in the J-band to leverage WINTER's J-band reference images and all-sky survey. In practice, we aim to conduct most of the search in J band but also follow up all interesting candidate events in the Y-band to study the Y — J color evolution. We will also study the g — J color pair, either with g-band images from ZTF follow-up or with the optical camera on WINTER's companion port. Time permitting, WINTER and its counterpart optical camera will also observe candidate events in the u, r, i, and H filters.

6.5   Conclusion

The BNS merger GW170817 brought about a new field of multimessenger astronomy, but despite extensive follow-up campaigns in O3, we have not observed a second multimessenger kilonova. In this study, we show infrared observations are a promising avenue for kilonova discovery, particularly for lanthanide-rich "red" kilonovae, as these are detectable to larger distances in the infrared than at optical wavelengths. We predict that infrared follow-up of GW triggers with WINTER could discover up to ten new kilonovae per year during O4. Furthermore, by employing more targeted follow-up strategies than those we have simulated, we can achieve a deeper sensitivity on a subset of interesting targets, therefore enhancing our ability to confirm new discoveries.

Moreover, we limit this study to BNS mergers and leave the study of NSBH kilonovae to a future work. Infrared follow-up of NSBH kilonovae is especially promising, as they are brighter in the infrared compared to BNS kilonovae [82, 340, 972], and observing both event types has the potential to increase the number of kilonovae discovered each year. Even just one new electromagnetic observation of a kilonova in O4 will double the number of known multimessenger kilonovae, helping to answer ongoing questions in the study of the Hubble tension, the neutron star equation of state, and r-process nucleosynthesis.

Chapter 7

Population properties and multimessenger prospects of neutron star-black hole mergers following GWTC-3

The content of this chapter was previously published in the Monthly Notices of the Royal Astronomical Society as Ref. [154] in Oct. 2022. ASB conceived the project, performed the analysis, and wrote the manuscript.

Abstract

Neutron star-black hole (NSBH) mergers detected in gravitational waves have the potential to shed light on supernova physics, the dense matter equation of state, and the astrophysical processes that power their potential electromagnetic counterparts. We use the population of four candidate NSBH events detected in gravitational waves so far with a false alarm rate ≤ 1 yr^{-1} to constrain the mass and spin distributions and multimessenger prospects of these systems. We find that the black holes in NSBHs are both less massive and have smaller dimensionless spins than those in black hole binaries. We also find evidence for a mass gap between the most massive neutron stars and least massive black holes in NSBHs at 98.6% credibility. Using an approach driven by gravitational-wave data rather than binary simulations, we find that fewer than 14% of NSBH mergers detectable in gravitational waves will have an electromagnetic counterpart. While the inferred presence of a mass gap and fraction of sources with a counterpart depend on the event selection and prior knowledge of source classification, the conclusion that the black holes in NSBHs have lower masses and smaller spin parameters than those in black hole binaries is robust. Finally, we propose a method for the multimessenger analysis of NSBH mergers based on the nondetection of an electromagnetic counterpart and conclude that, even in the most optimistic case, the constraints on the neutron star equation of state that can be obtained with multimessenger NSBH detections are not competitive with those from gravitational-wave measurements of tides in binary neutron star mergers and radio and X-ray pulsar observations.

7.1 Introduction

The detection of the neutron star-black hole (NSBH) mergers GW200105 and GW200115 [47] by the LIGO [3] and Virgo [53] gravitational-wave (GW) observatories confirmed the existence of this class of sources and has heralded the study of their population properties [52, 855, 330, 536, 969, 953]. Their inferred intrinsic properties and merger rates are consistent with the broad range of predictions for the astrophysical population of neutron-star black hole mergers [218, 178, 581], although isolated binary evolution likely dominates among the potential formation channels for NSBH systems based on rate arguments [581]. The chirp masses of these systems are expected to lie between $\sim 1.5 - 5\ M_\odot$ with a peak around 3 $M_\odot$, but the shapes and widths of the individual neutron star (NS) and black hole (BH) mass distributions for NSBH are much more uncertain theoretically [218, 178]. If the black hole forms second among the two compact objects, it can acquire spin through tidal spin-up of its progenitor, depending on the orbital separation prior to the second supernova [725, 128, 218, 475]. This is expected for up to $\sim 20\%$ of the intrinsic population of NSBH mergers [178, 217], while systems where the black hole forms first are expected to have negligible spin due to efficient angular momentum transport [832, 388, 387].

NSBH mergers are also potential multimessenger sources if the neutron star gets tidally disrupted outside the black hole innermost stable circular orbit [689, 367, 368, 687]. In this case, the disrupted material can power a range of electromagnetic counterparts including a kilonova [853, 854, 340, 505] and short gamma-ray burst (GRB) jet [630, 485, 692, 803, 775]. The detection of an electromagnetic counterpart to a NSBH merger observed in gravitational waves (or lack thereof) can be used to place multimessenger constraints on the neutron star equation of state (EoS), remnant mass, and the properties of the kilonova and GRB jet [94, 112, 462, 250, 245, 216, 728, 790]. Direct measurements of the EoS via gravitational-wave constraints on the tidal deformability of the neutron star in the binary are particularly difficult for NSBH systems—especially those with low signal-to-noise ratios (SNRs) and unequal mass ratios [531, 477]—so taking advantage of the multimessenger information encoded in these systems offers an alternative approach to constrain the EoS.

The latest catalog of gravitational-wave sources (GWTC-3) published by the LIGO-Virgo-KAGRA (LVK) collaboration includes 69 binary black hole (BBH) mergers, four NSBHs, and two binary neutron star (BNS) mergers detected with false alarm rate (FAR) less than 1 per year [44]. GW190814 also meets this FAR criterion, although its source characterization is uncertain as the secondary object is either the most massive neutron star or least massive black hole detected to date [39]. Previous works have sought both to compare the population properties of the black holes and neutron stars in NSBH mergers to those in BBH or BNS mergers and to constrain the properties of the compact object population as a whole. While the BBH primary mass distribution spans the mass range between $\sim 5 - 80\ M_\odot$, Ref. [969] finds that the masses of the black holes in NSBHs only extend out to $\sim 50\ M_\odot$ when considering a population of five NSBHs, including one source detected with FAR $> 1\,\mathrm{yr}^{-1}$, GW191219_163120. Similarly, the distributions of the effective aligned and precessing spins of NSBHs are found to favor smaller values than those of BBHs. Taking GW190814 to be a NSBH merger with a spinning neutron star, Ref. [953] finds evidence for a mass gap between the lightest black holes and heaviest neutron stars, although Refs. [52, 330] find that this evidence weakens when considering the full population of compact objects regardless of source type.

Previous works have also predicted the fraction of NSBH mergers that are expected to be accompanied by an electromagnetic counterpart, $f_{\mathrm{EM\text{-}bright}}$, although none have done so by simultaneously fitting for and marginalizing over the binary mass and spin distributions and by accounting for the uncertainty in the NS EoS. Refs. [303, 759,

178, 370] all find that NSBH mergers are unlikely sources of electromagnetic radiation based on population synthesis simulations of NSBH formation via isolated binary evolution; however, $f_{\text{EM-bright}}$ varies across their studies from $\sim 10^{-2} - 0.7$ depending on the binary evolution parameters. Higher $f_{\text{EM-bright}}$ is expected for higher black hole spin aligned to the orbital angular momentum and stiffer equations of state, both of which are qualitatively disfavored by current gravitational-wave observations. Ref. [227] constrains the contribution of NSBH mergers to r-process nucleosynthesis based on the observed populations of galactic neutron stars and binary black holes, finding that BNSs contribute at least twice as much to r-process element production.

In this work, we measure the population properties of NSBH sources first using gravitational-wave data alone, focusing on the pairing function determining the distribution of the mass ratio between the neutron star and the black hole. We use the posteriors on the population hyperparameters to take a data-driven approach to estimating $f_{\text{EM-bright}}$ by marginalizing over the uncertainty in the mass and spin distributions along with the EoS. We present the methods and results for the gravitational-wave-only analysis in Section 7.2. Since we find that the $f_{\text{EM-bright}}$ posterior peaks strongly at $f_{\text{EM-bright}} = 0$, we next extend the analysis to determine what constraints can be placed on the neutron star EoS under the best-case assumption that no electromagnetic counterpart was identified for any of the four NSBHs in our observed gravitational-wave population because none was produced, presenting the method and results in Section 7.3. We conclude with a discussion of the caveats and astrophysical implications of our results in Section 7.4.

7.2 Gravitational-wave-only analysis of NSBH population properties

7.2.1 Methods

We first employ the framework of hierarchical Bayesian inference to measure the mass and spin distributions of NSBH mergers using only gravitational-wave data.

Our population includes the four NSBH events reported in GWTC-3 detected with FAR ≤ 1 yr^{-1}: GW190426_152155, GW190917_114630, GW200105_162426, and GW200115_042309, listed in Table 7.1[1]. We do not include GW190814 as the secondary, less massive object in the binary is likely too massive to be a neutron star [39, 322], and it is a known outlier relative to observed NSBH (and BBH) systems [52, 49]. We seek to obtain posteriors for the hyperparameters $\mathbf{\Lambda}_{\mathrm{GW}}$ governing the population-level distributions of binary parameters like the masses and spins, $\boldsymbol{\theta}$, given a set of N events with data $\{d\}$,

$$p(\mathbf{\Lambda}_{\mathrm{GW}}|\{d\}) \propto p(\{d\}|\mathbf{\Lambda}_{\mathrm{GW}})\pi(\mathbf{\Lambda}_{\mathrm{GW}}), \tag{7.1}$$

where $\pi(\mathbf{\Lambda}_{\mathrm{GW}})$ is the prior on the hyperparameters, and the likelihood is given by multiplying the individual-event likelihoods marginalized over the binary parameters $\boldsymbol{\theta}$ [874],

$$p(\{d\}|\mathbf{\Lambda}_{\mathrm{GW}}) \propto \frac{1}{\alpha(\mathbf{\Lambda}_{\mathrm{GW}})^N} \prod_i^N \sum_j \frac{\pi_{\mathrm{pop}}(\boldsymbol{\theta}_{i,j}|\mathbf{\Lambda}_{\mathrm{GW}})}{\pi_{\mathrm{PE}}(\boldsymbol{\theta}_{i,j})}. \tag{7.2}$$

Here, $\pi_{\mathrm{PE}}(\boldsymbol{\theta})$ is the original prior applied for the binary parameters $\boldsymbol{\theta}$ during the individual-event parameter estimation step, while $\pi_{\mathrm{pop}}(\boldsymbol{\theta}|\mathbf{\Lambda}_{\mathrm{GW}})$ is the population-level distribution we assume describes the data characterized by hyperparameters $\mathbf{\Lambda}_{\mathrm{GW}}$. The index j represents the individual-event posterior samples, which we use to perform a Monte Carlo integral to marginalize over $\boldsymbol{\theta}$, and the index i indicates the event in our set of four NSBHs.

For the individual-event posterior samples, we use the results publicly released by the LVK including the effects of gravitational-wave emission from higher-order modes and spin precession,[2] where the prior on the dimensionless spin magnitude of both

---

[1]This is the same FAR threshold applied by the LVK in the binary black hole analyses presented in [52], although that work used a threshold of FAR ≤ 0.25 yr^{-1} for analyses with events containing neutron stars. We exclude the candidate NSBH event GW190531_023648 reported in [42] with maximum FAR $= 0.41$ yr^{-1}, as parameter estimates are not available. This is consistent with the treatment of this event in [52].

[2]This corresponds to the `PrecessingSpinIMRHM` samples for the 2019 events [33, 43], and the `C01:Mixed` samples for the 2020 events [46].

Table 7.1: Event names, FAR values, and references for the candidate signals included in our analysis.

Name	FAR (yr^{-1})	Reference
GW190426_152155	9.12×10^{-1}	[42]
GW190917_114630	6.56×10^{-1}	[44]
GW200105_162426	2.04×10^{-1}	[47]
GW200115_042309	$< 10^{-5}$	[47]

the black hole and neutron star covers the range $\chi \in [0, 1]$ and the spin tilts can be misaligned relative to the orbital angular momentum. We model the neutron star as a point mass, as there are no currently-available waveform models that include the effects of higher order modes, misalignment of the tilt of the black hole spin, and the tidal deformability of the neutron star. However, previous works have found that the effect of tides on the waveform for NSBH sources is the least significant of the three aforementioned processes [477].

The term $\alpha(\mathbf{\Lambda}_{\text{GW}})^N$ in the denominator of Eq. 7.2 represents the fraction of sources drawn from a population model with hyperparameters $\mathbf{\Lambda}_{\text{GW}}$ that would be detected. This correction accounts for the bias due to gravitational-wave selection effects in our chosen NSBH population and allows us to obtain unbiased estimates of the underlying astrophysical population, rather than the observed one [566, 585, 919]. We evaluate $\alpha(\mathbf{\Lambda}_{\text{GW}})$ by taking a Monte Carlo integral over detected events drawn from a simulated population following the method described in [334]. We use the sensitivity estimates for NSBH systems released at the end of the most recent LIGO-Virgo observing run (O3), obtained via a simulated injection campaign [45].

We assume the black hole mass distribution follows a truncated power-law [353],

$$\pi_{\text{pop}}(m_{\text{BH}}|\alpha, m_{\text{BH,min}}, m_{\text{BH,max}}) \propto \tag{7.3}$$

$$\begin{cases} m_{\text{BH}}^{-\alpha}, & m_{\text{BH,min}} \leq m_{\text{BH}} \leq m_{\text{BH,max}} \\ 0, & \text{otherwise} \end{cases}.$$

We explore two different possibilities for the pairing function governing the distribution of the mass ratio between the black hole and the neutron star, $q \equiv m_{\text{NS}}/m_{\text{BH}}$: a truncated Gaussian or another power law [355],

$$\pi_{\mathrm{pop}}(q|m_{\mathrm{BH}}, m_{\mathrm{NS,max}}, \mu, \sigma) \propto \tag{7.4}$$

$$\begin{cases} \mathcal{N}(q|\mu,\sigma), & q_{\mathrm{min}}(m_{\mathrm{BH}}) \leq q \leq q_{\mathrm{max}}(m_{\mathrm{BH}}, m_{\mathrm{NS,max}}) \\ 0, & \text{otherwise} \end{cases},$$

$$\pi_{\mathrm{pop}}(q|m_{\mathrm{BH}}, m_{\mathrm{NS,max}}, \beta) \propto \tag{7.5}$$

$$\begin{cases} q^{\beta}, & q_{\mathrm{min}}(m_{\mathrm{BH}}) \leq q \leq q_{\mathrm{max}}(m_{\mathrm{BH}}, m_{\mathrm{NS,max}}) \\ 0, & \text{otherwise} \end{cases},$$

emphasizing that the pairing function is a conditional distribution that depends on the value of the black hole mass. Because we assume that the black hole is always the more massive (primary) compact object in the binary, so that $q \leq 1$, this means that the range of allowed mass ratio values changes depending on the black hole mass. We fix the minimum neutron star mass to 1 $M_{\odot}$, so that $q_{\mathrm{min}} = 1/m_{\mathrm{BH}}$, and sample in the maximum neutron star mass, $m_{\mathrm{NS,max}}$, as a free parameter, such that $q_{\mathrm{max}} = \min(m_{\mathrm{NS,max}}/m_{\mathrm{BH}}, 1)$.

We fit the black hole spin with a Beta distribution with hyperparameters α_{χ} and β_{χ} [946],

$$\pi_{\mathrm{pop}}(\chi_{\mathrm{BH}}|\alpha_{\chi}, \beta_{\chi}) = \frac{\chi_{\mathrm{BH}}^{\alpha-1}(1 - \chi_{\mathrm{BH}})^{\beta-1}}{\mathrm{B}(\alpha_{\chi}, \beta_{\chi})}, \tag{7.6}$$

where $\mathrm{B}(\alpha_{\chi}, \beta_{\chi})$ is the Beta function. We do not explicitly fit the spin of the neutron star, but restrict it to lie within the breakup spin, χ_{Kep}, which represents the maximum neutron star spin at the mass-shedding limit. The exact value of the breakup spin depends on the EoS, but is about $\chi_{\mathrm{Kep}} \sim 0.7$ for most EoSs [802, 639]. We reweight the binary parameter posterior samples obtained under the precessing, high-spin prior for both the black hole and the neutron star so that the prior on the neutron star spin magnitude is uniform on $[0, 0.7]$. Choosing a prior that supports high spins avoids biases that could arise in the mass distribution due to mismodeling the neutron star

Table 7.2: Hyperparameters describing the mass and spin distributions and the maximum and minimum values allowed in the prior applied during hierarchical inference. The priors on all parameters are uniform.

Symbol	Parameter	Minimum	Maximum
α	black hole mass power-law index	-4	12
$m_{\mathrm{BH,min}}$	minimum black hole mass	$2\ M_\odot$	$10\ M_\odot$
$m_{\mathrm{BH,max}}$	maximum black hole mass	$8\ M_\odot$	$20\ M_\odot$
$m_{\mathrm{NS,max}}$	maximum neutron star mass	$1.97\ M_\odot$	$2.7\ M_\odot$
μ	mass ratio mean	0.1	0.6
σ	mass ratio standard deviation	0.1	1
β	mass ratio power-law index	-10	4
α_χ	black hole spin α	0.1	10
β_χ	black hole spin β	0.1	10

spin via the correlation between the component of the spin aligned with the orbital angular momentum and mass ratio [156, 953].

Our full population prior, $\pi_{\mathrm{pop}}(\boldsymbol{\theta}|\boldsymbol{\Lambda}_{\mathrm{GW}})$, is the product of the population distributions for the black hole mass, mass ratio, and black hole spin given in Eqs. 7.3-7.6. We reweight the publicly-released posterior samples into a redshift prior that is uniform in comoving volume and source-frame time. We use the DYNESTY sampler [828] as implemented in the GWPOPULATION [849] and BILBY [96] packages to draw samples from the posterior on $\boldsymbol{\Lambda}_{\mathrm{GW}}$ in Eq. 7.1. We apply uniform prior distributions for all the hyperparameters, $\pi(\boldsymbol{\Lambda}_{\mathrm{GW}})$, whose minimum and maximum values are given in Table 7.2. We note that these prior ranges by definition exclude GW190814 from our target population, as its primary mass is greater than the maximum black hole mass of $20\ M_\odot$ that we consider in our analysis.

7.2.2 Results

In Fig. 7-1 we show the inferred component mass, mass ratio, and black hole spin posterior population distributions (PPDs), which are the expected distributions for the binary parameters, $\boldsymbol{\theta}$, of the *astrophysical* population of new NSBH events irrespective of detectability inferred from the accumulated set of four detections:

$$p(\boldsymbol{\theta}|\{d\}) = \int \pi_{\text{pop}}(\boldsymbol{\theta}|\boldsymbol{\Lambda}_{\text{GW}})p(\boldsymbol{\Lambda}_{\text{GW}}|\{d\})d\boldsymbol{\Lambda}_{\text{GW}}. \qquad (7.7)$$

The results obtained under the Gaussian (top) and power-law (bottom) mass ratio models are qualitatively similar. The black hole mass distribution shown in the top left panels of each grid is constrained between $5.52^{+1.29}_{-2.66} - 9.96^{+8.54}_{-1.05}$ $M_\odot$ (maximum posterior value on the minimum and maximum black hole masses and 90% credible intervals calculated with the highest posterior density method) for the Gaussian pairing function, with discernible peaks at the maximum posterior values of both the minimum and maximum mass. This is due to a degeneracy between the maximum black hole mass, $m_{\text{BH,max}}$ and the black hole mass power-law index, α. The posterior on α has support for both positive and negative slopes; the branch with support for positive slopes strongly prefers a peak at $m_{\text{BH,max}} \sim 10$ $M_\odot$, while the negative-slope branch supports a wide range of maximum black hole masses. This can be understood in terms of the black hole mass posteriors for the individual events, shown in Fig. 7-2, as the posterior support for all four events falls off at $m_{\text{BH}} \gtrsim 10$ $M_\odot$. The maximum black hole mass inferred for the NSBH population using these four sources is significantly smaller than that inferred from the BBH population, which extends up to $80.60^{+18.68}_{-1.70}$ $M_\odot$ [45],[3] indicating that the black holes in NSBH systems are systematically lighter than those in BBH systems.

Our results also indicate that the black holes in NSBHs have systematically smaller spins than those in BBHs. Under both mass ratio models, the inferred spin distribution shown in the bottom left panels of each grid falls off steeply from $\chi_{\text{BH}} = 0$ and only extends up to a maximum spin magnitude of $\chi_{\text{BH,99}} \leq 0.68$ under both the Gaussian and power-law pairing functions at 90% credibility.[4] Support for high spin magnitudes is strongly suppressed relative to the 90% credible region encompassed by samples from the hyperparameter prior, shown in the dashed black line. Some of

---

[3] We use the publicly-released hyperparameter samples obtained under the Power Law + Peak mass model and Default spin model presented in [52] for all statements about the BBH population throughout the manuscript.

[4] $\chi_{\text{BH,99}}$ represents the spin at which 99% of the probability for the Beta distribution is enclosed.

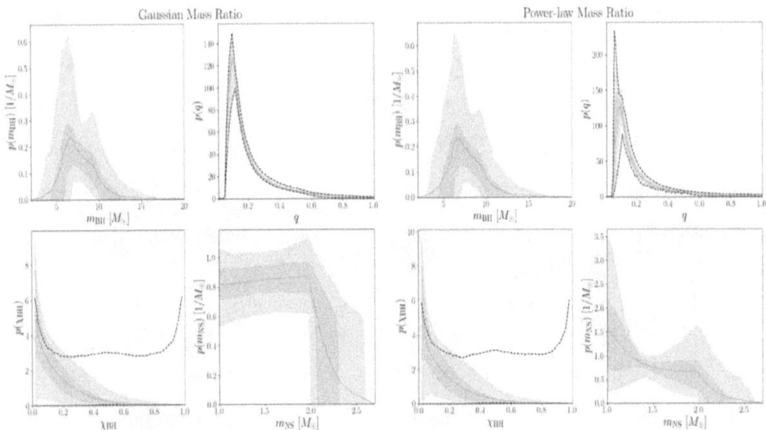

Figure 7-1: Posterior predictive distributions (solid blue) and 50% and 90% credible intervals (shaded blue) for the component masses, mass ratio, and black hole spin in the underlying, astrophysical NSBH population under the Gaussian mass ratio model (top) and power-law model (bottom). The black dashed lines show the 90% credible region enclosed by draws from the hyperparameter prior for the black hole spin and mass ratio.

the differences between our inferred spin distribution and the BBH spin distribution presented in [52] can be attributed to differences in the prior on the spin hyperparameters (see Appendix 7.5.2 for a detailed comparison), but the BBH spin distribution does not fall off as steeply and extends to higher spin magnitudes.

The upper right panels of each grid show the inferred mass ratio distribution under each mass ratio model. The shape of the mass ratio distribution is much more strongly determined by the priors on the maximum and minimum black hole and neutron star masses than by the functional form of the pairing function. Since the neutron star can only take on masses in the range $m_{\mathrm{NS}} \in [1, 2.7]\ M_\odot$ and the black hole mass distribution covers the range $m_{\mathrm{BH}} \in [2, 20]\ M_\odot$, this means that much of the prior probability is clustered around $q \sim 0.1$, with a lower bound of $q = 0.05$ set by the minimum neutron star and maximum black hole masses. Particularly for the Gaussian model, there is very little information gained in the posterior shown in blue relative to the 90% credible region allowed by the prior, shown in the dashed black line. We are only able to exclude narrow distributions peaked towards equal masses, with $\sigma \lesssim 0.2$ and $\mu \gtrsim 0.5$. Even for these distributions with higher values of μ, the nature of the priors on the component mass ranges is such that the Gaussian is truncated well below the peak for all but the lowest black hole masses, which means that most of the probability still lies around $q \sim 0.1$ The posterior on the power-law index under the power-law model is more informative, $\beta = -1.42^{+4.72}_{-3.32}$, although the shape of the mass ratio distribution under this model is similarly dominated by the choice of component mass ranges.

The implied neutron star mass distributions given the inferred black hole mass distributions and pairing function are shown in the remaining panel of each grid. While the shapes of the distributions obtained under the two mass ratio models are different, the constraints on the maximum neutron star mass are similar, $m_{\mathrm{NS,max}} = 2.07^{+0.59}_{-0.10}\ M_\odot$ for the power-law model and $m_{\mathrm{NS,max}} = 2.03^{+0.56}_{-0.06}\ M_\odot$ for the Gaussian model. The posterior on $m_{\mathrm{NS,max}}$ peaks at the lower bound of the prior, which is a conservative lower limit on the maximum NS mass observed electromagnetically [87, 258, 366]. The posterior support does extend up to higher masses rather than railing narrowly against the lower edge of the prior, which would indicate that the true value of the maximum neutron star mass may actually lie below the lower edge of the prior. This would hint that the maximum mass of neutron stars in NSBHs detected in gravitational waves is smaller than the maximum mass of neutron stars detected electromagnetically, on which the value of the lower edge of the prior on $m_{\mathrm{NS,max}}$ is based. Instead, we find that the two are consistent. The shape of the neutron star mass distribution is flatter under the Gaussian model, falling off sharply at the maximum mass, while for the power-law model the distribution peaks at the lower edge of the allowed mass range, but the posterior still supports a flat distribution. The effect of reweighting the individual-event posterior samples by the inferred population distribution, as shown on the right in Fig. 7-2, is to suppress the high-spin, equal-mass, and high-neutron-star-mass tails present under the original prior.

In order to determine if there is a statistical preference between the Gaussian and power-law pairing functions, we perform a posterior predictive check to compare the inferred detectable populations under each model to the observed population. To do

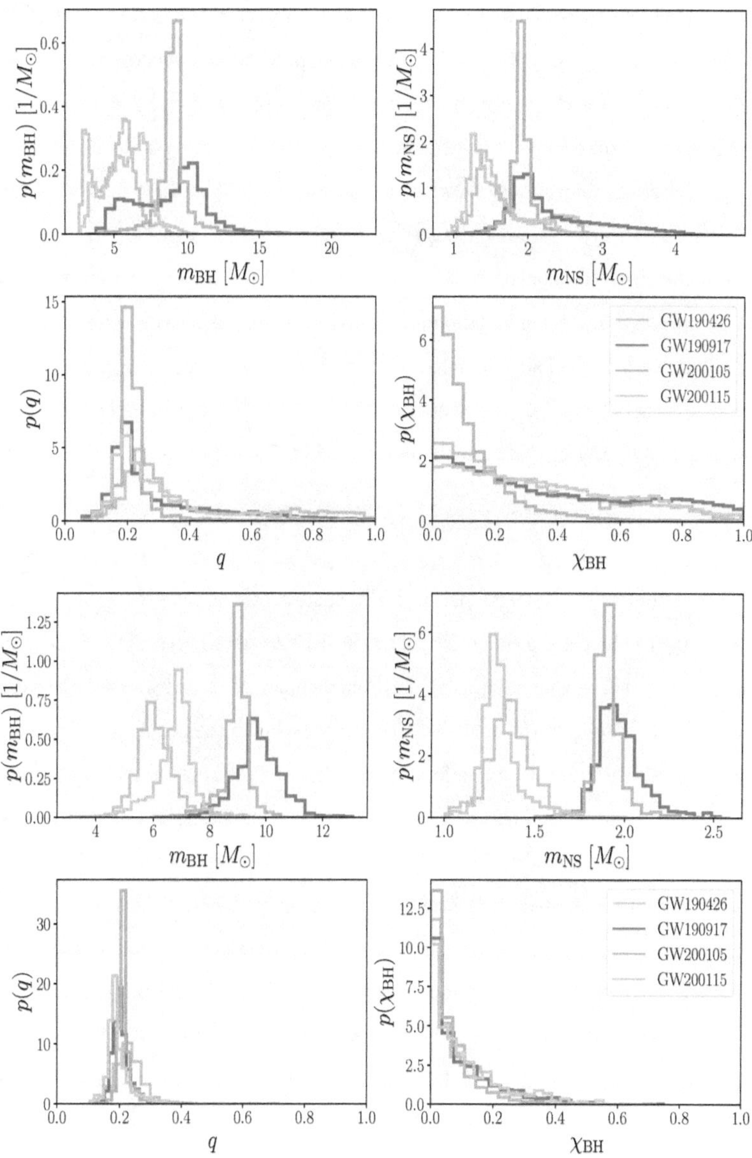

Figure 7-2: Distributions of the component masses, mass ratio, and black hole spin for each of the four individual events in our NSBH population under the original priors (left) and reweighted into the population prior inferred under the Gaussian mass ratio model (right).

this, we reweight the simulated detected events used to compute $\alpha(\Lambda_{\mathrm{GW}})$ in Eq. 7.2 by the full population distribution $\pi_{\mathrm{pop}}(\boldsymbol{\theta}|\Lambda_{\mathrm{GW}})$ implied by each hyperparameter posterior sample. For each hyperparameter posterior sample, we draw $N = 4$ simulated events from the reweighted set of detected events. We then compute the CDF of the mass ratio of the four events, showing the median and 90% credible interval of the CDFs across all hyperparameter posterior samples in blue in Fig. 7-3. To compare against the observed population, we draw one sample from the posterior on the mass ratio reweighted into the population prior implied by each hyperparameter posterior sample for each of the four real detected events (one sample from each histogram in the right panel of Fig. 7-2). We compute the CDF of these four samples and show the median and 90% credible interval in grey in Fig. 7-3.

The results for the Gaussian pairing function are shown on the left and the power-law pairing results are shown on the right. The discrete steps in the CDFs come from the fact that each is computed using only four samples. The CDF range for the observed events lies within the predicted CDF range based on the population results for both models. Because we have so few observations, there is considerable uncertainty in both the predicted and observed distributions, and the posterior predictive check does not lend more support to one model or the other. Both provide suitable fits to the data given the limited number of observations in our sample. This is consistent with the fact that the shapes of the mass ratio distributions for the Gaussian and power-law models are so similar due to their dependence on the component mass ranges via the minimum and maximum mass ratios. We note that while the observed CDFs peak around $q \sim 0.2$, the astrophysical mass ratio distribution shown in Fig. 7-1 instead peaks at lower mass ratios, around $q \sim 0.1$. At fixed chirp mass, binaries with more equal mass ratios take longer to merge and hence have more time to accumulate SNR in the detector, making them easier to detect. This selection effect explains the shift towards more equal mass ratios between the astrophysical and observed distributions.

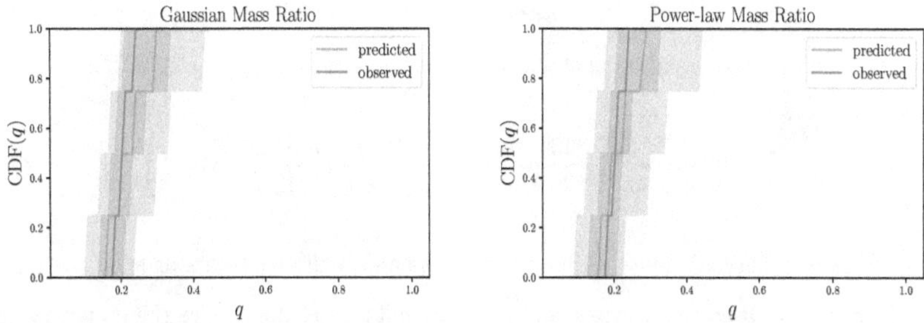

Figure 7-3: Posterior predictive check comparing the inferred population under the Gaussian (blue, left) and power-law (blue, right) mass ratio models to the observed population (grey).

7.2.3 EM-bright fraction

As an extension of the population measurements reported in the previous section, we can calculate the implied fraction of GW-detectable NSBH systems which can produce an electromagnetic counterpart due to tidal disruption of the neutron star outside the black hole's innermost stable circular orbit (ISCO) radius, R_{isco}, before the merger. We use the fitting formula for the remnant mass $\hat{M}_{\mathrm{rem}}$ presented in [368], which depends on the symmetric mass ratio, $\eta = q/(1+q)^2$, black hole spin aligned with the orbital angular momentum, $\chi_{\mathrm{BH},z}$, and neutron star EoS via the compactness, $C_{\mathrm{NS}} = Gm_{\mathrm{NS}}/(R_{\mathrm{NS}}c^2)$:

$$\hat{M}_{\mathrm{rem}} = m_{\mathrm{NS},b}\left[\max\left(\alpha\frac{(1-2C_{\mathrm{NS}})}{\eta^{1/3}} - \beta\frac{R_{\mathrm{isco}}(\chi_{\mathrm{BH},z})C_{\mathrm{NS}}}{\eta} + \gamma, 0\right)\right]^\delta. \qquad (7.8)$$

Here, $m_{\mathrm{NS},b}$ is the baryonic mass of the neutron star (the sum of its binding energy and gravitational mass m_{NS}), and $\alpha = 0.406, \beta = 0.139, \gamma = 0.255, \delta = 1.761$. Although this fitting formula was developed for NSBH mergers with nonrotating NSs, in the absence of a more general fit in the literature, we assume that it is also valid for spinning NSs if we take C_{NS} to be the compactness of the rotating NS.

To express the baryonic mass as an approximate function of the gravitational mass and spin for any equation of state, we use the empirical fitting formula from [236]:

$$\frac{m_{\mathrm{NS},b}}{M_\odot} = \frac{m_{\mathrm{NS}}}{M_\odot} + \frac{13}{20}\left(\frac{m_{\mathrm{NS}}}{M_\odot}\right)^2\left(1 - \frac{1}{130}\left(\frac{m_{\mathrm{NS}}}{M_\odot}\right)^{3.4}\chi_{\mathrm{NS}}^{1.7}\right). \tag{7.9}$$

While the baryonic mass itself does not depend on the neutron star spin, the relationship between the baryonic and gravitational masses does, since the properties of spinning neutron stars differ from those of their non-spinning counterparts. Spinning neutron stars are supported against gravitational collapse up to higher masses due to their rotation. We follow [953] and calculate the maximum neutron star mass accounting for rigid rotation using the universal relation presented in [176, 639],

$$m_{\mathrm{NS,crit}} = m_{\mathrm{TOV}}\left[1 + a_1\left(\frac{\chi_{\mathrm{NS}}}{\chi_{\mathrm{Kep}}}\right)^2 + a_2\left(\frac{\chi_{\mathrm{NS}}}{\chi_{\mathrm{Kep}}}\right)^4\right], \tag{7.10}$$

with $a_1 = 0.132$, $a_2 = 0.071$, and m_{TOV} is the maximum mass that can be supported against gravitational collapse for a non-spinning neutron star. The breakup spin introduced in the previous section, χ_{Kep}, can be expressed in terms of the compactness at the TOV mass, $C_{\mathrm{TOV}} = C_{\mathrm{NS}}(m_{\mathrm{TOV}})$, as

$$\chi_{\mathrm{Kep}} = \frac{\alpha_1}{\sqrt{C_{\mathrm{TOV}}}} + \alpha_2\sqrt{C_{\mathrm{TOV}}}, \tag{7.11}$$

with $\alpha_1 = 0.045$, $\alpha_2 = 1.112$ [176, 523, 640, 802].

The gravitational mass of a spinning neutron star is higher than the gravitational mass of the NS with the same central density at rest, and its equatorial radius is also larger. However, Ref. [524] finds that the compactness of a rotating neutron star is the same as the compactness of the non-rotating neutron star with the same central density to within a few percent for astrophysically-realistic values of the compactness up to the breakup spin. If we can compute the radius of the non-spinning neutron star with the same central density, this allows us to calculate $C_{\mathrm{NS}} \approx C_{\mathrm{NS},0}$. Because the gravitational mass is the quantity that we measure with gravitational waves but

Table 7.3: Descriptions of the various neutron star mass parameters used in our analysis.

Symbol	Definition	Equation
m_{NS}	gravitational mass including the effect of spin	-
$m_{\mathrm{NS},b}$	baryonic mass, binding energy + gravitational mass	7.9
$m_{\mathrm{NS},0}$	gravitational mass of the non-spinning neutron star with $m_{\mathrm{NS}}, \chi_{\mathrm{NS}}$	7.12
m_{TOV}	maximum mass of a non-spinning neutron star for a particular EoS	-
$m_{\mathrm{NS,crit}}$	maximum mass of a spinning neutron star for a particular EoS	7.10

the mass-radius relation for each EoS is given in terms of the non-spinning mass and radius, we need a way to map from the gravitational mass of the spinning neutron star to its rest mass. We assume that the expression in Eq. 7.10 holds for all neutron star masses, such that

$$m_{\mathrm{NS},0} = m_{\mathrm{NS}} \left[1 + a_1 \left(\frac{\chi_{\mathrm{NS}}}{\chi_{\mathrm{Kep}}} \right)^2 + a_2 \left(\frac{\chi_{\mathrm{NS}}}{\chi_{\mathrm{Kep}}} \right)^4 \right]^{-1}, \tag{7.12}$$

where we use m_{NS} to indicate the neutron star mass measured with gravitational waves and $m_{\mathrm{NS},0}$ to indicate the rest mass of the object with measured gravitational mass m_{NS} and spin χ_{NS}.[5] The various neutron star masses that we use in our analysis are summarized in Table 7.3.

To account for the uncertainty in the EoS in our calculation of $\hat{M}_{\mathrm{rem}}$, we marginalize over the publicly released non-parametric EoS posterior samples [547] conditioned on data from gravitational-wave observations of binary neutron star mergers and radio and X-ray pulsar observations obtained by [546]. We associate each mass and spin population hyperparameter posterior sample with an EoS sample, Λ_{EoS}, requiring that the critical mass in Eq. 7.10 for that EoS is greater than or equal to the value of $m_{\mathrm{NS,max}}$ for that hyperparameter sample, so that the maximum mass in the astrophysical population is within the maximum mass supported by that EoS.

For each sample from $\Lambda = (\Lambda_{\mathrm{GW}}, \Lambda_{\mathrm{EoS}})$, we reweight the simulated detected events used to calculate $\alpha(\Lambda_{\mathrm{GW}})$ in Eq. 7.2 by the implied population distribution.

---

[5]We opt to use the approximation in Eq. 7.12 rather than the universal relation between the spinning and non-spinning gravitational masses presented in [524], as the latter requires calculating the EoS-dependent moment of inertia to convert between dimensionless spin and angular rotation frequency.

For the neutron star in each of the reweighted NSBH binaries, we calculate the radius and compactness given by the EoS defined by that Λ sample by interpolating the non-spinning mass-radius relation given in the [547] dataset. If a particular neutron star sample is above the maximum neutron star mass supported by that EoS, $m_{\text{NS,crit}}$, we remove that sample from our population, which is by definition restricted only to valid NSBH systems.

With the compactness calculated for each neutron star, we then calculate $\hat{M}_{\text{rem}}$ for each of the binaries in the reweighted population. Because there is no universally accepted threshold for the remnant mass above which a NSBH merger will be electromagnetically bright (EM-bright)[6], we report our results for the EM-bright fraction in terms of the fraction of GW-detectable sources for each Λ sample for which $\hat{M}_{\text{rem}} \geq M_{\text{rem,min}}$, with $M_{\text{rem,min}} = 0, 10^{-2}, 10^{-1} \ M_{\odot}$. By reweighting the simulated detected events, we are quantifying $f_{\text{EM-bright}}$ for the population of NSBH mergers detectable in gravitational waves rather than the underlying astrophysical population.

The posteriors for $f_{\text{EM-bright}} = f(\hat{M}_{\text{rem}} \geq M_{\text{rem,min}})$ shown in Fig. 7-4 are strongly peaked at $f_{\text{EM-bright}} = 0$ with $f(\hat{M}_{\text{rem}} \geq 0 \ M_{\odot}) \leq 0.11$ at 90% credibility for the Gaussian pairing function and $f(\hat{M}_{\text{rem}} \geq 0 \ M_{\odot}) \leq 0.14$ for the power-law pairing. This means that at most 14% of GW-detectable sources will have a chance to be EM-bright. The posteriors on the EM-bright fraction using $M_{\text{rem,min}} = 10^{-4}, 10^{-3} \ M_{\odot}$ are statistically similar to the result obtained with $M_{\text{rem,min}} = 0$, leading us to conclude that we cannot distinguish the effect of values of the minimum remnant mass below 10^{-3} on the NSBH EM-bright fraction. The results under the power-law pairing function extend to larger values of $f_{\text{EM-bright}}$, which can be explained in terms of the more gradual drop-off from the peak of the mass ratio distribution relative to the Gaussian pairing shown in Fig. 7-1. More equal mass ratios lead to more significant neutron star tidal disruption and a larger remnant mass. A similar effect is responsible for the samples predicting the largest EM-bright fractions, as those correspond to the

---

[6]Precursor emission produced before the NSBH merger has also been studied as a potential electromagnetic counterpart, sourced either via magnetospheric interactions or neutron star tidal resonances [see, e.g., 340]. We do not consider the possibility of such precursor emission in the formulation of our analysis.

hyperparameter posteriors that favor black hole mass distributions that peak at small masses with a large negative slope, leading to more equal mass ratios.

We can use the same method to place constraints on the absolute contribution of NSBHs to heavy metal production in the universe via the ejection of r-process elements. Instead of sampling from the detectable population of NSBH mergers as we did for the calculation of $f_{\text{EM-bright}}$, we sample from the inferred underlying astrophysical population of sources (shown in Fig. 7-1) in proportion to our inferred rate of NSBH mergers, marginalizing over the uncertainty in the population properties, merger rate, and EoS using the [546] posterior samples. We obtain a merger rate in the range $1.24 - 62.3$ $\text{Gpc}^{-3}\text{yr}^{-1}$ ($1.15 - 57.2$ $\text{Gpc}^{-3}\text{yr}^{-1}$) under the Gaussian (power-law) pairing function assuming a prior $\pi(\mathcal{R}) \propto 1/\mathcal{R}$ on the merger rate. Our corresponding posterior on the total r-process ejecta mass contribution from the astrophysical NSBH population rails strongly at 0 $M_\odot$, with 90% of the probability lying within $\hat{M}_{\text{rem}}^{\text{tot}} \leq 0.72$ $M_\odot/\text{Gpc}^{-3}\text{yr}^{-1}$ ($\hat{M}_{\text{rem}}^{\text{tot}} \leq 1.03$ $M_\odot/\text{Gpc}^{-3}\text{yr}^{-1}$).

To put this number in context, GW170817 produced at least 0.01 $M_\odot$ of ejecta (see, e.g., Table 1 of [242] and references therein). Assuming this is typical for BNS mergers, and adopting the preferred GWTC-3 BNS rate estimate of 170 Gpc^{-3} yr^{-1} [52], the total BNS ejecta mass yield is ~ 1.70 $M_\odot/\text{Gpc}^{-3}\text{yr}^{-1}$. Thus, this back-of-the envelope estimate indicates that NSBHs contribute at most $\sim 40\%$ of the r-process ejecta in the universe, consistent with the simulation-based estimates of [227], although we emphasize that the exact fraction depends sensitively on the assumed BNS population properties.

7.3 Multimessenger constraints on the EoS and NSBH population properties

7.3.1 Methods

We find that our population hyperparameter posteriors obtained using gravitational-wave data alone under the Gaussian (power-law) pairing function predict that there

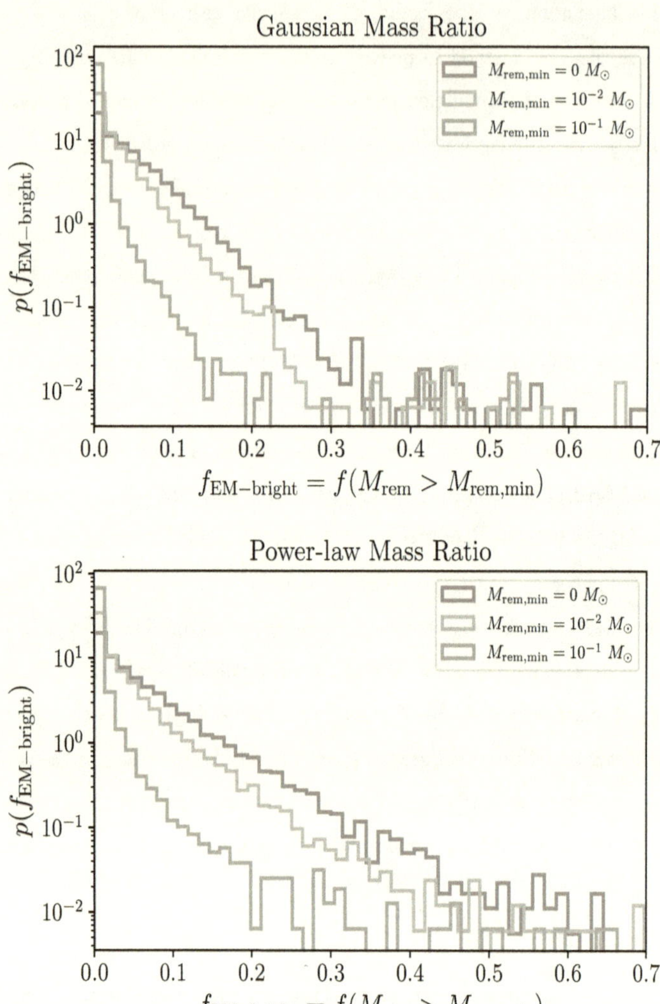

Figure 7-4: Posterior on the fraction of GW-detectable NSBH systems that will be electromagnetically bright with remnant mass $\hat{M}_{\rm rem} \geq M_{\rm rem,min}$, with the different colors indicating different values of $M_{\rm rem,min}$. The posterior is marginalized over the uncertainty in the neutron star equation of state and in the population hyperparameters for both the Gaussian (top) and power-law (bottom) pairing functions.

is a $98.8^{+1.2}_{-33.5}\%$ ($98.8^{+1.2}_{-43.6}\%$) probability that none of the four detected NSBH systems were EM-bright. This probability is obtained by applying a binomial distribution to each of the posterior samples in $f_{\mathrm{EM-bright}}$ shown in Fig. 7-4 for $M_{\mathrm{rem,min}} = 0 \ M_\odot$ with $p(\mathrm{success}) = f_{\mathrm{EM-bright}}$ for zero successes and four failures. Using this bound, we seek to determine if a meaningful multimessenger constraint can be placed on the NSBH population properties and neutron star EoS using the lack of detection of any electromagnetic counterpart for our four observed sources. We make the assumption that no counterpart was detected for any of the observed gravitational-wave events because there was no counterpart to detect, namely $M_{\mathrm{rem}} = 0$ for all four of the detected sources. We use $\hat{M}_{\mathrm{rem}}$ to indicate the predicted remnant mass given the NSBH binary parameters and EoS using Eq. 7.8, and M_{rem} to indicate a measured value of the remnant mass for an observed system.

We emphasize that the assumption we make represents the best-case scenario for EoS constraints from nondetection; in reality a number of factors contribute to the detectability of the EM counterpart, such as the intrinsic brightness of the emission if $M_{\mathrm{rem}} > 0$ depending on the lightcurve model, the sensitivity and coverage of the available telescope network, and the promptness and accuracy of the sky localization of the event released by the LVK [see, e.g. 617, 339]. We do not attempt to carefully account for these electromagnetic selection effects, as their contributions to the lack of counterpart detection are highly uncertain, and instead make the maximally optimistic assumption that $M_{\mathrm{rem}} = 0$ for each of the four NSBH systems detected in gravitational waves: i.e., we assume that the EM surveys would have detected any $M_{\mathrm{rem}} > 0$. As such, the results we present based on this assumption represent a conservative upper limit on the constraining power of nondetection in the multimessenger analysis of NSBH sources.

In addition to the gravitational-wave data for each event, we now also include the measurement of the remnant mass of each event as an independent data point, assuming $M_{\mathrm{rem}} = 0$. The incorporation of the remnant mass data into the hierarchical inference method amounts to the introduction of an additional term in the numerator of Eq. 7.2—the likelihood of observing remnant mass $M_{\mathrm{rem}} = 0$ given the parameters

$q, m_{\rm NS}, \chi_{\rm NS}, \chi_{\rm BH,z}$ and $\Lambda_{\rm EoS}$,

$$p(M_{\rm rem}|q, m_{\rm NS}, \chi_{\rm BH,z}, \Lambda_{\rm EoS}) \tag{7.13}$$

$$= \delta(M_{\rm rem} - \hat{M}_{\rm rem}(q, m_{\rm NS}, \chi_{\rm NS}, \chi_{\rm BH,z}, \Lambda_{\rm EoS})) \tag{7.14}$$

$$= \delta(\hat{M}_{\rm rem}(q, m_{\rm NS}, \chi_{\rm NS}, \chi_{\rm BH,z}, \Lambda_{\rm EoS})). \tag{7.15}$$

The full derivation of the multimessenger likelihood is given in Appendix 7.5.1. In practice, this means that in addition to the mass and spin hyperparameters, $\Lambda_{\rm GW}$, we also now sample in a set of hyperparameters describing the EoS, which enter the likelihood via the compactness that goes into the calculation of $\hat{M}_{\rm rem}$.

For the multimessenger analysis, we adopt the piecewise polytrope EoS characterized by four parameters introduced in [739], where the pressure as a function of density is given by

$$p(\rho) = K_i \rho^{\Gamma_i} \tag{7.16}$$

in each of three different regions, $i = 1, 2, 3$, with transition densities of $\rho_1 = 10^{14.7}$ g/cm^3 and $\rho_2 = 10^{15}$ g/cm^3. Requiring that the pressure be continuous across the transition densities means the full EoS can be determined by four parameters—the three Γ_i adiabatic indices and an overall pressure scale, $p_1 = p(\rho_1)$. Instead of using the [547] posteriors on the EoS as was done in Section 7.2.3, we now directly sample in $\Lambda_{\rm EoS} = (\log(p_1/[{\rm dyne/cm}^2]), \Gamma_1, \Gamma_2, \Gamma_3)$ to obtain an independent posterior from NSBH observations alone, applying a uniform prior on these EoS hyperparameters.

We follow [457] in setting the prior ranges on $\Lambda_{\rm EoS}$, shown in Table 7.4, and in imposing two additional constraints. We require the maximum mass of a non-spinning neutron star supported by the EoS determined by a particular draw from $\Lambda_{\rm EoS}$ to be $m_{\rm TOV} \geq 1.97 \, M_\odot$. This value is chosen to match the lower bound of the prior on the maximum neutron star mass hyperparameter described in Section 7.2.1, $m_{\rm NS,max}$. We also require that the EoS does not violate causality, so that the speed of sound in the neutron star is less than the speed of light. Because of the accuracy limitations of the

Table 7.4: Hyperparameters describing the piecewise polytrope neutron star equation of state and the maximum and minimum values allowed in the prior. The priors on all parameters are uniform.

Symbol	Parameter	Minimum	Maximum
$\log(p_1/[\mathrm{dyne/cm^2}])$	log-pressure at ρ_1	33.6	34.8
Γ_1	first adiabatic index	2	4.5
Γ_2	second adiabatic index	1.1	4.5
Γ_3	third adiabatic index	1.1	4.5

piecewise polytrope fit to the EoS, in practice we only enforce the causality constraint when the speed of sound exceeds $1.1c$. We calculate the radius and compactness given by the EoS defined by each Λ_{EoS} sample using the EOSINFERENCE package [530, 532] in conjunction with Eq. 7.12 to take into account the neutron star spin.

We do not impose additional EoS prior information, e.g., from the neutron star tidal deformability derived from gravitational-wave observations of the binary neutron star mergers GW170817 [17] and GW190425 [32], so our prior, shown in the dashed black line in Fig. 7-6, includes very stiff equations of state.[7] We make this choice to obtain an independent posterior on the EoS parameters using the NSBH data alone in order to compare the constraining power of this multimessenger analysis against the tidal information that can be extracted from GW observations of BNS mergers without any multimessenger observations.

For the multimessenger analysis, we only use the Gaussian pairing function model, since the gravitational-wave-only analysis revealed that the two pairing functions give statistically similar results. We no longer sample directly in $m_{\mathrm{NS,max}}$, and instead use the spin-dependent critical neutron star mass calculated using Eq. 7.10 for each sample from Λ, i.e., we impose that the maximum mass in the astrophysical NS population is the spinning maximum mass supported by the EoS, $m_{\mathrm{NS,max}} = m_{\mathrm{NS,crit}}$. This means that the maximum mass ratio, q_{max}, defined in Eq 7.4 is now a function of $(m_{\mathrm{BH}}, \chi_{\mathrm{NS}}, \Lambda_{\mathrm{EoS}})$ rather than just m_{BH} and the single hyperparameter $m_{\mathrm{NS,max}}$. We also impose that the neutron star spin should lie within the specific value of the

---

[7]Our prior and model choice for the EoS parameterization is different than that used in [546], although we do not expect the model differences to qualitatively affect the comparison of the two results in Fig. 7-6.

breakup spin, $\chi_{\rm Kep}$, defined by each EoS sample rather than setting a universal value of $\chi_{\rm Kep} = 0.7$ as was done in the GW-only analysis. For this multimessenger analysis, we generate samples from the posterior on Λ using the NESTLE sampler [111].

Since we allow the NS and BH mass ranges to overlap in order to probe the presence of a mass gap between them (see prior ranges in Table 7.2), it is possible that a binary in our population with two equal-mass components of $2\ M_\odot$ each, for example, could also be either a BNS or BBH by our own definitions of the NS and BH mass ranges. By insisting that all four of our events are NSBHs, we are imposing prior knowledge of source classification, which can lead to an overestimation of the information provided by a nondetection. If the population actually includes some low-mass BBHs, they should be discarded since they do not contribute any information to the constraint on the NS EoS. For now we include the simplifying assumption that all sources in the population are NSBHs, in the spirit of being maximally optimistic about the constraining power of our multimessenger analysis.

7.3.2   Results

The population distributions inferred for the NSBH binary parameters folding in the nondetection of any electromagnetic counterparts are shown in Fig. 7-5. The results are very similar to those obtained using the GW data only, shown in the top panel of Fig. 7-1, particularly for the black hole mass and spin distributions. The implied distribution for the neutron star mass tapers off more gradually as the upper limit is EoS-dependent under the multimessenger model, unlike the universal, hard cutoff at $m_{\rm NS,max}$ under the GW-only model. Because we are assuming $m_{\rm NS,max} = m_{\rm NS,crit}$ in the multimessenger case, the maximum neutron star mass depends on the spin and the EoS following Eq. 7.10. The posterior on the mass ratio distribution slightly prefers more extreme mass ratios for the multimessenger analysis compared to the GW-only analysis. This extra information on the mass ratio comes from the effect of q on the remnant mass. More equal mass ratios lead to more remnant mass, so enforcing that $M_{\rm rem} = 0$ pushes the posterior on the mean of the mass ratio distribution, μ, towards lower values.

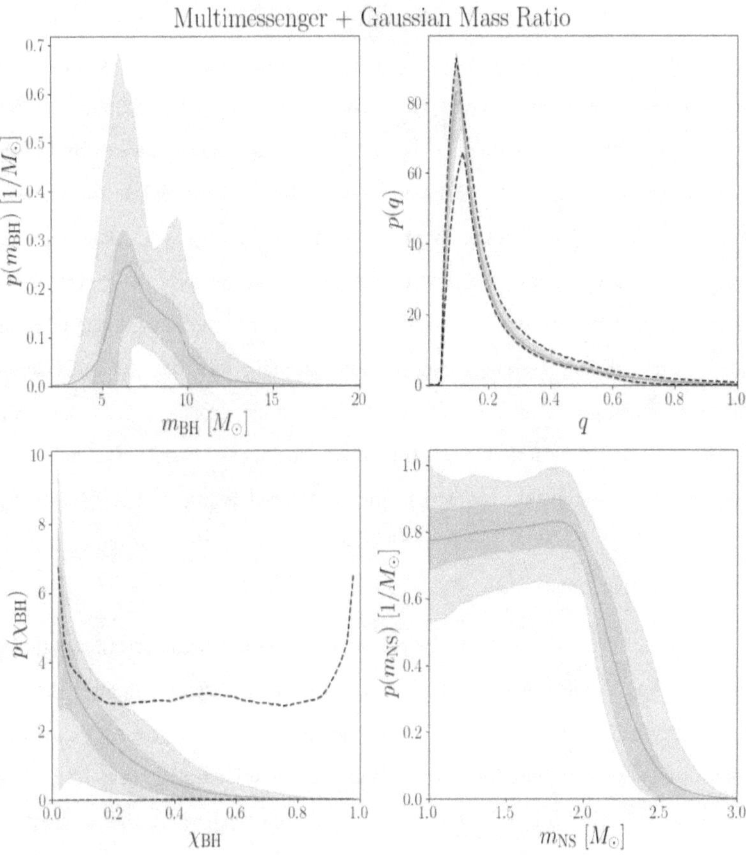

Figure 7-5: Posterior predictive distributions (solid blue) and 50% and 90% credible intervals (shaded blue) for the component masses, mass ratio, and black hole spin in the underlying, astrophysical population under the Gaussian mass ratio model from the multimessenger analysis that assumes none of the four detected NSBHs produced ejecta. The black dashed lines show the 90% credible region enclosed by draws from the hyperparameter prior for the black hole spin and mass ratio.

In Fig. 7-6, we show the constraints obtained on the neutron star mass-radius relation and EoS from the multimessenger NSBH analysis (blue) and the constrains from GW observations of BNS mergers and radio and X-ray pulsar observations from [546] (red). Our result in blue represents a first step towards an EoS constraint from gravitational waves that self-consistently accounts for neutron star spin. Compared to the mass-radius relations allowed by the prior on Λ_{EoS} (dashed black), our posterior rules out the stiffest EoSs yielding the largest radii for a given mass. Stiffer EoSs support more significant tidal disruption of the neutron star, which leads to enhanced remnant mass ejection. Thus, enforcing that there should be no remnant mass left over after the merger rules out this part of the EoS parameter space. Moreover, as can be seen in the EoS posterior in the right panel of Fig. 7-6, the constraining power of this hierarchical multimessenger nondetection method is relatively consistent across all plausible neutron star densities, rather than being more concentrated at a specific density scale as is typical for the constraint from tidal measurements of an individual BNS merger. We find that the upper limit on the radius from the NSBH analysis is comparable to the [546] analysis, while their lower limit on the radius is more constraining than ours.

Finally, in Fig. 7-7, we show the posterior on the EM-bright fraction obtained using our NSBH multimessenger analysis. Rather than marginalizing over the EoS uncertainty using the [547] posteriors, we use the posterior on the Λ_{EoS} piecewise polytrope parameters obtained using the method described in Section 7.3.1. The posterior on the EM-bright fraction is similar to the one shown in the top panel of Fig. 7-4 but is more narrowly peaked at $f_{\mathrm{EM\text{-}bright}} = 0$. We find $f(\hat{M}_{\mathrm{rem}} \geq 0) \leq 0.075$ at 90% credibility. Because our EoS constraints are comparable to those of [546], the source of this difference must be the improved measurement of the mass ratio distribution under the multimessenger model, which favors more extreme mass ratios and hence less remnant mass following the merger. In this case we find that the samples corresponding to the largest EM-bright fractions are driven by the hyperparameter posteriors that favor large black hole spins and stiff EoSs, consistent with the prediction that the NSs in these systems are more easily disrupted.

247

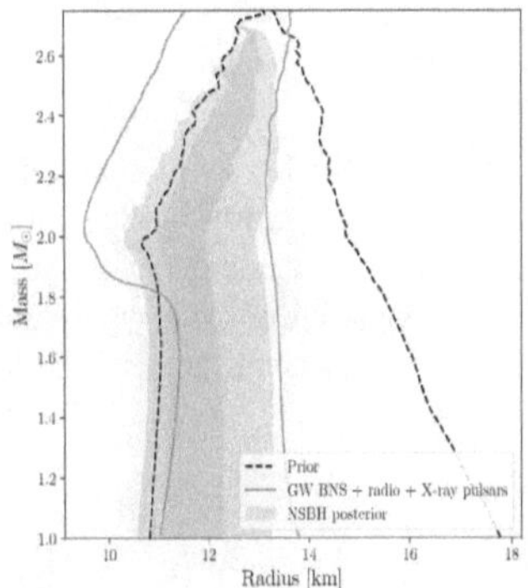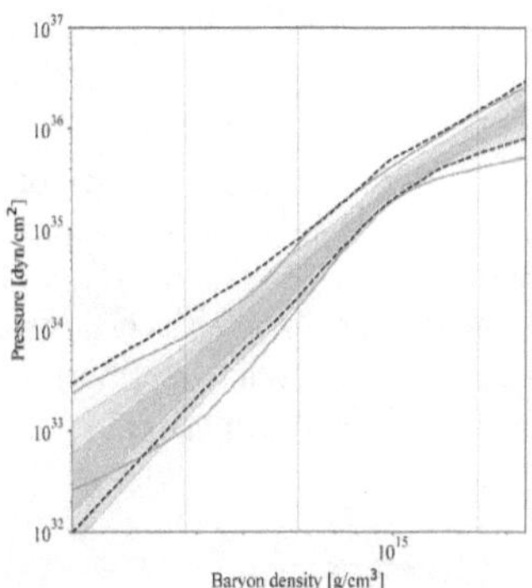

Figure 7-6: Constraints on the neutron star mass-radius relation (left) and equation of state (right) from our NSBH multimessenger analysis (blue) compared to those inferred using GW observations of BNS mergers and radio and X-ray pulsar observations obtained in [546] (red, 90% credible interval). The shaded regions show the 50% and 90% credible intervals, while the dashed black lines enclose the 90% credible region spanned by the prior on the Λ_{EoS} parameters. The faint blue lines show individual mass-radius relation or EoS posterior draws from our NSBH analysis. The grey vertical lines in the right panel indicate once, twice and six times the nuclear saturation density of 2.8×10^{14} g/cm^3.

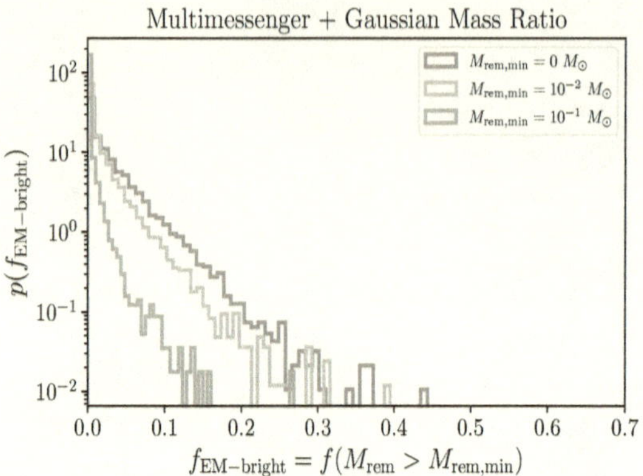

Figure 7-7: Posterior on the fraction of GW-detectable NSBH systems that will be electromagnetically bright with remnant mass $\hat{M}_{\rm rem} \geq M_{\rm rem,min}$, with the different colors indicating different values of $M_{\rm rem,min}$. The posterior is obtained by marginalizing over the uncertainty in both the binary population and equation of state hyperparameters under the NSBH-only multimessenger analysis that assumes none of the four detected NSBHs produced ejecta.

7.4   Discussion

In this work, we have analyzed the population properties and multimessenger prospects of four neutron star-black hole merger events detected in gravitational waves with a false alarm rate less than 1 per year [44], GW190426_152155, GW190917_114630, GW200105_162426, and GW200115_042309. We exclude GW190814 from our analysis as the classification of the secondary object as a neutron star is unlikely. We first measured the population distributions of the black hole mass and spin and binary mass ratio using gravitational-wave data alone, marginalizing over the uncertainty in these distributions along with realistic uncertainty in the neutron star EoS to obtain a posterior on the fraction of NSBH sources detectable in gravitational waves that will be electromagnetically bright. We then developed a new Bayesian multimessenger analysis method that folds in the nondetection of any electromagnetic counterpart for these four sources to obtain constraints on the mass and spin distributions, the neutron star EoS, and the EM-bright fraction.

We find that the maximum black hole mass in NSBH systems is much lower than the maximum mass for binary black hole systems, with a distance of 24 standard deviations of the NSBH $m_{\mathrm{BH,max}}$ measurement between them. Our measurement of the maximum mass is lower than that obtained in [969] due to their inclusion of an additional event with a more massive primary that was detected with lower FAR. Our measurement of the minimum black hole mass, on the other hand, is consistent with [969], [953], and the minimum mass inferred from the BBH population, which all peak around $\sim 5.5\ M_\odot$. When taken together with the inference on the maximum neutron star mass, we measure the width of the lower mass gap between the most massive neutron stars and the least massive black holes to be $3.52^{+1.15}_{-2.56}\ M_\odot$, with no mass gap (width $\leq 0\ M_\odot$) disfavored at 98.6% credibility under the Gaussian pairing function, consistent with the results of [953]. This inference is inherently contingent on the exclusion of GW190814 from our analysis. Our inferred neutron star mass distribution is qualitatively consistent with the result of [52]; the differences in minimum and maximum neutron star mass are largely attributable to different prior choices between our analyses.

Our inference on the mass ratio distribution is dominated by the prior ranges imposed on the component masses, which result in a peak at $q \sim 0.1$. We find no statistical preference between a Gaussian or power-law pairing function, as may be expected given our limited sample size of four detections. The distribution obtained under the power-law pairing function is more informative, however, and falls off more gradually from the peak. When we fold in the nondetection of any electromagnetic counterparts for the four NSBH systems in our population, the multimessenger analysis rules out a slightly larger region of the mass ratio hyperparameter space corresponding to large values of the mean and small values of the standard deviation for the Gaussian pairing function.

Unlike [953] which do not explicitly fit the black hole spin distribution and [969] which fit only the effective aligned and precessing spin distributions, we fit the black hole spin magnitude distribution directly, assuming that the spin orientations are isotropically distributed. We find that the spin magnitude distribution is strongly peaked at $\chi_{\rm BH} = 0$, with a maximum spin smaller than the BBH population, although some of the differences between the NSBH and BBH spin magnitude distribution can be attributed to different prior choices (see Appendix 7.5.2).

We have verified that our results are robust when using just the two events with the lowest FAR in the population, GW200105_162426 and GW200115_042309, as indicated in Table 7.1. This is expected due to the fact that the two lower-significance events have very similar masses to the two higher-significance events, as shown in Fig. 7-2. As such, the population parameters we infer with just two events are similar to those inferred with all four events, just less well-constrained (see Fig. 7-9). However, some of our results naturally depend on how we have defined the selected population and imposed prior knowledge of source classification. This is a feature of hierarchical inference when all analyzed candidates are assumed to be real astrophysical events belonging to a single population. For example, when we include GW190425—most likely a binary neutron star merger—in our population along with the four events previously analyzed, we find no evidence for a mass gap, with $m_{\rm BH,min} \leq 2.40\ M_\odot$ at 90% credibility. Correspondingly, the posterior on the EM-bright fraction shifts to higher values, with $f(\hat{M}_{\rm rem} > 0\ M_\odot) = 0.16^{+0.13}_{-0.16}$. More details on the analysis including GW190425 are included in Appendix 7.5.4. An alternative approach that we leave to future work would be to simultaneously infer the properties and classification of individual events into multiple sub-populations [330].

The differences between the mass and spin distributions of the black holes in NSBHs and BBHs may indicate that the two populations draw from different stellar progenitors and potentially form via different channels. While it is likely that some fraction of BBHs form dynamically [e.g. 762, 963, 173], potentially via hierarchical mergers [514, 642] in order to explain the support for masses in the "upper mass gap" predicted by isolated binary evolution models [453, 132] along with population-level evidence for spin precession [49, 52], the lower masses and smaller spin magnitudes observed in the NSBH population do not as strongly suggest a dynamical origin [954]. Concrete statements about the formation channels of NSBH systems are difficult to make with so few observations, however. A meaningful constraint on the distribution of black hole spin tilts would be particularly useful for this purpose; the posteriors for several of the candidate events we consider individually support tilts anti-aligned to the orbital angular momentum [422], although Ref. [588] finds that this is driven by the prior choice. They conclude that a more astrophysically-motivated prior [177, 217] yields posteriors consistent with small black hole spin, as we find in this work. We leave a full hierarchical analysis of the black hole tilt distribution for NSBHs to future work when more detections can be included in the inference.

In addition to constraining the distributions of the binary mass and spin parameters, we present a data-driven estimate of the fraction of NSBH sources detectable in gravitational waves that may also have an electromagnetic counterpart. When considering the gravitational-wave NSBH data alone in conjunction with recent constraints on the neutron star EoS from gravitational-wave observations of BNS mergers and radio and X-ray pulsar observations, we find that at most 14% of detectable sources will be left with any remnant mass outside the black hole ISCO radius that can potentially power an electromagnetic counterpart. Consistent with the small fraction of sources that have the potential to be EM-bright, we find that the maximum contribution of the astrophysical population of NSBH mergers to heavy element production in the universe is $\leq 1.03\ M_\odot/\mathrm{Gpc}^{-3}\mathrm{yr}^{-1}$. When we factor in the nondetection of any electromagnetic counterparts to the four NSBH mergers in our population, the EM-bright fraction drops to at most 7.5%. Unlike previous estimates of the EM-bright fraction based on population synthesis simulations [303, 370, 178], our result accounts for the measured black hole spin and mass ratio distributions, which favor small spins and extreme mass ratios and hence a lower neutron star disruption probability. We also account for neutron star spin in modeling the EoS that goes into the calculation of the remnant mass. This data-driven approach leads to a more pessimistic outlook on the likelihood of observing electromagnetic counterparts to NSBH mergers detected in gravitational waves.

Finally, our multimessenger analysis that includes the nondetection of any NSBH electromagnetic counterparts allows us to place independent constraints on the neutron star EoS. Our results represent a conservative upper limit on the constraining power of nondetection in such a multimessenger NSBH analysis, as we ignore highly uncertain electromagnetic selection effects and instead make the maximally optimistic assumption (in the case of nondetection) that no counterpart was observed because there was exactly no remnant mass left after the merger. Even in this most optimistic scenario that assumes perfect surveys with no selection effects, our EoS constraints are comparable to those obtained using existing gravitational-wave measurements of tides in BNS and radio and X-ray pulsar observations. As such, we conclude that multimessenger analyses of NSBH mergers are not a promising method for measuring the neutron star EoS. More realistic constraints for NSBH would require considerably more complicated modeling of EM selection effects and will be even less constraining than the ones we present here. Since the results obtained with the current set of NSBH detections are less constraining than the joint BNS and pulsar measurements, we expect the disparity in the constraining power of the two methods to persist or even grow as the relative number of NSBH and BNS observations remains constant due to detection rate scaling arguments. Even if an EM counterpart were detected for a NSBH merger, EM selection effects must still be accounted for to obtain an accurate joint multimessenger constraint on the EoS from multiple such detections. Taken together, our results suggest that detections of electromagnetic counterparts to NSBH mergers are likely to be rare and that the lack of detections is relatively uninformative about the EoS compared to other means of probing neutron star matter[8].

7.5 Supplementary material

7.5.1 Derivation of the multimessenger likelihood

The joint likelihood of observing a particular NSBH merger event with gravitational-wave data, d, and remnant mass measurement $M_{\rm rem}$, is the product of the likelihoods of making each of those observations individually. We use $\boldsymbol{\theta}$ to refer to the full set of binary parameters needed to characterize the gravitational-wave emission and $\mathbf{x}$

---

[8]Our hierarchical inference results including hyperparameter posterior samples are publicly available on Zenodo.

to refer to the subset of those parameters that are also needed to characterize the remnant mass measurement, $\mathbf{x} = (q, m_{\mathrm{NS}}, \chi_{\mathrm{NS}}, \chi_{\mathrm{BH},z})$. We do not include the neutron star tidal deformability among the $\boldsymbol{\theta}$ parameters, as the waveform models we choose for $h_k(\boldsymbol{\theta})$ do not include this effect.

$$p(d, M_{\mathrm{rem}}|\boldsymbol{\theta}, \boldsymbol{\Lambda}_{\mathrm{EoS}}) = p(d|\boldsymbol{\theta})p(M_{\mathrm{rem}}|\mathbf{x}, \boldsymbol{\Lambda}_{\mathrm{EoS}}) \qquad (7.17)$$
$$= p(d|\boldsymbol{\theta})\delta(\hat{M}_{\mathrm{rem}}(\mathbf{x}, \boldsymbol{\Lambda}_{\mathrm{EoS}})).$$

The remnant mass likelihood is a delta function at $M_{\mathrm{rem}} = 0\ M_\odot$ because we explicitly make the assumption that there was precisely no remnant mass left after the merger due to the nondetection of any electromagnetic counterpart. A more realistic analysis would relax this assumption and take into account the uncertainty in the remnant mass due to various electromagnetic selection effects including telescope sensitivity and uncertainty in the brightness of the emission. We emphasize that we make this simplifying assumption in order to present the most optimistic multimessenger constraints on the neutron star EoS.

Instead of measuring the binary parameters $\boldsymbol{\theta}$ for individual events, we are interested in measuring the hyperparameters, $\boldsymbol{\Lambda} = (\boldsymbol{\Lambda}_{\mathrm{GW}}, \boldsymbol{\Lambda}_{\mathrm{EoS}})$ governing the distributions of $\boldsymbol{\theta}$ across a population of sources. The likelihood of observing d, M_{rem} given $\boldsymbol{\Lambda}$ is obtained by marginalizing over $\boldsymbol{\theta}$, and the likelihood of observing a set of events with $\{d, M_{\mathrm{rem}}\}$ is the product of the individual-event likelihoods:

$$p(d, M_{\mathrm{rem}}|\boldsymbol{\Lambda}_{\mathrm{GW}}, \boldsymbol{\Lambda}_{\mathrm{EoS}}) = \int p(d, M_{\mathrm{rem}}|\boldsymbol{\theta}, \boldsymbol{\Lambda}_{\mathrm{EoS}})\pi_{\mathrm{pop}}(\boldsymbol{\theta}|\boldsymbol{\Lambda}_{\mathrm{GW}})d\boldsymbol{\theta} \qquad (7.18)$$

$$p(\{d, M_{\mathrm{rem}}\}|\boldsymbol{\Lambda}_{\mathrm{GW}}, \boldsymbol{\Lambda}_{\mathrm{EoS}})$$
$$= \prod_i \int p(d_i|\boldsymbol{\theta}_i)\delta(\hat{M}_{\mathrm{rem}}(\mathbf{x}_i, \boldsymbol{\Lambda}_{\mathrm{EoS}}))\pi_{\mathrm{pop}}(\boldsymbol{\theta}_i|\boldsymbol{\Lambda}_{\mathrm{GW}})d\boldsymbol{\theta}. \qquad (7.19)$$

The gravitational-wave likelihood in Eq. 7.19 can be replaced via Bayes' Theorem with the ratio of the posterior to the prior, which allows the integral to be evaluated using a sum over individual-event posterior samples, j:

$$p(\{d, M_{\text{rem}}\}|\Lambda_{\text{GW}}, \Lambda_{\text{EoS}})$$

$$\propto \prod_i \int \frac{p(\boldsymbol{\theta}_i|d_i)}{\pi_{\text{PE}}(\boldsymbol{\theta}_i)} \delta(\hat{M}_{\text{rem}}(\mathbf{x}_i, \Lambda_{\text{EoS}}))\pi_{\text{pop}}(\boldsymbol{\theta}_i|\Lambda_{\text{GW}})d\boldsymbol{\theta} \tag{7.20}$$

$$\propto \prod_i \sum_j \frac{\delta(\hat{M}_{\text{rem}}(\mathbf{x}_{i,j}, \Lambda_{\text{EoS}}))\pi_{\text{pop}}(\boldsymbol{\theta}_{i,j}|\Lambda_{\text{GW}})}{\pi_{\text{PE}}(\boldsymbol{\theta}_{i,j})} \tag{7.21}$$

Because we are neglecting electromagnetic selection effects, the joint likelihood in Eq. 7.21 can be amended to account for gravitational-wave selection effects in the usual way,

$$p(\{d, M_{\text{rem}}\}|\Lambda_{\text{GW}}, \Lambda_{\text{EoS}})$$

$$\propto \frac{1}{\alpha(\Lambda_{\text{GW}})^N} \prod_i^N \sum_j \frac{\delta(\hat{M}_{\text{rem}}(\mathbf{x}_{i,j}, \Lambda_{\text{EoS}}))\pi_{\text{pop}}(\boldsymbol{\theta}_{i,j}|\Lambda_{\text{GW}})}{\pi_{\text{PE}}(\boldsymbol{\theta}_{i,j})} \tag{7.22}$$

where N is the number of observed NSBH mergers in the population being analyzed. This is the same expression as Eq. 7.2 with the addition of the remnant mass likelihood to the numerator.

7.5.2 Black hole spin comparison

In our analysis, we use uniform priors on the α_χ, β_χ hyperparameters governing the black hole spin magnitude Beta distribution and include values of $\alpha_\chi, \beta_\chi \leq 1$, which correspond to singular Beta distributions. This choice allows the spin distribution to peak at $\chi_{\text{BH}} = 0$, as we might expect if NSBHs form via isolated binary evolution and the black hole is the first compact object to form [725, 128, 218, 388, 832], or $\chi_{\text{BH}} = 1$, consistent with measurements of spin in black hole X-ray binaries [e.g., 745]. However, in order to obtain an apples-to-apples comparison between the black hole spin distribution we infer for NSBH and the one measured in [52], we need to apply the same prior. The LVK analysis does not allow for singular Beta distributions, and

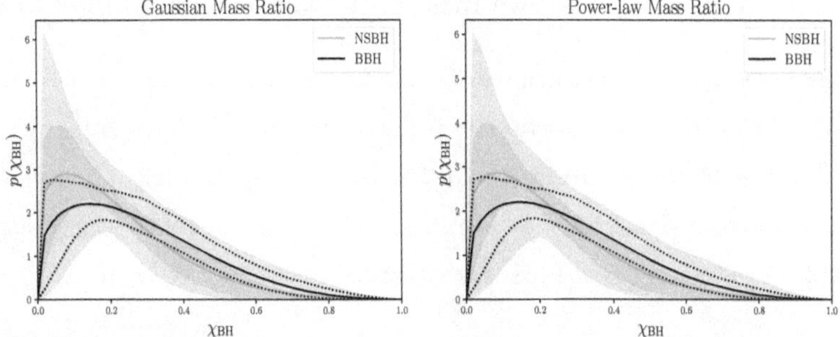

Figure 7-8: Posterior on the black hole spin magnitude distribution for the NSBH population obtained using gravitational-wave data alone (blue) using the same prior assumptions that go into the BBH spin magnitude distribution inference in [52], shown in black. The shaded blue region shows the NSBH 50% and 90% credible intervals, while the dotted black lines enclose the 90% credible region for BBH. The top (bottom) shows the result under the Gaussian (power-law) pairing function.

instead applies uniform priors on the mean and variance of the Beta distribution,

$$\mu = \frac{\alpha}{\alpha + \beta} \tag{7.23}$$

$$\sigma^2 = \frac{\alpha\beta}{(\alpha + \beta)^2(\alpha + \beta + 1)}. \tag{7.24}$$

In Fig. 7-8, we show in blue the posterior on the NSBH black hole spin magnitude distribution obtained using GW data alone (bottom left panels of Fig. 7-1) reweighted to match the prior choices made in the LVK analysis of BBH spins, while the BBH spin distribution is shown in black. This direct comparison demonstrates that our conclusion that the black holes in NSBHs have smaller spins still holds when equivalent prior assumptions are made. The NSBH result is considerably more uncertain than the BBH result, consistent with the fact that there are only four NSBH events included in our analysis but 69 BBH events going into the LVK result. With this choice of prior, we obtain $\chi_{\mathrm{BH},99} = 0.57^{+0.27}_{-0.17}$ ($\chi_{\mathrm{BH},99} = 0.53^{+0.34}_{-0.10}$) for the Gaussian (power-law) pairing function, while for the BBH analysis, $\chi_{\mathrm{BH},99} = 0.76^{+0.06}_{-0.06}$.

7.5.3 Analysis of the two most significant events

Here we present the population distributions for the component masses, mass ratio, and black hole spin magnitude inferred using only the two lowest-FAR events, GW200105_162426 and GW200115_042309. By comparison with Fig. 7-1, the distributions obtained with just these two events are similar to those obtained including the less significant candidates, just more uncertain.

7.5.4 Including GW190425

In order to verify the effect of our event selection and imposed prior knowledge of source classification on our results, we repeat our analysis of only the GW data to include GW190425, widely considered to be a binary neutron star merger given its inferred masses. This system has posterior support in the region of parameter space covered by our hyperparameter priors in Table 7.2, so it is not a priori excluded from our population, as is the case for GW190814.

Because this source is believed to be a binary neutron star rather than a NSBH merger, all of the publicly-available posterior samples for this event use waveform models that assume the tidal deformabilities of the two components are free parameters but do not include the effect of higher-order modes. This is in contrast to the posteriors we had analyzed for the analysis in the main text obtained with waveforms that assume both components are point masses with no tidal deformability but include the effects of higher-order modes. We choose to use the `PublicationSamples` posteriors for GW190425, where the spin magnitude prior extends up to $\chi < 0.89$ and the directions can be misaligned to the orbital angular momentum. We cannot self-consistently perform the analysis that models the NS EoS using the electromagnetic counterpart nondetection including GW190425, because there are no available posterior samples assuming the two components are point masses.

The results of the analysis including GW190425 are presented in Figs. 7-10-7-12. As might be expected given the low primary mass inferred individually for GW190425, the posterior predictive distribution for the black hole mass changes significantly when

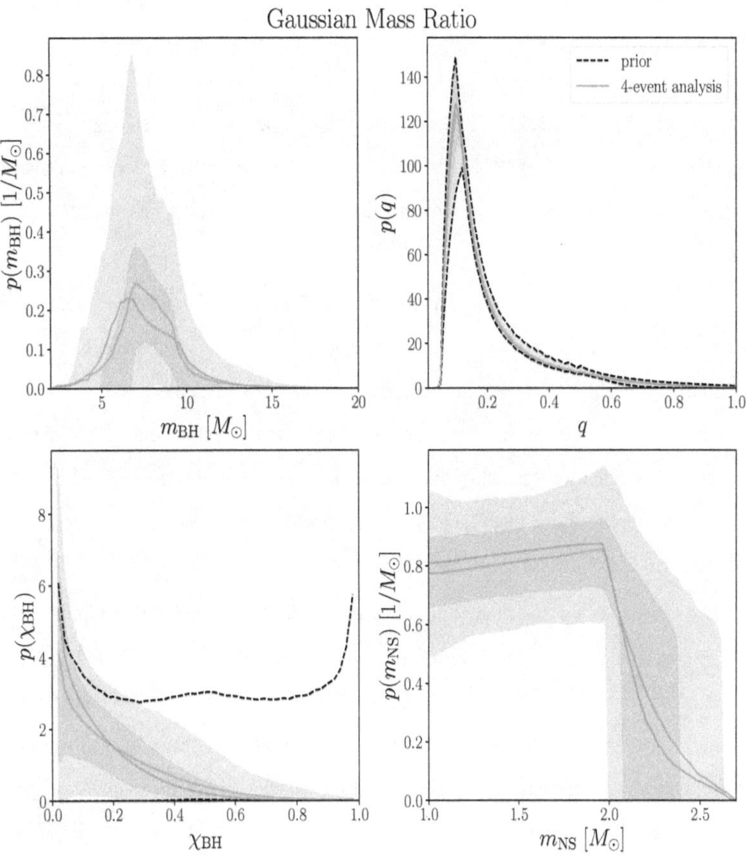

Figure 7-9: Posterior predictive distributions (solid blue) and 50% and 90% credible intervals (shaded blue) for the component masses, mass ratio, and black hole spin in the underlying, astrophysical NSBH population under the Gaussian mass ratio model using a population consisting of only GW200105_162426 and GW200115_042309. The black dashed lines show the 90% credible region enclosed by draws from the hyperparameter prior for the black hole spin and mass ratio, and the red lines show the PPDs inferred using the original event selection for Fig. 7-1.

this event is included in the analysis, compared to the distribution presented in Fig. 7-1. The posterior on the minimum black hole mass rails against the lower edge of the prior, shifting the PPD to peak at 2 $M_\odot$ rather than ~ 6 $M_\odot$. The upper tail of the distribution does not change substantially when this event is included, consistent with the fact that the posteriors on the maximum black hole mass are qualitatively similar with or without GW190425. The posterior on α, however, more strongly disfavors negative values when GW190425 is included, as the peak at low black hole masses required by GW190425 must be accommodated by a power-law with a negative slope (positive α).

The marginalized mass ratio distribution does not vary significantly upon including GW190425 in the analysis, as it is prior-driven, similar to the result obtained without GW190425. However, the posteriors on the mass ratio mean and width more strongly prefer values at the lower edges of their respective priors. This may be due to the information gained by increasing the number of events analyzed by 25%. The distributions of the black hole spin and neutron star mass also do not change significantly when GW190425 is included.

The posteriors for the binary parameters of GW190425 under the original prior and reweighted into the new inferred population prior are shown in Fig. 7-11. The effect of assuming that this event is an NSBH rather than a binary neutron star merger is to upweight the posterior support at unequal mass ratios. All the support at equal mass is removed, and the mass ratio posterior under the population prior peaks staunchly at $q = 0.49$. The spin distribution also shifts towards lower values.

Finally, in Fig. 7-12, we show the posteriors on the fraction of GW-detectable NSBH systems that will be EM-bright for different values of the threshold remnant mass obtained including GW190425. As expected given the shift in the black hole mass distribution towards lower masses which leads to more significant disruption of the neutron star outside the ISCO radius, there is a commensurate increase in the EM-bright fraction relative to the result without GW190425 shown in Fig. 7-4. When GW190425 is included, the posterior peaks away from $f_{\mathrm{EM-bright}} = 0$, and we infer $f(\hat{M}_{\mathrm{rem}} > 0\ M_\odot) = 0.16^{+0.13}_{-0.16}$.

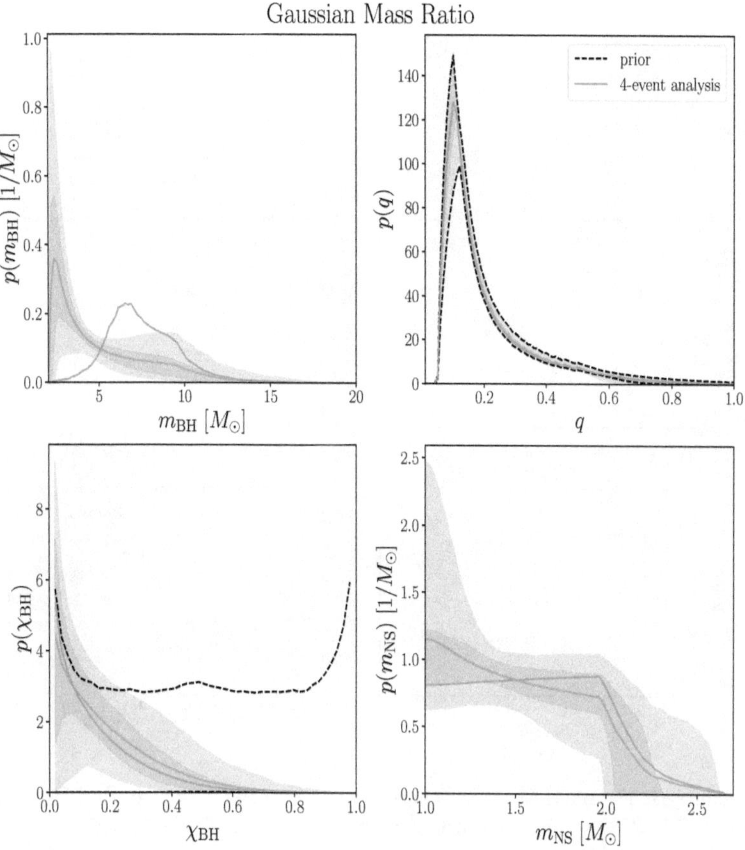

Figure 7-10: Posterior predictive distributions (solid blue) and 50% and 90% credible intervals (shaded blue) for the component masses, mass ratio, and black hole spin in the underlying, astrophysical NSBH population under the Gaussian mass ratio model using a population consisting of the four candidates analyzed in the main text *and* GW190425. The black dashed lines show the 90% credible region enclosed by draws from the hyperparameter prior for the black hole spin and mass ratio, and the red lines show the PPDs inferred using the original event selection for Fig. 7-1.

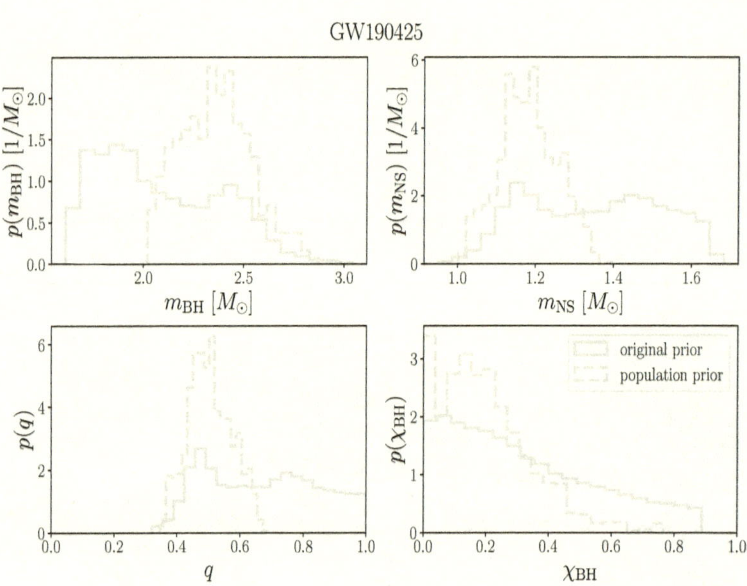

Figure 7-11: Distributions of the component masses, mass ratio, and black hole spin for GW190425 assuming it belongs to our population NSBH mergers under the original priors (solid lines) and reweighted into the population prior inferred under the Gaussian mass ratio model (dashed line).

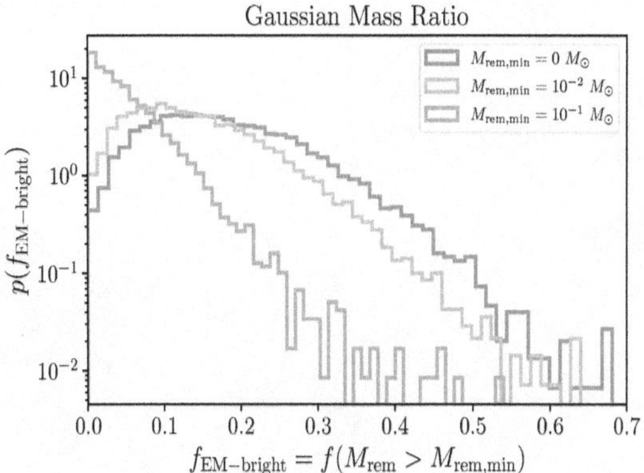

Figure 7-12: Posterior on the fraction of GW-detectable NSBH systems that will be electromagnetically bright with remnant mass $\hat{M}_{\rm rem} \geq M_{\rm rem,min}$ obtained by analyzing a population consisting of GW190425 along with the four candidate events presented in the main text. The different colors indicate different values of $M_{\rm rem,min}$. The posterior is marginalized over the uncertainty in the neutron star equation of state and in the population hyperparameters for the Gaussian pairing function.

While the posteriors for the minimum black hole mass and EM-bright fraction change significantly when GW190425 is included in the analysis, our main conclusions that the black holes in NSBHs are both less massive and more slowly spinning than those in BBH remain robust. The change in these posteriors highlights the importance of event selection and prior knowledge of source classification in population analyses. As suggested in the main text, this dependence can be removed by simultaneously fitting and sorting individual events into multiple sub-populations based on distinct distributions [330]. We leave this analysis to future work.

Part II

Binary black holes

Chapter 8

Measuring the spins of heavy binary black holes

The content of this chapter was previously published in Physical Review D as Ref. [152] in Nov. 2021. ASB performed the analysis, wrote the manuscript, and contributed to project development.

Abstract

An accurate and precise measurement of the spins of individual merging black holes is required to understand their origin. While previous studies have indicated that most of the spin information comes from the inspiral part of the signal, the informative spin measurement of the heavy binary black hole system GW190521 suggests that the merger and ringdown can contribute significantly to the spin constraints for such massive systems. We perform a systematic study into the measurability of the spin parameters of individual heavy binary black hole mergers using a numerical relativity surrogate waveform model including the effects of both spin-induced precession and higher-order modes. We find that the spin measurements are driven by the merger and ringdown parts of the signal for GW190521-like systems, but the uncertainty in the measurement increases with the total mass of the system. We are able to place meaningful constraints on the spin parameters even for systems observed at moderate signal-to-noise ratios, but the measurability depends on the exact six-dimensional spin configuration of the system. Finally, we find that the azimuthal angle between the in-plane projections of the component spin vectors at a given reference frequency cannot be well-measured for most of our simulated configurations even for signals observed with high signal-to-noise ratios.

8.1   Introduction

Gravitational-wave observations offer a unique way to directly measure the masses and spins of astrophysical black holes, providing insights into their formation channels [404, 922, 836, 758, 835, 850, 353, 333, 966, 115, 862, 91, 769, 946, 798, 105, 515, 963]. The latest observing run of the advanced LIGO [3] and Virgo [53] gravitational-wave interferometers resulted in the discovery of tens of new binary black hole systems [41] leading to the best constraints on the population-level mass and spin distributions of these objects to date [49]. Reliable measurements of the masses and spins of individual black holes are crucial ingredients for population analyses and of particular interest for exceptional events.

GW190521 is one such exceptional event, representing the first direct detection of an intermediate-mass black hole [37]. The inferred component masses of this system are above the theoretical upper limit for black holes formed directly from stellar collapse due to the effects of pair-instability supernovae [453, 676, 132, 596, 940] (although see [131, 354, 657]). Together with some evidence for general-relativistic spin-induced precession, this suggests that this binary could have formed in a dynamical environment [706, 708, 624, 666, 785, 636, 371] as the result of a hierarchical merger [753, 401, 757, 172, 372]. Previous works have claimed evidence for eccentricity in this system [763, 398, 196], another signature of dynamical formation.

Intermediate-mass black holes, with masses approximately $10^2 - 10^5 \; M_\odot$, occupy the mass range in between stellar-mass and supermassive black holes, lying above the putative pair-instability mass gap described above and potentially serving as the seeds for the supermassive black holes at the centers of galaxies [311, 623, 621, 436]. Such objects can form directly in the early Universe via the collapse of massive Population III stars [384, 452, 829, 560] or low-angular-momentum gas clouds [565, 179, 564, 130], or dynamically in dense environments via runaway collisions [708, 705, 101] or hierarchical mergers [624, 666, 415], as suggested above. Given the unique formation channels of binaries involving these objects and the fact that the sensitivity of ground-based gravitational-wave detectors increases with the total mass of the binary (up to a few hundred solar masses) [353, 325], such systems are of particular interest.

Because gravitational-wave frequency scales inversely with the total mass, heavy binaries merge at lower frequencies, leaving few orbital cycles in the sensitive band of ground-based detectors. This would imply that our ability to constrain the spin parameters, which is traditionally thought to be driven by the inspiral part of the waveform, is limited for such heavy systems. However, the informative spin measurement and evidence for precession found in GW190521 seem to indicate that meaningful inferences can be made for the spins of heavy binary black holes (BBHs). In this work, we seek to systematically investigate how well spins can be measured for such heavy systems.

To leading post-Newtonian order, the influence of the spins can be parameterized in terms of a single mass-weighted effective aligned spin parameter, χ_{eff} [271, 64, 62, 786, 720]:

$$\chi_{\mathrm{eff}} = \frac{\chi_1 \cos\theta_1 + \chi_2 q \cos\theta_2}{1+q}, \tag{8.1}$$

where $q \equiv m_2/m_1$, $m_2 \leq m_1$ is the binary mass ratio, χ_i is the component spin magnitude with $i = 1, 2$, and θ_i is the component spin tilt relative to the orbital angular momentum. For binaries where the component spins are aligned or antialigned with the orbital angular momentum, the spin modifies the inspiral rate relative to the equivalent non-spinning system, but the waveforms are otherwise morphologically similar [200].

However, when the spins of the black holes are misaligned to the orbital angular momentum, the orbital plane and the individual spins precess about the total angular momentum [90, 513]. The waveforms for these systems are morphologically richer, leading to characteristic modulations in the gravitational-wave amplitude and phase. Even in this generic case, however, the average influence of the remaining four spin degrees of freedom can be reduced to a single effective precessing spin parameter,

χ_p[1] [795]:

$$\chi_p = \max\left(\chi_1 \sin\theta_1, \left(\frac{4q+3}{4+3q}\right) q\chi_2 \sin\theta_2\right). \tag{8.2}$$

Because these parameters encode the leading order effect of spins on the inspiral phasing, χ_{eff} and χ_p are typically much better measured than the individual component spins for signals dominated by the inspiral [921, 800]. Most previous studies investigating the measurability of spin parameters with ground-based gravitational-wave detectors have focused on systems in this inspiral-dominated regime [900, 737, 231, 668, 923, 408, 224, 712, 433]. Such work finds that the secondary spin is particularly hard to measure [737], as are high component spins when χ_{eff} is small [224]. However, unequal masses significantly improve the resolution of the spin parameters [923, 688], particularly when higher-order multipoles are included in the waveform model [231, 668, 712]. Precession effects can further serve to break degeneracies and improve spin constraints [900, 231, 668, 712].

While χ_{eff} and χ_p encapsulate the spin effects for the inspiral part of the waveform, heavier binary black hole systems will have significant contributions to the detectable signal from the merger and ringdown. The morphology of the latter is determined by the mass and spin of the final black hole. Most previous studies investigating the ability to characterize such heavy systems have focused on nonspinning [427, 450, 907] or aligned-spin [721, 800, 608] BBHs. Ref. [921] included precessing systems in their analysis, but used a waveform model that only includes the contribution from the dominant $\ell = |m| = 2$ mode to the signal. This phenomenological waveform model, IMRPhenomPv2 [445, 482, 509], approximates the effects of precession in two important ways: (i) only four spin degrees of freedom are included, and the azimuthal spin angles are ignored; and (ii) this model is not informed by numerical relativity (NR) simulations of precessing binaries, which are critical to accurately model the merger process. Instead, an equivalent nonprecessing waveform is "twisted up" to account

---

[1]See [865, 405] for alternative definitions of an effective precessing spin parameter that more robustly capture the effects of higher-order modes and variations occurring on the precessional timescale, respectively.

for some effects of orbital precession [794]. Furthermore, IMRPhenomPv2 and other such phenomenological waveform models rely on heuristic recipes for modeling the ringdown of the remnant black hole and attaching it to the rest of the waveform. Accurate ringdown modeling is especially important for high-mass systems since this part of the waveform contributes significantly to the detectability of the signal.

Considering the importance of precession, ringdown modeling, and higher-order modes for heavy binary black hole systems [800], we seek to expand upon previous studies to determine how well the spins can be measured for such systems using a numerical relativity surrogate waveform that includes the effects of the full six-dimensional spin degrees of freedom and higher harmonics, NRSur7dq4 [903]. Trained directly on precessing NR simulations, this model captures the effects of precession at an accuracy level comparable to the simulations themselves.

The rest of the chapter is organized as follows. In Section 8.2, we describe the Bayesian inference formalism we employ for our analysis. Section 8.3 presents an investigation into the measurability of precession in highly precessing systems comparable to GW190521. A systematic study of the measurability of the spin degrees of freedom is presented in Section 8.4 for the component and effective spins, and in Section 8.5 for the azimuthal angles. We find that the spin information is driven by the merger and ringdown parts of the signal for GW190521-like systems, but that the measurability of the spin parameters depends sensitively on the exact six-dimensional spin configuration of the binary.

8.2 Parameter Estimation

A quasicircular binary black hole merger is completely characterized by 15 parameters: eight intrinsic parameters comprising of the masses and spin vectors for each component black hole, and seven extrinsic parameters. The extrinsic parameters include the sky location (α, δ), the luminosity distance to the source, d_L, the inclination angle between the total angular momentum and the line of sight of the observer, θ_{JN}, the polarization angle ψ, and the time and orbital phase ϕ at coalescence. The spin

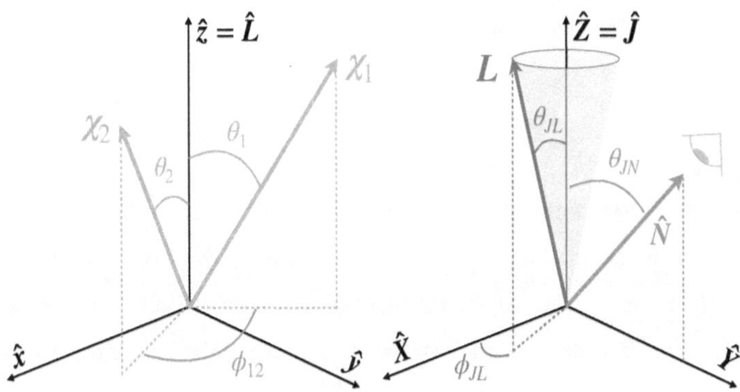

Figure 8-1: *Left:* Diagram of the component spin vectors relative to the orbital angular momentum pointing along the $\hat{z}$ axis. ϕ_{12} is the azimuthal angle between the projections of the component spins onto the orbital plane. *Right:* Diagram demonstrating the definition of ϕ_{JL}, which is the azimuthal angle between the $\hat{X}$ axis and the projection of the orbital angular momentum onto the $\hat{X} - \hat{Y}$ plane in the frame where the total angular momentum points along the $\hat{Z}$ axis and the line-of-sight vector, $\hat{N}$, lies in the $\hat{Y} - \hat{Z}$ plane.

degrees of freedom are typically parameterized in terms of the component magnitudes, χ_i, tilt angles relative to the orbital angular momentum θ_i, and two azimuthal angles, ϕ_{12} and ϕ_{JL}, defined in the diagrams in Fig. 8-1. The component black holes are usually sorted according to their masses, with χ_1 indicating the spin magnitude of the more massive black hole, but they can also be sorted according to their spins, with subscripts A, B [153].

In order to obtain estimates of these 15 binary parameters from gravitational-wave strain data, we employ the framework of Bayesian inference. The posterior probability distribution for the binary parameters $\boldsymbol{\theta}$ given the observation of data d is:

$$p(\boldsymbol{\theta}|d) \propto \mathcal{L}(d|\boldsymbol{\theta})\pi(\boldsymbol{\theta}). \tag{8.3}$$

The likelihood of observing the data d is [906, 761]

$$\mathcal{L}(d|\boldsymbol{\theta}) \propto \exp\left(-\sum_k \frac{2|d_k - h_k(\boldsymbol{\theta})|^2}{TS_k}\right), \tag{8.4}$$

where $h(\boldsymbol{\theta})$ represents the gravitational waveform which depends on the binary parameters, T is the duration of the analyzed data, S is the power spectral density (PSD) characterizing the noise in the interferometer, and the subscript k indicates the frequency dependence of the data, waveform, and PSD. The prior probability distributions for the binary parameters are represented by $\pi(\boldsymbol{\theta})$.

We employ the formalism described above to perform parameter estimation for high-mass binary black hole merger simulations. Unless otherwise stated, the following settings are used for all the simulations described in this manuscript. We use a three-detector network of the advanced LIGO [3] and Virgo [53] interferometers operating at design sensitivity and do not add Gaussian noise to the simulated data in each detector so that the analysis is performed in zero-noise [895]. We analyze 8 s segments of data sampled at a rate of 2048 Hz. The signals are simulated and recovered with the NRSur7dq4 waveform [903] for $h(\boldsymbol{\theta})$, and samples from the posterior distribution given in Eq. (8.3) are obtained using the LALInference software [906].

NRSur7dq4 is trained on numerical relativity simulations with mass ratios $q \geq 1/4$ and spin magnitudes $\chi_1, \chi_2 \leq 0.8$, but extrapolates reasonably well to $q \leq 1/6$ and $\chi_1, \chi_2 = 1$ [2]. We therefore allow our priors to extend to $q_{\min} = 1/5$ and $\chi_{\max} = 0.99$. Unless otherwise stated, the reference frequency at which the spin parameters are specified is 25 Hz, and this is also the minimum frequency included in the likelihood. While advanced LIGO and Virgo are sensitive to lower frequencies, NRSur7dq4 only includes about ~ 20 orbits before merger [903], which can lead to the signal abruptly starting within the detector band for small total masses or small mass ratios. To avoid this, we use a conservative minimum frequency of 25 Hz so that NRSur7dq4 waveforms

---

[2] While the number of precessing NR simulations with spins $\gtrsim 0.8$ are limited, we have checked that for the ~ 30 such available waveforms, the surrogate agrees with NR to within NR resolution errors. We have also verified that parameter estimation results within the training region are unaffected when using priors extending into the extrapolation region relative to those restricted to the training region.

can be successfully generated for the smallest total mass ($M_{\text{tot}} = 70\ M_\odot$) and mass ratio we allow in our prior. We use priors that are uniform in the component masses, spin magnitudes, azimuthal angles and $\cos\theta_i$,[3] and $\propto d_L^2$ in luminosity distance. Standard priors are used for all other parameters [906].

8.3 Measurability of spins in GW190521-like systems

GW190521 is the heaviest binary black hole system detected in gravitational waves to date, and its χ_p posterior shows a noticeable deviation from the prior, peaking at around $\chi_p \approx 0.7$ [37]. This suggests that general-relativistic spin-induced precession is measurable in high-mass signals with moderate signal-to-noise ratios (SNRs). We therefore seek to verify this measurement using three simulated systems of varying inclination angle and mass ratio, q, (and hence varying χ_{eff}) in order to determine how the measurability of χ_p depends on these parameters. The distance for each system is adjusted so that the network optimal SNR is fixed to $\rho_{\text{opt}}^{\text{net}} = 15$ to match the SNR of GW190521. The true value of $\chi_p = 0.90$ for all three simulations, and the remaining parameters which are kept the same are presented in Table 8.1. For these systems we use minimum and reference frequencies of 11 Hz, as was done in the analysis of GW190521 [37]. The three choices of inclination and mass ratio are:

1. Nearly edge-on, nearly equal-mass: $\theta_{JN} = 1.4$, $q = 0.9$, $\chi_{\text{eff}} = 0.074$

2. Nearly face-on, nearly equal-mass: $\theta_{JN} = 0.5$, $q = 0.9$, $\chi_{\text{eff}} = 0.074$

3. Nearly edge-on, more unequal masses: $\theta_{JN} = 1.4$, $q = 0.7$, $\chi_{\text{eff}} = 0.067$

GW190521 is consistent with having been observed face-on [37], so it is most similar to simulation number 2 above. The results of our simulations are shown in Fig. 8-2. We are unable to recover a significant measurement of χ_p matching the measurement obtained for GW190521 with any of the three systems. In fact, the posteriors for the component spin magnitudes prefer low values, peaking at $\chi_i = 0$

---

[3]See [920, 478, 964] for discussions on the effects of prior choices on spin measurements.

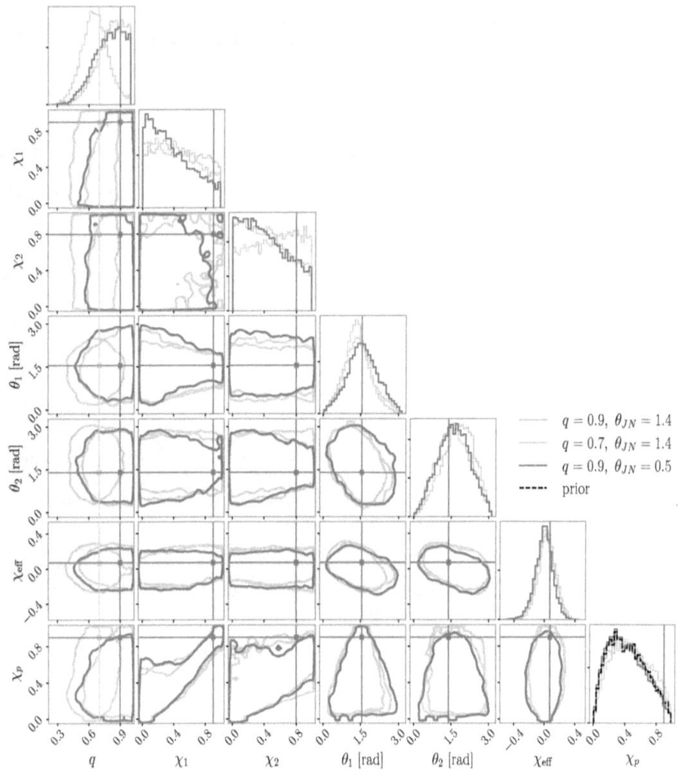

Figure 8-2: Comparison corner plot for the three simulated systems with $\chi_p = 0.9$ at SNR=15 showing the posteriors for mass ratio, spin magnitudes, χ_{eff}, and χ_p. The prior for χ_p is shown in the dashed black line.

Table 8.1: True values for the parameters which are kept the same for all three simulations with SNR=15 shown in Fig. 8-2.

Parameter	Symbol	Value
Detector-frame total mass	M_{tot}	270 $M_\odot$
Primary spin magnitude	χ_1	0.9
Secondary spin magnitude	χ_2	0.8
Primary tilt[4]	θ_1	1.55
Secondary tilt	θ_2	1.40
Azimuthal inter-spin angle	ϕ_{12}	6.28
Azimuthal precession cone angle	ϕ_{JL}	6.28
Effective precessing spin	χ_p	0.90
Coalescence phase	ϕ	4.70
Polarization angle	ψ	0.28
Coalescence GPS time	t_c	1126259642 s
Right ascension	α	1.09
Declination	δ	0.47

for both of the nearly-equal mass simulations. The spin magnitude posteriors for the more unequal-mass system show less of a preference for low spin magnitudes, although they only exhibit a minor deviation from the uniform prior. The spin tilt posteriors are also not informative enough to drive the measurement of χ_p to high values; while the true values of the tilt angles do fall at the peak of the posterior, the posterior distributions are largely unconstrained relative to the prior. The resulting posterior for χ_p is also driven by the prior, shown in the dashed black line in Fig. 8-2. The fact that we are unable to recover an informative posterior for χ_p—even for the system observed nearly edge on, for which the effects of precession should be most apparent—suggests that obtaining a more informative measurement, such as the one for GW190521, requires tuning to a specific configuration among the spin degrees of freedom. It could also be an indication that the posterior for GW190521 is driven to high values of χ_p due to the properties of the detector noise at the time of the event.

To investigate these possibilities, we conduct parameter estimation for a simulated system with parameters specified by the maximum-likelihood point from the publicly-released posterior samples obtained for GW190521 by LIGO-Virgo [38, 48] using the NRSur7dq4 waveform. This ensures that any parameter correlations that led to a more informative measurement of χ_p for GW190521 are included in our simulation.

Table 8.2: Maximum likelihood parameter values used for the simulated system drawn from the GW190521 posterior.

Parameter	Symbol	Value
Mass ratio	q	0.82
Detector-frame total mass	M_{tot}	268.83 $M_\odot$
Primary spin magnitude	χ_1	0.09
Secondary spin magnitude	χ_2	0.90
Primary tilt[5]	θ_1	0.92
Secondary tilt	θ_2	1.41
Azimuthal inter-spin angle	ϕ_{12}	1.04
Azimuthal precession cone angle	ϕ_{JL}	2.48
Effective aligned spin	χ_{eff}	0.094
Effective precessing spin	χ_p	0.70
Coalescence phase	ϕ	0.004
Polarization angle	ψ	2.38
Coalescence GPS time	t_c	1242442967.41 s
Right ascension	α	0.16
Declination	δ	-1.14
Luminosity distance	d_L	2941.03 Mpc
Inclination angle	θ_{JN}	0.95

We add this BBH signal to ten different realizations of Gaussian noise colored by the power spectral densities of the two advanced LIGO detectors and advanced Virgo calculated for the segment of data containing GW190521 [38]. We also use minimum and reference frequencies of 11 Hz for analyzing this maximum-likelihood system. The binary parameters are given in Table 8.2.

The posterior probability distributions obtained for χ_p for the ten different realizations of Gaussian noise are shown in Fig. 8-3, along with the posterior obtained in the absence of Gaussian noise. Even though the true value of $\chi_p = 0.70$ is lower than that of the previous three simulations, the measurement is much more informative and peaks closer to the true value for the comparable zero-noise case. While one of the noise realizations leads to a posterior peaked at low values of χ_p and another returns a flatter posterior, these deviations are consistent with Gaussianity. Because real interferometer data is known to include non-Gaussian excursions, we repeat the experiment by adding the maximum-likelihood system into ten different segments of real data from the third observing run of advanced LIGO and Virgo instead of simu-

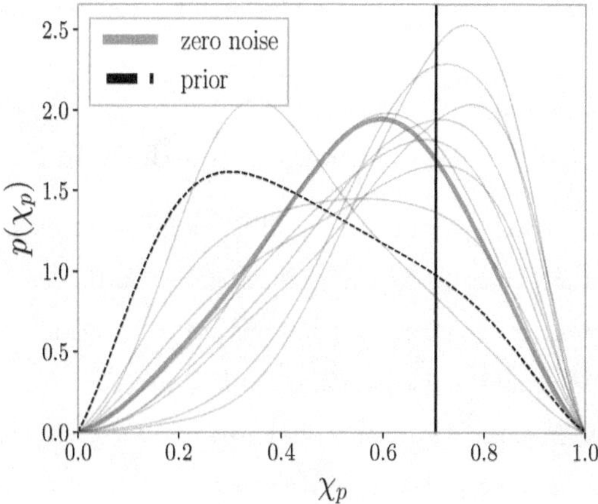

Figure 8-3: Probability density for χ_p for the GW190521 maximum-likelihood simulation added to ten different realizations of Gaussian noise colored by the PSDs estimated for the time around GW190521. The blue line shows the result for the simulation without Gaussian noise. The prior is shown in the dashed black line.

lated Gaussian noise. We find a similar spread in the χ_p posteriors to that shown in Fig. 8-3; further details can be found in Appendix 8.7.1.

These results indicate that even the spin degrees of freedom that are not usually measurable can have a significant effect on the detectability of precession. Only specific values of the spin angles yield a system where χ_p is measured as well as it was for GW190521. The χ_p posteriors shown in Fig. 8-2 are qualitatively similar to those found in [223], where it is suggested that high values of χ_p are more difficult to recover in systems with low values of χ_{eff}. However, the systems shown in Figs. 8-2 and 8-3 have similar true values of χ_{eff}, all < 0.1, but very different posteriors for χ_p, suggesting that the measurability of χ_p depends on more than just χ_{eff}. While we have found that certain configurations can produce systems with informative χ_p posteriors, such configurations might be uncommon. If this is the case, more exotic, but also rare, explanations could be invoked to explain the highly precessing nature of this event, such as residual eccentricity at merger [398, 763], which can be confused with precession when analyzed with waveforms assuming the binary system is circular [763, 195].

8.3.1   Effect of the cutoff frequency on the spin measurement

The information on the spin parameters in BBH sources is traditionally thought to come predominantly from the inspiral phasing, and yet in the previous section we have demonstrated that an informative measurement of χ_p can be made for a heavy GW190521-like system with few inspiral cycles in the sensitive band of the LIGO-Virgo detectors. In order to determine which part of the frequency band drives the measurement of the spin parameters for high-mass, highly-precessing systems, we perform a study where the maximum frequency that is included in the likelihood in Eq. 9.1 for the system in Table 8.2 is incrementally increased. We begin with a maximum frequency of 16 Hz, so that the total analyzed bandwidth is only 5 Hz, but rescale the distance of the system so that the total SNR is kept constant at 15 for all the choices of f_{max}. Gaussian noise is not added to these simulations. The time evolution of the gravitational-wave frequency in the coprecessing frame [903] is shown in the top panel of Fig. 8-4. The gravitational-wave frequency is calculated as $f_{\mathrm{GW}} = \Omega_{\mathrm{orb}}/\pi$, where Ω_{orb} is the time derivative of the orbital phase as defined in Eq. (3) of [903]. The gravitational-wave amplitude peaks at $f_{\mathrm{GW}} = 43.05$ Hz, which we treat as the merger frequency. The analysis with the full band up to 512 Hz is the same as in the dark blue line in Fig. 8-3.

The posterior probability distributions obtained using each of the different maximum frequencies are shown in Fig. 8-5. The posteriors for all the parameters shown remain largely unchanged as the maximum frequency is increased until an f_{max} of 65 Hz is reached. This can be explained due to the fact that the SNR of each simulated system is kept constant. Even though more inspiral cycles are added as we increase the analyzed bandwidth, each cycle becomes less informative since the amplitude of the signal decreases to keep the overall SNR the same.

Even for the choices of $f_{\mathrm{max}} = 46, 50$ Hz, which include frequency content after the merger, the posteriors are qualitatively similar to those obtained when analyz-

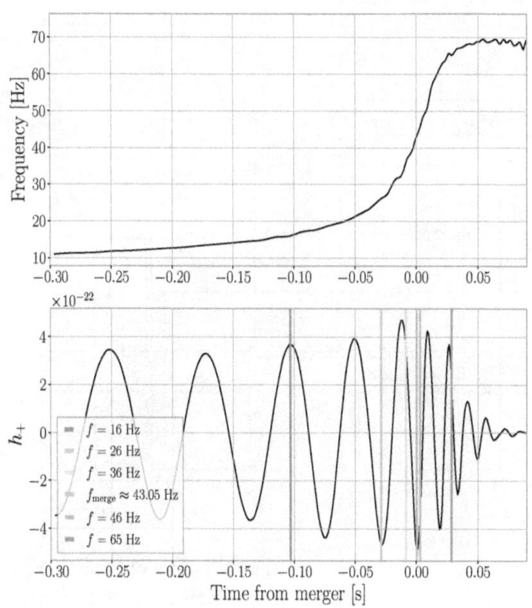

Figure 8-4: *Top panel:* Evolution of the gravitational-wave frequency in the coprecessing frame as a function of time for the GW190521 maximum-likelihood simulation, with the merger time set to $t = 0$. *Bottom panel:* Time-domain waveform with the varying maximum frequencies we use in our analysis marked in time as vertical lines. The frequency at the merger time is also shown.

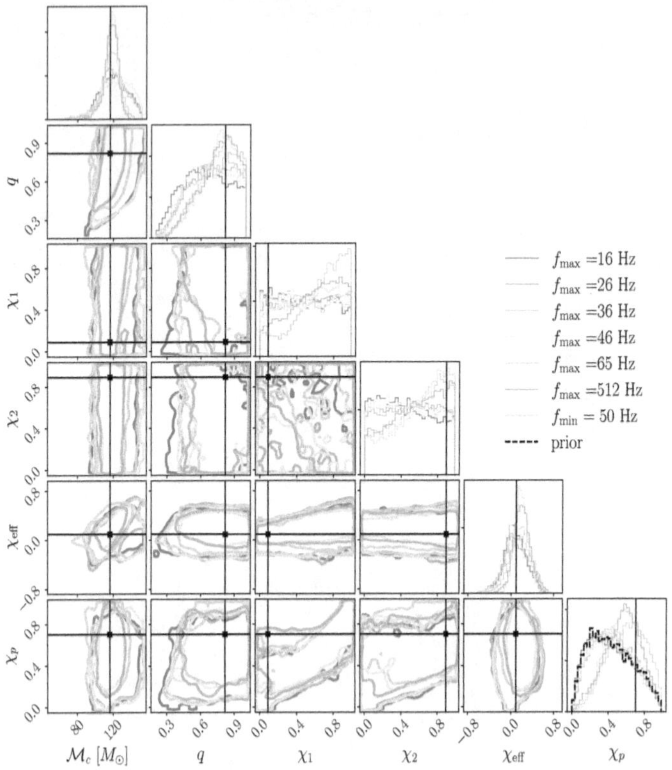

Figure 8-5: Comparison corner plot for the posteriors obtained for the chirp mass, mass ratio, component spin magnitudes, and effective aligned and precessing spin parameters for the GW190521 maximum-likelihood simulation using different maximum and minimum frequencies.

ing only the inspiral frequency content. However, the posterior constraints improve dramatically for all parameters when $f_{\mathrm{max}} = 65$ Hz is used, and become further constrained when the full frequency band is analyzed. Allowing for frequency content up to 65 Hz includes significantly more of the postmerger waveform than either 45 or 50 Hz. This indicates that the ringdown part of the signal, where the frequency as a function of time changes concavity in the top panel of Fig. 8-4, contributes the most to the measurement of both the mass and spin parameters for such heavy BBH systems.

As a verification of the contribution of the postmerger part of the signal to the ability to constrain the spin parameters, we repeat the analysis of the system above but instead of changing the maximum frequency, we limit the minimum frequency to 50 Hz. This ensures that the inspiral has been completely removed from the analyzed signal, and any information on the spins must come entirely from the merger and ringdown. The posteriors for the spin parameters obtained with $f_{\mathrm{min}} = 50$ Hz are shown in gold in Fig. 8-5. These posteriors are nearly identical to those obtained when analyzing the full frequency band. This demonstrates that the strong dependence of the measurability of the spin parameters on the postmerger part of the signal is not artificially due to the limited sensitivity of the detectors at low frequencies. Rather, the physical richness of the merger and ringdown alone is sufficient to provide a spin constraint equal to that obtained when also including the inspiral in the analysis [480, 555].

8.4 Accuracy and precision of spin measurements for high-mass systems

In the previous section we have shown that certain spin configurations can lead to informative measurements driven by the postmerger part of the signal for heavy BBH systems. Next, we seek to extend this analysis to generic systems to determine how the measurability of the spin parameters depends on the total mass of the binary. We investigate the effect of systematically varying the total mass for different choices of primary spin tilt, mass ratio, and inclination angle. We consider systems with two different inclination angles, $\theta_{JN} = 0.5, 1.4$ (which we refer to as nearly face-on and nearly edge-on, respectively) and two mass ratios, $q = 1, 1/4$. We also use three different primary tilt angles, $\theta_{1/A} = 0, \pi/3, \pi/2$ rad, corresponding to systems with aligned, moderate, and in-plane tilt angles (see the first set of simulations presented in Table 10.4).

Table 8.3: True parameter values for the four different sets of simulations performed for Sections 8.4 and 8.5 of this work. In the text, we refer to the two different choices of mass ratio as equal and unequal, and to the two different choices of inclination angle for Simulation 1 as nearly face-on and nearly edge-on.

Set	1	2	3	4
SNR	30	30	30	60
$M_{\mathrm{tot}}\ [M_\odot]$	$75-250$	$75-250$	$75-300$	$75-300$
q	$1, 0.25$	$1, 0.25$	$1, 0.25$	$1, 0.25$
χ_1	0.8	0.8	0.8	0.8
θ_1 [rad]	$0, \pi/3, \pi/2$	$\pi/3$	$\pi/3$	$\pi/3$
χ_2	0	0.2	0.2	0.2
θ_2 [rad]	$-$	$\pi/6$	$\pi/6$	$\pi/6$
ϕ_{12} [rad]	$-$	$0, \pi/2, \pi$	$0, \pi/2, \pi$	$0, \pi/2, \pi$
θ_{JN} [rad]	$0.5, 1.4$	0.5	0.5	0.5

In this section, we sort the binary components by their dimensionless spin magnitudes instead of their masses, as is customary. We use the subscript A (B) to indicate the highest (lowest)-spinning black hole [153]. Because we include equal-mass systems in our simulations, this sorting helps break the degeneracy between the spins of the two black holes. In the unequal-mass case, the heavier black hole is also the one with higher spin, so there is no distinction between the two sorting methods. The true values for the remaining parameters which are kept the same across all the simulations presented in this section are listed in Table 8.4.

While the simulated systems presented in this section are not necessarily tuned to the optimal configuration found in the previous section, the accuracy and precision of the spin measurements are comparable to those previously presented. The conclusions we draw in this section—particularly with respect to comparisons between different inclination and primary tilt angles—are thus robust across a broad range of systems.

8.4.1   Equal-mass systems

Figure 8-6 shows the bias and uncertainty in the recovered component spin magnitudes, χ_A and χ_B, tilt angle of the highest-spinning black hole, θ_A, and effective aligned and precessing spins, χ_{eff} and χ_p, as a function of the true total mass of the system for all simulated systems with $q = 1$. We define the bias as the difference

Table 8.4: True values for the parameters common to all four simulation sets described in Sections 8.4 and 8.5.

Parameter	Symbol	Value
Azimuthal precession cone angle	ϕ_{JL}	6.28
Polarization angle	ψ	0.28
Coalescence phase	ϕ_c	4.70
Right ascension	α	1.09
Declination	δ	1.47
Coalescence GPS time	t_c	1126259642 s

between the median of the posterior and the true value, and the uncertainty is represented by the width of the 90% credible interval (CI) of the posterior. Thus, the bias is a measure of the accuracy of the posterior, and the 90% CI width is a measure of the precision. The bias is ill-defined for θ_B since this quantity is undefined for a nonspinning component; consequently, this parameter is unconstrained, and its posteriors just return the prior for all simulations.

For equal-mass spin-aligned systems ($\theta_{1/A} = 0$; light blue lines in Fig. 8-6), there is no significant trend in the accuracy or precision of the measurement of the component spin magnitudes or tilts with the mass of the system for either of the two inclination angles we simulated. The large positive bias in the θ_A posterior for these systems is due to the fact that the measurement of θ_A is not very informative and is hence driven by the uniform-in-$\cos\theta_A$ prior, which peaks at $\theta_A = \pi/2$, so the median is always higher than the true value of $\theta_A = 0$. The posteriors for both χ_A and θ_A are more constrained and less biased for systems with nearly edge-on inclinations (dashed lines in Fig. 8-6), however. This is a generic feature of all our simulations, and can be explained due to the fact that, for a fixed SNR, viewing the system nearly edge-on enhances the effect of precession on the observed waveform. In some configurations where $\cos\theta_{JL} < \sin\theta_{JN}$ (see the right panel of Fig. 8-1), the observer can see both sides of the orbital plane as it oscillates [923].

For systems with moderate primary spin tilt, $\theta_{1/A} = \pi/3$, shown in the medium-blue lines in Fig. 8-6, the constraints on χ_A improve slightly at higher total masses, although the same trend is not observed in θ_A for this choice of tilt angle. The width of the posterior is largest at about $M_{\text{tot}} \approx 160\ M_\odot$, corresponding to a local minimum

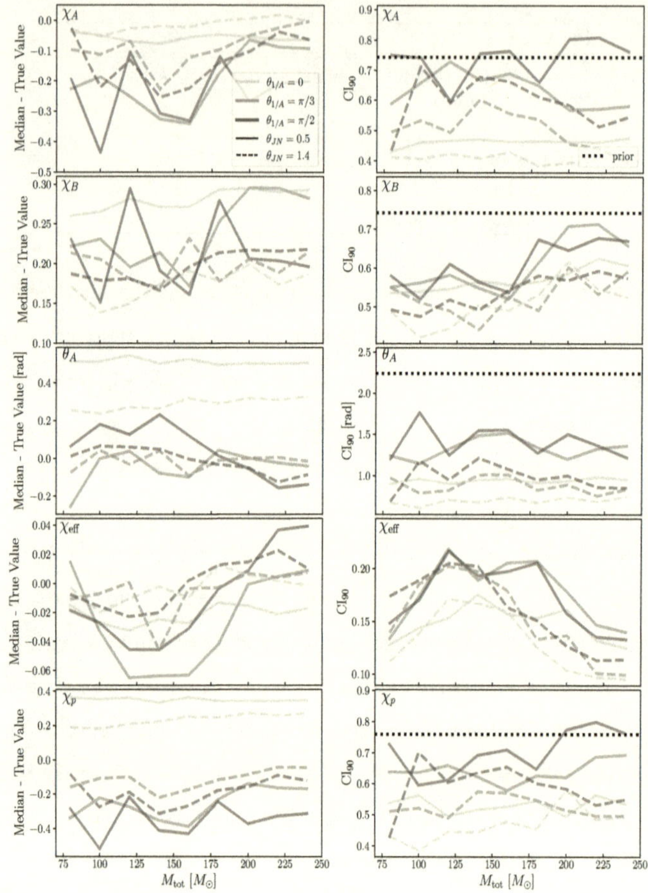

Figure 8-6: The bias and width of the spin-sorted spin magnitudes, tilt of the highest-spinning black hole, and effective aligned and precessing spins as a function of the total mass for systems with equal mass. The bias is defined as the difference between the median of the posterior and the true value, while the width is the symmetric 90% credible interval. The secondary black hole is nonspinning, and the remaining true parameter values are given in Tables 10.4 and 8.4. Aligned spin systems with $\theta_{1/A} = 0$ are shown in light blue, those with $\theta_{1/A} = \pi/3$ in medium-blue, and those with in-plane primary tilt, $\theta_{1/A} = \pi/2$, in dark blue. Systems observed nearly face-on with $\theta_{JN} = 0.5$ rad are represented with a solid line, and those observed nearly edge-on with $\theta_{JN} = 1.4$ rad are represented with a dashed line. The dotted black lines show the width of the 90% CI for the prior for all parameters except $\chi_{\rm eff}$, for which the prior width of 0.84 is much larger than the posterior widths.

in the bias as a function of mass for χ_A. This trend is observed for both inclination angles, although the posteriors are again more constrained for the nearly edge-on simulations. For systems with in-plane primary spin, this trend is only observed at nearly edge-on inclination angles (dark blue dashed line), while the width of the 90% credible interval for the nearly face-on systems (dark blue solid line) remains largely constant with mass.

The posteriors of the lowest-spinning object do not exhibit a turn-over in the width of the 90% CI for either $\theta_{1/A} = \pi/3$ or $\theta_{1/A} = \pi/2$. Instead, the width increases slightly as a function of mass. The bias for χ_B is roughly constant as a function of both mass and primary tilt angle. On the other hand, both the accuracy and the precision of the χ_A posteriors improve as the primary spin tilt decreases (as the line color gets lighter in the top panel of Fig. 8-6). This is indicative of the fact that that it is easier to measure the component spins for aligned-spin systems, where the absence of precession in the waveform is informative, as seen in Ref. [921], since only the well-measured aligned-spin component, $\chi_{A,z}$, contributes.

Comparing the scale of the first and second panels on the right side of Fig. 8-6 shows that χ_B is better constrained than χ_A for systems with nonzero primary tilt. The inference for these sources consistently recovers lower values for χ_A than the true value, as indicated by the negative values of the bias. This is due to a degeneracy with χ_p, which is shown in the corner plot in Fig. 8-7. The low-χ_p tail driven by the prior (black dashed line) pulls the spin magnitude to lower values. This preference for low spins translates into a better constraint on the spin magnitude of the lowest-spinning black hole, since high spin values can be entirely ruled out for that object but not for the highest-spinning object.

This bias in the magnitude of χ_A towards low values is also present in the posteriors for χ_{eff} and χ_p, as shown in the lower two panels on the left side of Fig. 8-6. The width of the 90% CI for χ_{eff} decreases at the highest masses regardless of primary tilt or inclination angle. We observe no significant trend in the width of the posteriors as a function of mass for χ_p. The large positive bias in χ_p for aligned-spin systems follows from the corresponding bias in θ_A, and is due to the fact that the true value

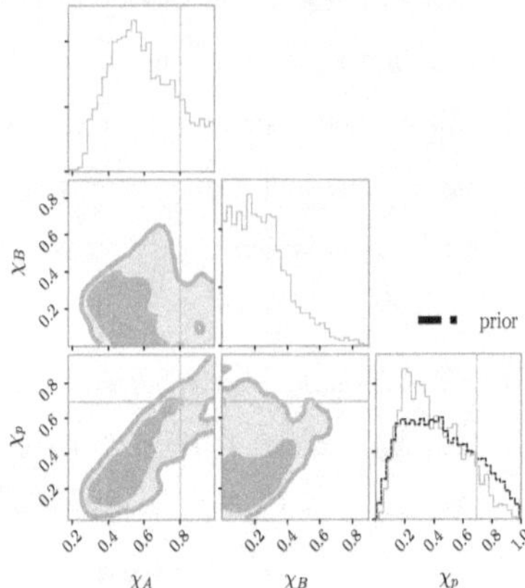

Figure 8-7: Corner plot showing the χ_A, χ_B, and χ_p posteriors for an equal-mass system with nonzero primary tilt. The light shaded region shows the 90% CI, and the darker region shows the 50% CI. The prior for χ_p is shown in the black dashed line, and the true parameter values are given by the vertical orange lines.

of $\chi_p = 0$ for those systems, so the median is always larger since the prior for χ_p is peaked away from $\chi_p = 0$. The posteriors for the systems observed nearly edge-on are less biased and better constrained for all choices of primary tilt compared to those for systems observed nearly face-on for both χ_{eff} and χ_p. We emphasize that the biases we comment on above are driven by the effects of the prior shape and projecting the posteriors into one dimension to calculate the median. The maximum likelihood point in the full 15-dimensional parameter space should coincide with the true parameter values since we are not adding Gaussian noise to the signal. Hence, any bias in the posterior median away from the true value occurs when the impact of the prior exceeds that of the likelihood.

8.4.2 Unequal-mass systems

The bias and uncertainty for the same spin parameters presented above are shown for our $q = 1/4$ simulations in Fig. 8-8. The most salient features of our analysis of unequal-mass systems are the apparent outliers in the width of the 90% credible interval of χ_A and θ_A at ~ 180 $M_\odot$ for systems with in-plane primary tilt observed nearly face-on (dark red solid line) and at ~ 160 $M_\odot$ for systems with in-plane primary tilt observed nearly edge-on (dark red dashed line). We verify that these apparent outliers are not due to sampling issues by repeating the inference with a different random seed for the stochastic sampler and by performing more simulations with a finer mass spacing (5 $M_\odot$) for the region with width 40 $M_\odot$ centered on the outlier value in each case. These additional simulations, which are included among the systems plotted in Fig. 8-8, demonstrate that the outliers represent local maxima in the width of the 90% CI as a function of mass, with values $\gtrsim 0.2$ in excess of those for the lowest and highest masses. As we will describe in Section 8.4.3, these apparent outliers are a consequence of the choice to define the true spin parameter values at a fixed reference frequency of 25 Hz for all systems regardless of total mass. This results in the compared systems having different values of the spin angles at the same point in their evolution as the total mass increases. We emphasize, however, that the presence of these apparent outliers does not affect the conclusions drawn when comparing the results across different values of inclination, primary tilt, and mass ratio, which are discussed in the rest of this section.

Both the accuracy and the precision of the measurements of the spin components for the highest-spinning object improve considerably for unequal-mass systems compared to those with equal mass. This is due to the fact that the unequal mass serves to break parameter degeneracies in the system [513]. The constraints for the lowest-spinning object conversely suffer due to this degeneracy breaking, as all the spin information obtained pertains to the highest-spinning object in this case, and the secondary spin measurements are completely uninformative.

The unequal-mass aligned-spin systems, represented by the light red lines in Fig. 8-8, show a gradual broadening of the posteriors on the magnitude and tilt of the highest-spinning black hole as a function of mass. A similar trend is observed for those systems with moderate primary spin tilt, shown in medium-red. While the results presented in Section 8.3.1 suggest that the spin information is dominated by the merger and ringdown parts of the signal for heavy BBH, it is possible that having

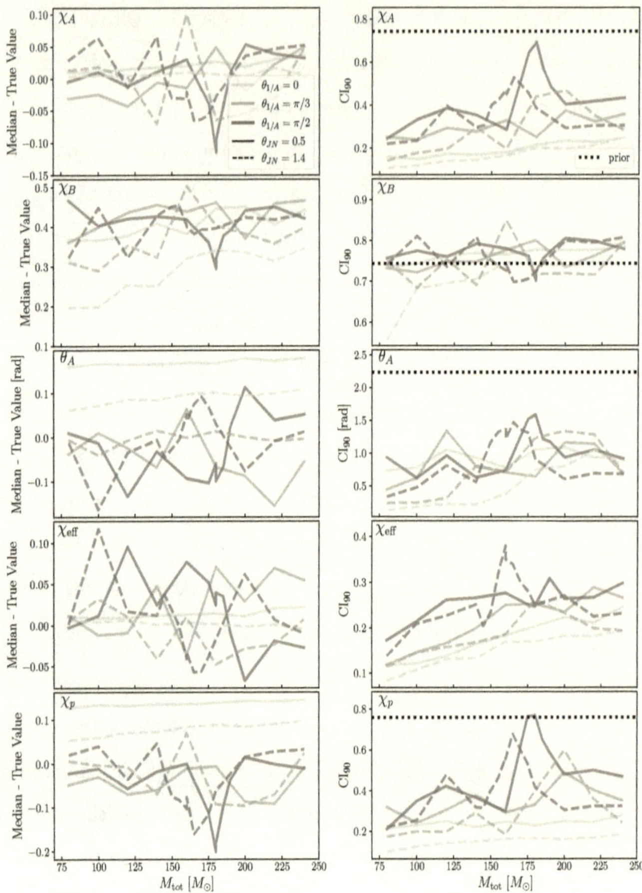

Figure 8-8: The bias and width of the spin-sorted spin magnitudes, tilt of the highest-spinning black hole, and effective aligned and precessing spins as a function of the total mass for systems with unequal mass, $q = 1/4$. The bias is defined as the difference between the median of the posterior and the true value, while the width is the symmetric 90% credible interval. The secondary black hole is nonspinning, and the remaining true parameter values are given in Tables 10.4 and 8.4. Aligned spin systems with $\theta_{1/A} = 0$ are shown in light red, those with $\theta_{1/A} = \pi/3$ in medium-red, and those with in-plane primary tilt, $\theta_{1/A} = \pi/2$ in dark red. Systems observed nearly face-on with $\theta_{JN} = 0.5$ rad are represented with a solid line, and those observed nearly edge-on with $\theta_{JN} = 1.4$ rad are represented with a dashed line. The dotted black lines show the width of the 90% CI for the prior for all parameters except χ_{eff}, for which the prior width of 0.84 is much larger than the posterior widths.

access to a longer inspiral still leads to improvements in the measurability of spin relative to a system with a shorter inspiral. Thus, the degradation of the component spin measurements for the highest-spinning black hole as a function of increasing total mass could be explained due to the presence of fewer inspiral cycles in the band of the gravitational-wave detector as the mass increases. That is, as the total mass increases, even though the spin information gained from the merger-ringdown becomes relatively more significant, the absolute spin information gain still decreases because fewer cycles fall in the detector band. The same explanation would apply for the increase in the width of the 90% CI for χ_B as a function of mass for equal-mass systems. We leave detailed investigation of this hypothesis to future work.

The bias in the χ_{eff} posteriors for unequal-mass systems varies significantly as a function of mass, although the scale of the bias is smaller than that of the component spin magnitudes. χ_p is much less biased for unequal compared to equal masses, indicating that it is easier to accurately capture precession via χ_p when the parameter degeneracies in the system are broken. This is due to the fact that there are post-Newtonian terms proportional to the difference in component masses that disappear for equal-mass systems, making the spin parameters harder to measure in this case [513].

The χ_{eff} posteriors for unequal-mass systems do not show the same decrease in the width of the 90% credible interval at the highest masses as the equal-mass systems, with the exception of the source with in-plane primary spin observed edge-on that exhibits the outlier in χ_A (dark red dashed line). The same outlier and trend is visible in χ_{eff} for this system, but not for the source observed nearly face-on (dark red solid line in the penultimate panel on the right side of Fig. 8-8). For the nearly face-on source, the increased uncertainties in the parameters going into the numerator and denominator of Eq. (10.1) cancel out, since the mass ratio suffers from a similar increase in the width of the 90% credible interval (see Fig. 8-17 in Appendix 8.7.2). The outliers in the systems with in-plane primary spin observed at both inclination angles are present in the width of the 90% credible interval for χ_p, however. There is another smaller outlier in χ_p at around $M_{\text{tot}} = 180\ M_\odot$ for the unequal-mass systems with moderate primary tilt observed nearly edge-on (medium-red dashed line), that can also be seen in the widths of the posteriors for χ_A and θ_A. With the exception of these outliers, the width of the 90% credible interval for both the $\chi_{\rm eff}$ and χ_p posteriors gradually increases with the total mass of the system.

8.4.3 Outliers

The apparent $q = 1/4$ outliers mentioned in the previous section show a deterioration in the precision of the measurements of χ_A, $\chi_{\rm eff}$, and χ_p at masses in the range of $\sim 160-180\ M_\odot$, and then a subsequent improvement at the highest masses. This goes against the naive expectation that the spin measurements should get worse for higher total masses since fewer inspiral cycles are accessible in the band of the interferometers as the total mass of the system increases. The constraints on the mass parameters for these systems, shown in Fig. 8-17 in Appendix 8.7.2, exhibit a similar trend.

The local maxima in the widths of the χ_A posteriors correspond to local minima in the widths of the χ_B posteriors, indicating that one is measured better at the expense of the other, although the peaks are still visible in the posteriors for the effective spins since the scale of the upward deviation is much larger than that of the downward deviation. These features are therefore not an artifact of the spin sorting, as the definitions of $\chi_{\rm eff}$ and χ_p do not change between sortings. The peaks in the widths as a function of mass also correspond to minima in the bias for χ_A and χ_p, indicating that for these systems the inference recovers spins that are systematically lower than the true values and not well constrained. We note that the two sets of simulations that exhibit this outlier behavior correspond to sources with the same intrinsic parameters just viewed at different inclination angles.

In Fig. 8-9, we plot the cumulative SNR in the Hanford interferometer as a function of frequency for the different total masses we simulated. For the nearly face-on case (top panel), the apparent outlier at 180 $M_\odot$ occurs when the inspiral contributes most to the total SNR of the signal (with the exception of the lowest mass sources). However, looking instead at the nearly edge-on case, the opposite is true; the outlier at 160 $M_\odot$ occurs when the merger contributes most to the total SNR of the signal

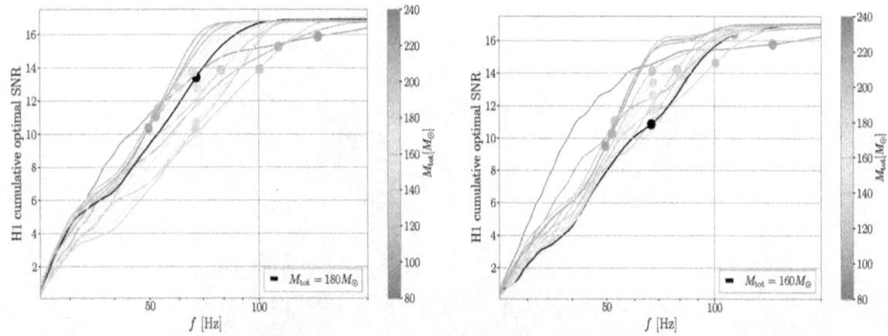

Figure 8-9: Cumulative SNR as a function of frequency in the Hanford interferometer for different total masses for the simulations observed nearly face-on (top) and nearly edge-on (bottom). The marker shows the merger frequency for each total mass, and the outlier mass is shown in black.

(except for the highest-mass sources). This feature is consistent across the SNR distributions for all three interferometers, and indicates that the apparent outliers cannot simply be attributed to one part of the signal contributing most of the spin information for these systems.

Instead, these plots highlight that the merger frequency does not monotonically decrease as a function of total mass for these systems, since the simulations were performed at a fixed reference frequency of 25 Hz. This reference frequency corresponds to different stages in the evolution of the source for systems of different masses, so the primary spin tilt angle will be different at a fixed dimensionless frequency, $f_{\mathrm{ref}} M_{\mathrm{tot}}$, with M_{tot} in units of seconds. This means the simulated binaries are not the same up to a total mass scaling, but rather correspond to different physical systems with distinct spin parameters. The tilt angles of some of the systems with masses in the vicinity of the two outliers are shown in Fig. 8-10 at $f_{\mathrm{ref}} M_{\mathrm{tot}} = 0.0263$, approximately one cycle before merger. That the two orientations have overlapping tilt angles in the common mass range is expected since they represent the same intrinsic system viewed at different inclination angles. While the overall scale of the variation in the primary tilt angle is less than a tenth of a radian, the apparent outliers (indicated with vertical lines) occur at inflection points in the tilt as a function of mass. This suggests that

290

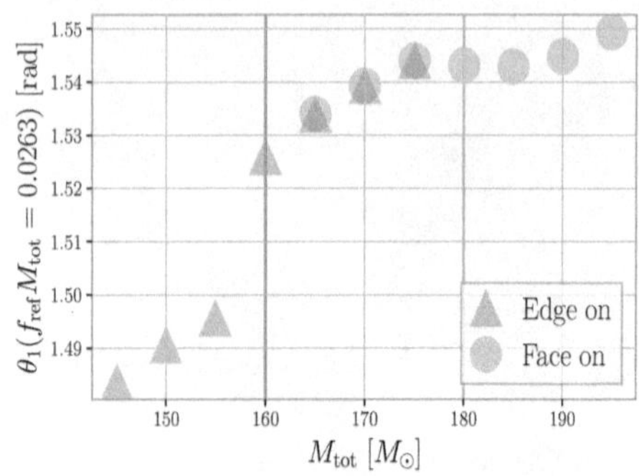

Figure 8-10: Tilt angle at a fixed dimensionless reference frequency, $f_{\mathrm{ref}}M_{\mathrm{tot}} = 0.0263$, as a function of mass for some of the systems with masses in the vicinity of the two outliers. The nearly edge-on systems are shown in blue, and the nearly face-on systems in red. The two overlap for the common masses because they correspond to the same intrinsic system viewed at different angles by the observer. The masses of the outlier systems are indicated with vertical lines.

the exact spin configuration can have a significant effect on the measurability of the spin parameters for high-mass systems, reinforcing the finding from Section 8.3.

To verify that the apparent outliers are due to the choice of fixed reference frequency, we repeat the simulations for the configurations with outliers—namely, unequal-mass systems with in-plane primary tilt observed both nearly edge-on and nearly face-on—but, instead of using a fixed reference frequency of 25 Hz, we use a fixed dimensionless reference frequency of $f_{\mathrm{ref}}M_{\mathrm{tot}} = 0.0296$. This value was chosen so that the waveform for the highest-mass system we simulated, $M_{\mathrm{tot}} = 240\ M_{\odot}$, could be generated with $f_{\mathrm{ref}} = f_{\mathrm{min}} = 25$ Hz. The widths of the 90% credible intervals for the component spin magnitudes, effective spins, and masses are shown in Fig. 8-11 for both the original fixed reference frequency (shown in dark red) and the new fixed dimensionless reference frequency (shown in orange). While the trend in the posterior width as a function of mass is not completely smoothed out when using the fixed dimensionless reference frequency, the outliers are no longer present. Instead, the width increases nearly monotonically as a function of total mass for all parameters except χ_B. For this parameter, the width is instead roughly constant when comparing the scale of the variation to that of the other parameters. The posteriors for θ_A with the new fixed dimensionless reference frequency exhibit similar behavior. The large excursions in the bias at the outlier mass for the component spin magnitudes and effective spins in Fig. 8-8 are also smoothed out.

By using a fixed dimensionless reference frequency, increasing the total mass of the system does not change the relative phasing of the waveform, but only the overall amplitude and time-frequency scaling. This results in the expected behavior that for a fixed spin configuration, more massive systems have fewer cycles in-band and hence the constraint on the spin parameters gradually worsens as the total mass of the system increases. However, these apparent outliers demonstrate that even small changes in the spin configuration, of the order of a tenth of a radian in the tilt angle, can lead to substantial differences in the posteriors for the spin parameters.

To determine whether these apparent outliers are unique to the NRSur7dq4 waveform, we repeat the parameter estimation for the original systems with a fixed reference frequency of 25 Hz with two additional waveform models, IMRPhenomPv2 [445, 482, 509] and IMRPhenomXPHM [710, 711, 397]. Detailed results and comparisons of the three waveform models are presented in Appendix 8.7.3. In general, we find that these three waveform models yield different results for both the accuracy and precision of the spin and mass parameters.

8.5   Measurement of azimuthal spin angles

Measurements of the azimuthal spin angles for binary black hole systems can carry important information about their formation channels and evolution [404, 406]. For example, binaries formed in isolation may become locked in a "spin-orbit resonance", where the orbital angular momentum and component spin vectors are coplanar and jointly precess around the total angular momentum [796]. In this configuration, the projections of the two component spins onto the orbital plane are either aligned or antialigned, so that the angle between them $\phi_{12} = 0, \pm\pi$. For stellar binaries with significant supernova natal kicks and stellar tides, an increasing fraction of systems

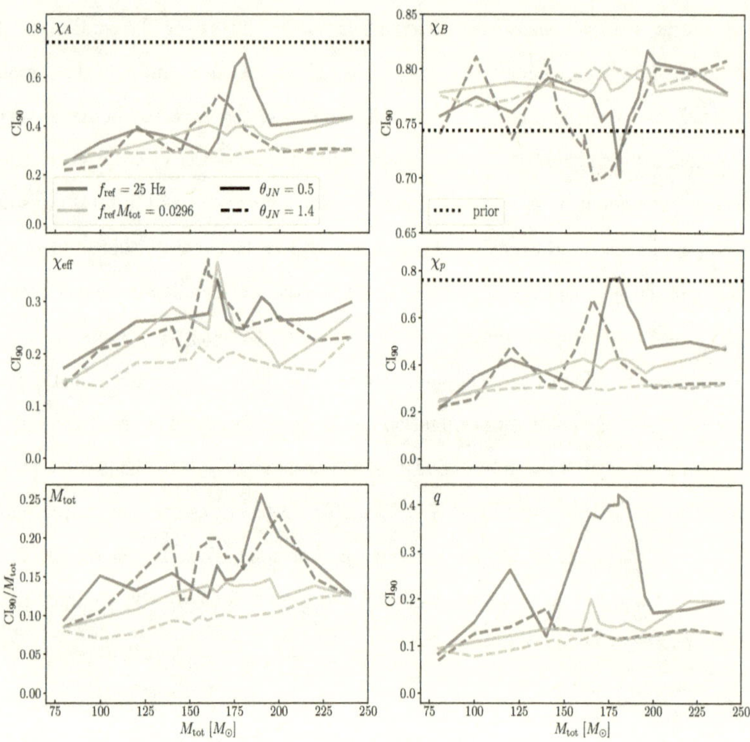

Figure 8-11: Width of the 90% credible interval as a function of total mass for the original systems with outliers at fixed $f_{\rm ref} = 25$ Hz shown in dark red compared to the simulations with the same parameters but at fixed dimensionless reference frequency $f_{\rm ref} M_{\rm tot} = 0.0296$ in orange. The systems viewed nearly edge-on are indicated with dashed lines and those observed nearly face-on with solid lines. For the total mass, the credible interval width is normalized by the value of the total mass of each system so that this quantity is dimensionless.

will get captured into this resonant configuration as the binary orbital separation decreases, where ϕ_{12} librates about 0 or $\pm\pi$ instead of circulating over the full allowable range of values as the binary evolves. Whether the in-plane spin components are aligned or antialigned depends on the efficiency of mass transfer between the first-born black hole and its remaining stellar companion [404].

The detectability of such spin-orbit resonances has previously been studied in [438, 406, 882, 60]. While Refs. [438, 406, 60] computed the overlap between the waveforms of different systems to determine if the two resonant configurations are detectable and distinguishable from each other and from non-resonant configurations, Ref. [882] conducted full parameter estimation on low-mass resonant BBH sources using the SpinTaylorT4 model, which includes the effects of spin precession but only in the inspiral regime relying on the post-Newtonian approximation [186, 190]. Here we investigate the measurability of ϕ_{12} defined at 25 Hz for massive BBH systems using the NRSur7dq4 waveform model including the full spin evolution and higher-order modes through the merger and ringdown.

The 90% credible intervals for the ϕ_{12} posteriors obtained for the equal-mass simulations presented in the previous section are shown in Fig. 8-12. For all the unequal-mass systems (not shown in Fig. 8-12) and all but one of the equal-mass configurations, the 90% CI includes nearly the entire prior range. However, we find that the posterior can be better constrained for systems with aligned primary spin and equal mass when observed nearly edge-on (light blue dashed line) compared to all the other simulated configurations. A corner plot showing the posteriors for the spin parameters for one such binary where ϕ_{12} is well-measured is shown in Fig. 8-13. In this case, we recover a high primary spin magnitude with a small primary tilt angle and a small secondary spin magnitude with a large tilt angle, so the secondary spin is reconstructed to lie almost entirely in the orbital plane (even though the true value is $\vec{\chi}_B = 0$).

If the perpendicular components of the spin vectors cancel, such that $\vec{S}_{1,\perp} + \vec{S}_{2,\perp} = 0$, then the total angular momentum lies entirely along the orbital angular momentum, so there is no precession. This can be achieved when $\phi_{12} = \pi$. Be-

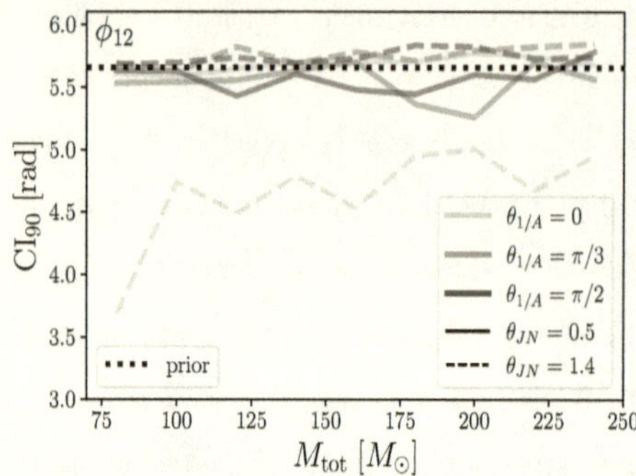

Figure 8-12: 90% credible interval for the ϕ_{12} posteriors obtained for the first set of simulations presented in Table 10.4.

cause the system is observed nearly edge-on, any modulations in the waveform due to orbital-plane precession would be more easily observable, so the lack of precession in this system means that the in-plane spin components should cancel, leading to the preference for $\phi_{12} = \pm\pi$. The lack of precession is not as clear when the system is observed nearly face-on, leading to a worse constraint on ϕ_{12}. A similar constraint on ϕ_{12} is not obtained for the equivalent unequal-mass systems, since in this case the primary magnitude and tilt are very well-measured, but the secondary spin is relatively unconstrained. This leaves no degeneracy between the primary and secondary in-plane spin vectors that needs to accommodate the lack of precession in the signal by preferring $\phi_{12} = \pi$.

Because ϕ_{12} does not have a physical meaning for systems with only one spinning component, we perform three additional sets of simulations, described by Simulation Sets 2–4 in Table 10.4. In these new simulations, we try two different values of the secondary spin and two different values of the SNR, to verify whether higher-SNR events will improve the measurability of ϕ_{12}. We also choose three different values of ϕ_{12}, although the systems we simulate are *not* necessarily in resonant configurations, which would require also choosing the spin tilt angles so that the value of ϕ_{12} is constant in time. The 90% credible intervals for the ϕ_{12} posteriors for Simulation Sets 2, 3, and 4 are shown in Fig. 8-14.

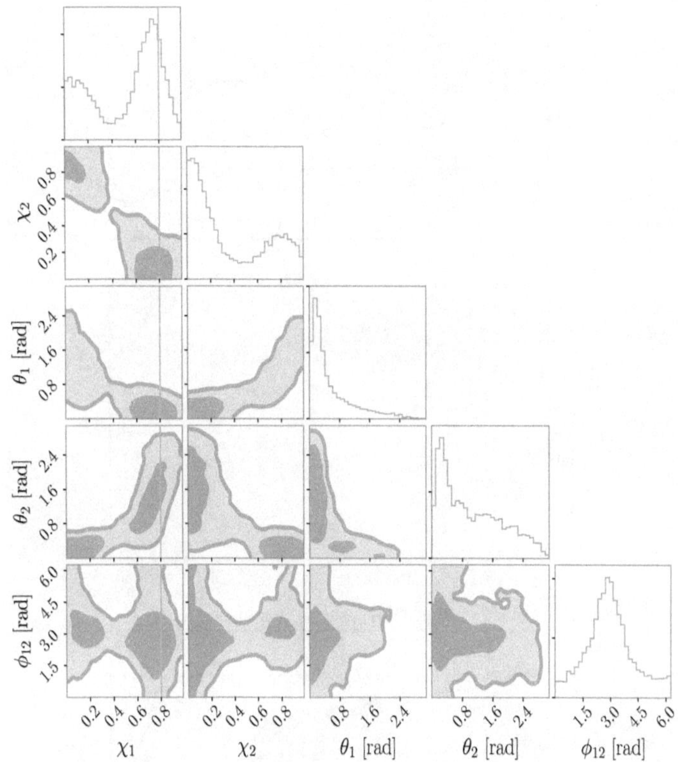

Figure 8-13: Corner plot showing the posteriors for the component spin parameters and ϕ_{12} for a simulated system with an aligned primary spin and nonspinning secondary component. The orange lines show the true parameter values, and hence are omitted for θ_2 and ϕ_{12} as these parameters do not have a physical meaning when only one object in the binary is spinning.

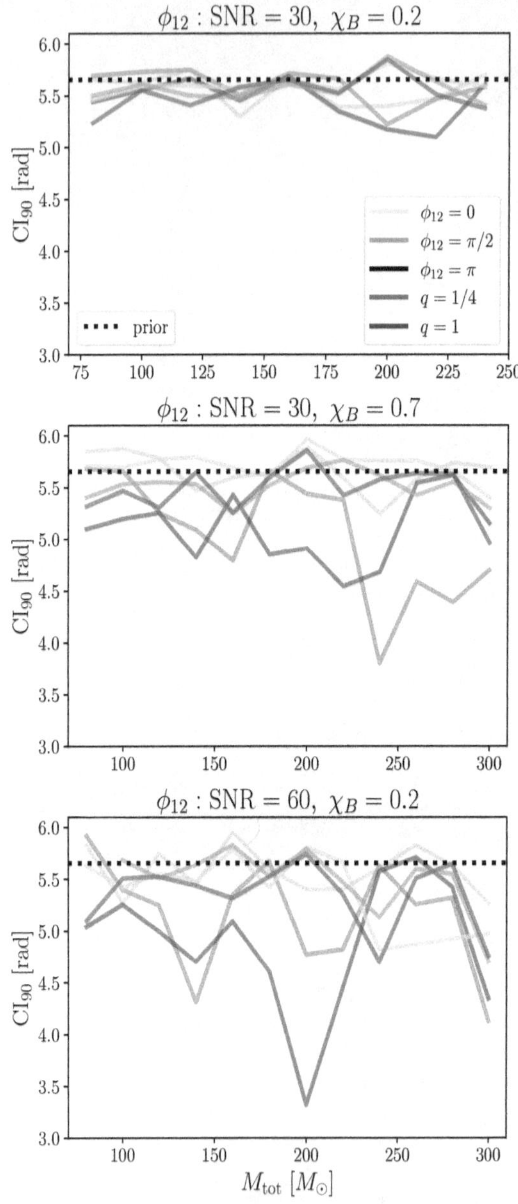

Figure 8-14: 90% credible intervals for the ϕ_{12} posteriors for Simulation Sets 2 (top; SNR=30, $\chi_2 = 0.2$), 3 (middle; SNR=30, $\chi_2 = 0.7$) and 4 (bottom; SNR=60, $\chi_2 = 0.2$). The other parameters of the simulated systems are given in Tables 10.4 and 8.4.

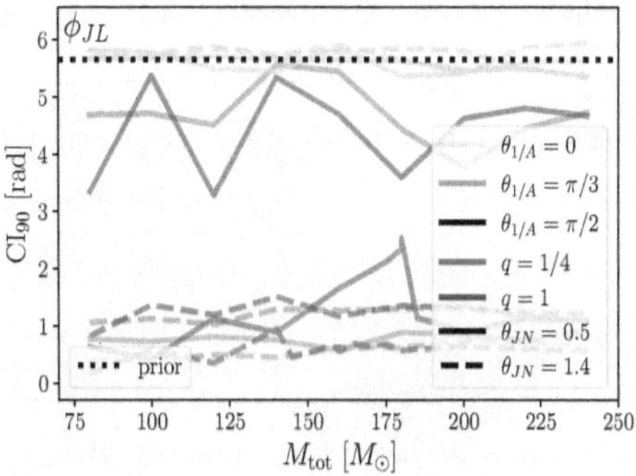

Figure 8-15: 90% credible interval for the ϕ_{JL} posteriors obtained for the first simulation set after wrapping the posterior samples around $\phi_{JL} = 2\pi$.

For the systems with SNR $= 30$ and low secondary spin, $\chi_2 = 0.2$, shown in the top panel of Fig. 8-14, ϕ_{12} is unconstrained relative to the prior regardless of the value of ϕ_{12} or the mass ratio of the system across all the total masses we analyzed. While the measurements improve slightly for equal-mass systems when the secondary spin is increased to $\chi_2 = 0.7$ in the middle panel and when the SNR is increased to 60 in the bottom panel, ϕ_{12} remains generally poorly constrained at 25 Hz for most systems we analyze.

The remaining spin degree of freedom is the azimuthal phase of the total angular momentum, $\vec{L}$, in its precession cone around the total angular momentum, $\vec{J}$, called ϕ_{JL} (see right panel of Fig 8-1). Along with the other five spin parameters and the binary mass ratio, this angle determines the recoil velocity and direction of the remnant black hole, which can have implications for the rate of hierarchical mergers in dynamical environments like globular clusters [904]. Because the true value of ϕ_{JL} used in our simulations is at the edge of the prior (see Table 8.4), the posterior probability density is bimodal with peaks at $\phi_{JL} = 0, 2\pi$. In order to meaningfully calculate the width of the 90% credible interval, we map the posterior samples with $\phi_{JL} < \pi$ to $\phi_{JL} + 2\pi$. The resulting 90% credible interval for ϕ_{JL} as a function of the total mass of the simulated system is shown in Fig. 8-15.

The worst constraint on ϕ_{JL} comes from systems with aligned primary spin, consistent with the fact that this parameter is not physically meaningful for aligned-spin systems where the total and orbital angular momentum vectors point in the same direction. The measurement of ϕ_{JL} improves for system with unequal mass ratios compared to those of equal mass and for systems observed nearly edge-on compared to those observed nearly face-on. Systems with unequal mass ratios have fewer degeneracies and enhanced higher-order mode emission, making the presence of precession—and hence an offset between the total and orbital angular momenta—easier to measure. This effect is similarly easier to discern for systems viewed nearly edge-on, when the effects of precession are enhanced. The primary tilt angle does not have a significant effect on the width of the ϕ_{JL} posterior for systems with the same mass ratio and inclination. We conclude that, while ϕ_{JL} can be well-measured for systems where the effects of precession are apparent, ϕ_{12} is generally poorly constrained even for highly spinning systems and those with high SNR, but leave further investigation of the measurability of ϕ_{12} for systems locked in spin-orbit resonance to future work [905].

8.6 Conclusions

We have presented a systematic analysis of the measurability of spin in heavy binary black hole sources including both aligned and precessing spins. We used the NR-Sur7dq4 waveform [903], which models the effects of all six spin degrees of freedom on the signal, as well as higher-order modes up to $\ell \leq 4$. Motivated by the informative measurement of χ_p for GW190521 [37], we performed several simulations of systems with parameters consistent with the GW190521 posterior, but were unable to recover an informative measurement of χ_p until we used the parameters of the maximum-likelihood posterior sample as the true values for our simulated system, indicating that the correlations between parameters and the exact six-dimensional spin configuration have a significant effect on the measurability of χ_p. For this maximum-likelihood system, we find that the spin measurement hinges on the merger and ringdown part of the signal.

We extended our study to generic configurations to investigate how the bias and precision of the spin posteriors scale with the total mass of the system. We find that the spin magnitude of the highest-spinning black hole [153], χ_A, can be accurately constrained to a 90% credible interval width of $\lesssim 0.5$ for systems with aligned spin or unequal masses (compared to a prior width of 0.89). Because of the degeneracies in the system, the spin of the lowest-spinning black hole, χ_B, is marginally better-constrained for equal-mass systems, but the posteriors for this parameter are uninformative for unequal-mass systems. The constraint on χ_B worsens as the total mass of the system increases for equal-mass systems, and the same is true for χ_A for the unequal-mass systems. No significant trend is observed for the width of the χ_A posterior as a function of mass for equal-mass systems.

The tilt of the highest-spinning black hole, θ_A, can be measured with a posterior 90% CI width of $\lesssim 1.5$ rad, and is in general better-measured for unequal-mass systems [231, 668, 923, 688, 712]. The corresponding prior width for θ_A is 2.21 rad. Both the spin magnitude and tilt of the highest-spinning black hole, χ_A and θ_A, are more accurately and precisely measured for aligned-spin systems, since the absence of precession is informative. We also find a generic improvement in the posterior bias and width for systems viewed nearly edge-on compared to those observed face-on [923], as the observer can see both sides of the orbital plane as it oscillates for some orientations when $\cos\theta_{JL} < \sin\theta_{JN}$, enhancing the visibility of precession for a fixed SNR.

Interestingly, we find that the effective aligned spin, χ_{eff}, is in general measured to a posterior 90% credible interval width of $\lesssim 0.35$ for all the high-mass systems we simulate (prior width of 0.84), despite the limited number of inspiral cycles available in the analysis. While the width increases as a function of mass for most of the unequal-mass systems, it actually narrows at the highest masses for equal-mass systems. On the other hand, χ_p is not well-constrained for equal-mass systems, but is in general measured to a posterior 90% credible interval width of $\lesssim 0.5$ for unequal-mass systems (compared to a prior width of 0.75). As was the case for the component spins, the effective spins are more narrowly constrained for aligned-spin systems, although the median of the χ_p posterior is biased away from the true value by $\sim$0.4 for equal-mass systems regardless of the primary tilt angle.

The trends in the width of the 90% credible interval of the χ_p posterior as a function of mass appear to exhibit significant outliers for unequal-mass systems with in-plane primary tilt at $\sim$160 $M_\odot$ for systems observed nearly edge-on and at $\sim$180 $M_\odot$ when observed nearly face-on. These apparent outliers are due to the variation in the exact spin configuration at a fixed dimensionless reference frequency. Changing the total mass of the simulated systems but keeping the reference frequency fixed to 25 Hz results in variations in the true simulated primary tilt angle at fixed dimensionless reference frequency of the order of a tenth of a radian. When repeating the analysis with a fixed dimensionless reference frequency, these apparent outliers disappear, re-inforcing the previous indication that small changes in the exact spin configuration can have a significant effect on the measurability of the spin parameters. Reanalyzing the systems with a fixed reference frequency of 25 Hz with two different waveform models, IMRPhenomPv2 [445, 482, 509] and IMRPhenomXPHM [710, 711, 397], led to different results, as shown in detail in Appendix 8.7.3.

We also investigate the measurability of the azimuthal spin angles that complete the six spin degrees of freedom. The ϕ_{12} angle can encode information about the system's formation history via spin-orbit resonances [796, 404, 406], and both ϕ_{12} and ϕ_{JL} affect the recoil kick of the remnant black hole [904]. While ϕ_{JL} is generally measured to a posterior 90% credible interval width of $\lesssim 1.5$ (prior width 5.65) for systems with some misalignment with unequal masses or viewed nearly edge-on, ϕ_{12} is generally unconstrained regardless of the SNR or the spin magnitude of the secondary for the configurations we simulated. The lack of constraint for ϕ_{12} could potentially be explained by the choice of reference frequency, as Ref. [905] recently showed that measuring the spins close to merger improves the constraint on this parameter. We do find, however, that some aligned-spin systems preferentially recover posteriors peaked at $\phi_{12} = \pm\pi$, since this configuration allows the effects of precession, which are absent in the signal, to cancel in the posterior.

To summarize, we conclude that:

1. It is possible to capture the effects of precession via the posterior on χ_p for highly-precessing, heavy systems with moderate SNR like GW190521, but the measurability of χ_p depends on the exact six-dimensional spin configuration.

2. For a highly precessing system including the parameter correlations that led to the informative χ_p measurement for GW190521, the inference of the component spins and hence χ_p depends strongly on the merger and ringdown parts of the signal, rather than the inspiral.

3. Consistent with previous studies, the spin of the highest-spinning black hole and χ_p are better measured for systems with unequal masses compared to those with equal masses. All spin parameters are better measured when the system is observed at higher inclination angles for systems with total masses in the range $80 - 240 \ M_\odot$.

4. The spin parameters are better constrained for systems with aligned spins compared to those in generically precessing configurations, since the lack of precession leaves a characteristic imprint on the waveform in these cases. Even when allowing for precession in the recovery of these systems, the lack of obvious signs of precession in the data leads to improved constraints on the spin parameters.

5. χ_{eff} is well-measured for all the masses we considered, with a 90% posterior credible interval width of $\lesssim 0.35$, demonstrating that the effective spin can be inferred in the absence of a prominent inspiral. For comparison, the prior width for χ_{eff} is 0.84. Furthermore, the constraint improves as the mass increases for some configurations, particularly those with equal mass.

6. Variations of even a tenth of a radian in the primary tilt angle at a fixed dimensionless reference frequency can lead to significant differences in the ability to accurately and precisely measure the spins of a particular system.

7. The azimuthal angle between the two component spin vectors is in general unconstrained at a fixed reference frequency of 25 Hz, even for systems detected at high SNR. However it can be recovered with a preference towards $\phi_{12} = \pm\pi$ for some aligned-spin systems, even though the true value of this parameter is not physically meaningful in these cases.

Our study raises several questions, particularly about the optimal configuration for measuring the effects of precession. While we find that the constraint on χ_p varies significantly depending on the exact six-dimensional spin configuration of the system, a more detailed exploration of the parameter space would be necessary to reveal which configurations lead to the best measurement of χ_p. We find that the true value of $\chi_{\rm eff}$ or χ_p alone is not enough to explain these variations. Furthermore, despite the fact that we find that the spin information is primarily driven by the merger and ringdown parts of the signal for the GW190521 maximum-likelihood simulation, the uncertainty in the component spin parameters continues to increase as the total mass of the system increases, even for those systems simulated at a fixed dimensionless reference frequency. This suggests that losing inspiral cycles leads to a loss of spin information, but further investigations are required to discern the effects of the inspiral vs. postmerger parts of the signal on the measurability of spin in generic systems. Finally, while we find that ϕ_{12} is unconstrained for most systems explored in this work, these systems are not necessarily locked in spin-orbit resonances. We leave the investigation of the measurability of ϕ_{12} for resonant systems to future work.

8.7 Supplementary material

8.7.1 GW190521 maximum-likelihood simulation

For the simulations where the GW190521 maximum-likelihood system is added to real detector noise from O3, we choose ten distinct 4 s data segments beginning 8 s after the coalescence times of ten of the higher-mass BBH systems included in GWTC-2 and assume the data in each segment is well-characterized by the PSDs computed

Event	Time
GW190413_134308	1239198214.76
GW190421_213856	1239917962.27
GW190503_185404	1240944870.30
GW190517_055101	1242107487.84
GW190527_092055	1242984081.81
GW190602_175927	1243533593.11
GW190706_222641	1246487227.35
GW190719_215514	1247608540.95
GW190909_114149	1252064535.73
GW190915_235702	1252627048.71

Table 8.5: Events from O3 whose PSDs were used and the corresponding coalescence times at which we added the GW190521 maximum-likelihood simulation into real detector data.

for each of those events (publicly available at [33]). The events and true coalescence times we use are detailed in Table 8.5. For these analyses, we use a bandwidth of $16 - 256$ Hz and a sampling rate of 512 Hz. We only include the data from the Hanford and Livingston detectors in this set of simulations, since Virgo data was not available for all of our chosen events. Because we allow the PSDs to vary, we adjust the distances of the ten simulated systems to keep the network matched filter SNR fixed to 15 for consistency with the Gaussian noise tests. The results of the real noise simulations are shown in the left panel of Fig. 8-16. The variation in the χ_p posterior is similar to that observed in Fig. 8-3.

We also repeat the experiment using Gaussian noise colored by the PSDs calculated for the data segment containing GW190521 but instead add a system with parameters drawn from the GW190521 posterior with a low value of χ_p. The resulting χ_p posteriors are shown in the right panel of Fig. 8-16, and the true values of the parameters are given in Table 8.6. Although the true value of $\chi_p = 0.18$, these posteriors much more closely resemble those obtained for the three initial highly-precessing systems shown in Fig. 8-2. They are less informative and peak at lower values of χ_p, driven by the prior. Although there is one noise realization that results in a posterior peaking at high χ_p, this is consistent with Gaussianity. Taken together, the results for all three experiments using draws from the posterior of GW190521 as the true

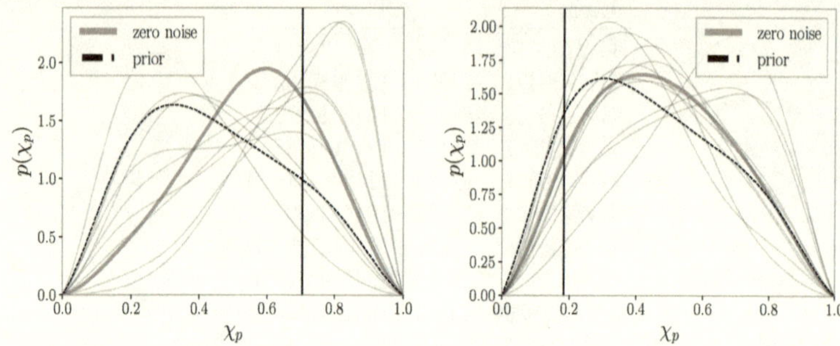

Figure 8-16: *Left:* Probability density for χ_p for the GW190521 maximum-likelihood simulation added to ten different segments of real detector noise from O3. *Right:* Probability density for χ_p for the simulated system with low-χ_p drawn from the GW190521 posterior added to ten different realizations of Gaussian noise colored by the PSDs calculated for the data segment containing GW190521. The thick, dark blue line shows the result of the corresponding simulation without Gaussian noise for comparison.

simulation parameters indicate that while it is not possible to rule out the hypothesis that the informative posterior on χ_p for GW190521 is driven by noise fluctuations at the time of the event, it is more likely due to the exact spin configuration that leads to χ_p being easier to measure.

8.7.2 Mass measurements for simulated single-spin systems

The bias and uncertainty in the total mass and mass ratio posteriors for the simulations discussed in detail in Section 8.4 are shown in Fig. 8-17. For the total mass, we normalize the bias and credible interval width by the value of the total mass of each system so that these quantities are dimensionless, consistent with the other parameters. Several trends observed for the spin parameters are also present in the measurements of the mass parameters. The width of the 90% CI increases with increasing total mass for the total mass of the system for all configurations, and for the mass ratio for equal-mass systems observed nearly face-on (solid blue lines). Both mass parameters are also generally more accurately and precisely recovered for aligned-spin systems (lightest line color) compared to those with some primary spin

Parameter	Symbol	Value
Mass ratio	q	0.76
Detector-frame total mass	M_{tot}	278.63 $M_\odot$
Primary spin magnitude	χ_1	0.18
Secondary spin magnitude	χ_2	0.52
Primary tilt[6]	θ_1	1.62
Secondary tilt	θ_2	0.19
Azimuthal inter-spin angle	ϕ_{12}	5.99
Azimuthal precession cone angle	ϕ_{JL}	4.73
Coalescence phase	ϕ	5.92
Polarization angle	ψ	2.48
Coalescence GPS time	t_c	1242442967.41 s
Right ascension	α	0.13
Declination	δ	-1.17
Luminosity distance	d_L	4735.54 Mpc
Inclination angle	θ_{JN}	0.56

Table 8.6: Parameter values drawn from the GW190521 posterior for the simulation with low χ_p.

tilt. While the total mass constraints are similar for systems observed nearly edge-on and nearly face-on, the mass ratio measurement improves considerably for systems observed nearly edge-on in terms of both the bias and the posterior width, particularly for equal-mass systems. This can be explained due to the enhanced observability of higher-order-mode emission at higher inclinations. The equal-mass systems are always biased towards lower values of q by definition, since the median of the posterior will always fall below the true value of $q = 1$, which is at the upper edge of the prior.

The outliers discussed in Section 8.4.3 for the spin parameters are also present in the mass parameters. The location of the outliers in total mass and mass ratio for the system observed nearly face-on (dark red solid line) matches with that of the outliers in the spin parameters for this system. However, the outlier is absent in the mass ratio posterior for the system observed nearly edge-on, and the peak in the width of the 90% credible interval for the total mass posterior shifts towards higher masses than the corresponding peak in the posteriors of the spin parameters. Comparing with the results for the systems analyzed at a fixed dimensionless reference frequency

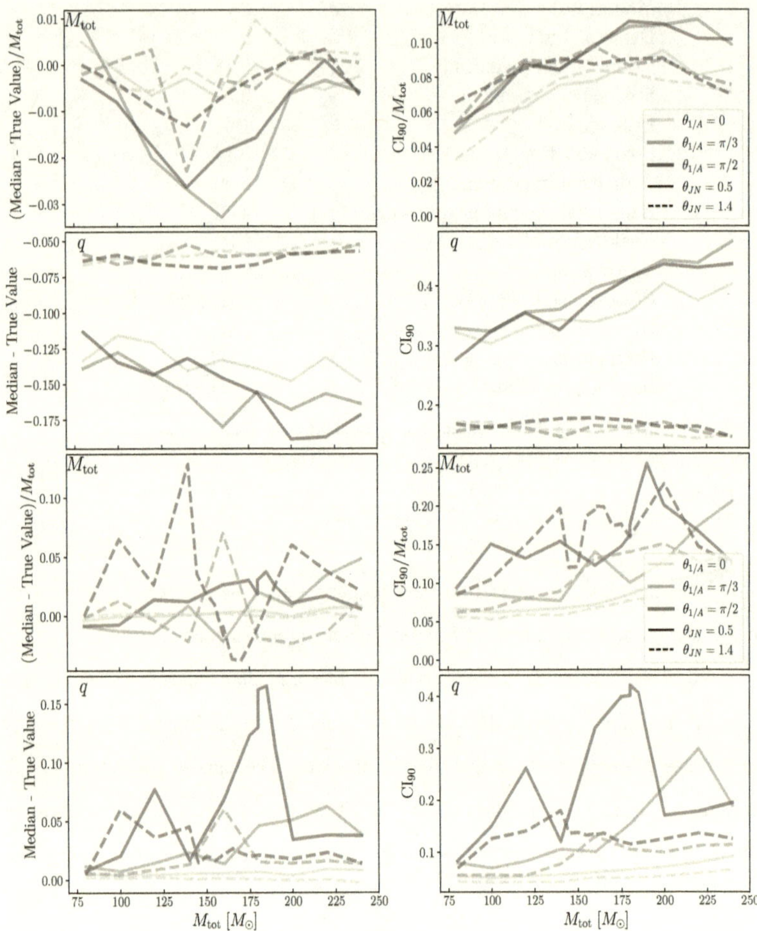

Figure 8-17: The bias and width of the total mass and mass ratio posteriors as a function of the total mass of the system for the first simulation set described in Table 10.4. The bias is defined as the difference between the median of the posterior and the true value, while the width is the symmetric 90% credible interval. For the total mass, these quantities are normalized by the value of the total mass of each system so that they are dimensionless. Equal-mass systems are shown in blue, while unequal-mass systems are shown in red. Systems observed nearly face-on are shown with a solid line, while those observed nearly edge-on are represented by a dotted line. The shading of the line corresponds to the tilt of the primary spin.

307

shown in Fig. 8-11, we conclude that these outliers in the mass parameters are also caused by differences in the spin angles of the underlying systems due to using a fixed reference frequency of 25 Hz for the analysis.

8.7.3 Waveform systematics

In order to verify whether the observed outliers described in Section 8.4.3 are unique to the NRSur7dq4 waveform model, we repeat the original simulations with a fixed reference frequency of 25 Hz for the systems demonstrating the outlier behavior with two additional waveform models, IMRPhenomPv2 [445, 482, 509] and IMRPhenomX-PHM [710, 711, 397]. These are both frequency-domain models, in contrast to the numerical relativity surrogate, which is a time-domain model. IMRPhenomPv2 includes only the dominant $\ell = |m| = 2$ multipoles, while IMRPhenomXPHM includes a subset of the higher-order modes included by the surrogate. Both of these phenomenological models incorporate the effects of precessing spins by approximating the full waveform as an underlying non-precessing waveform in the co-precessing frame that gets "twisted up" into the inertial frame via a rotation encoding the evolution of the orbital plane [794]. IMRPhenomPv2 uses the next-to-next-to-leading (NNLO) order, single spin, post-Newtonian approach described in [795], where $\vec{\chi}_2 = 0$, and only includes spin-orbit (not spin-spin) interactions. IMRPhenomXPHM adds the option of another implementation based on the multiple-scale analysis (MSA) introduced in [223] and previously applied in [508, 510], which enables the inclusion of double-spin effects and includes some contributions from spin-spin coupling. We use the MSA prescription for the Euler angles for our analysis, falling back to the NNLO presciption if the MSA system fails to initialize.

A comparison of the accuracy and precision of the spin and mass measurements for the three different waveform models is shown in Figs. 8-18–8-19. We use the same waveform model for both the simulation and recovery of these systems, so the purpose of this study is not to determine which model is the best at recovering the parameters of the same "true" underlying system. Rather, we want to determine if the trends we observe in the measurability of spins for high-mass systems are universal, particularly

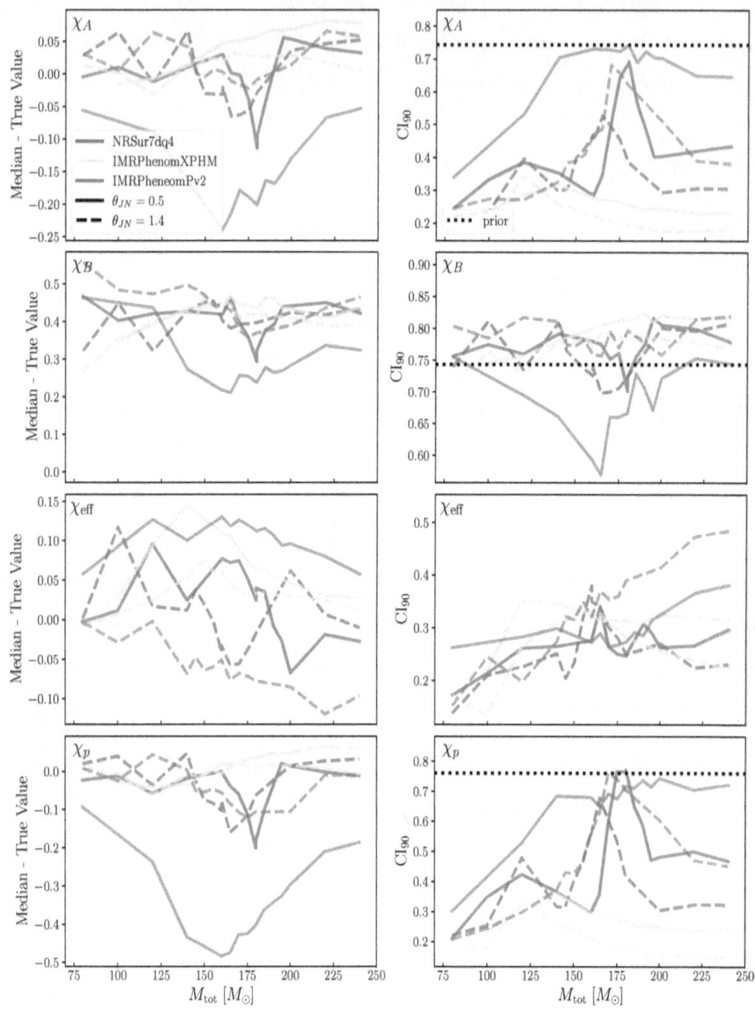

Figure 8-18: Bias and width of the 90% credible interval of the posteriors for component spin magnitudes and effective spins as a function of the true total mass for the original systems with outliers analyzed with three different waveform models: NRSur7dq4 (red), IMRPhenomXPHM (gold), and IMRPhenomPv2 (purple). The systems viewed nearly edge-on are indicated with dashed lines and those observed nearly face-on with solid lines. The dotted black lines show the width of the 90% CI for the prior for all parameters except $\chi_{\rm eff}$, for which the prior width of 0.84 is much larger than the posterior widths.

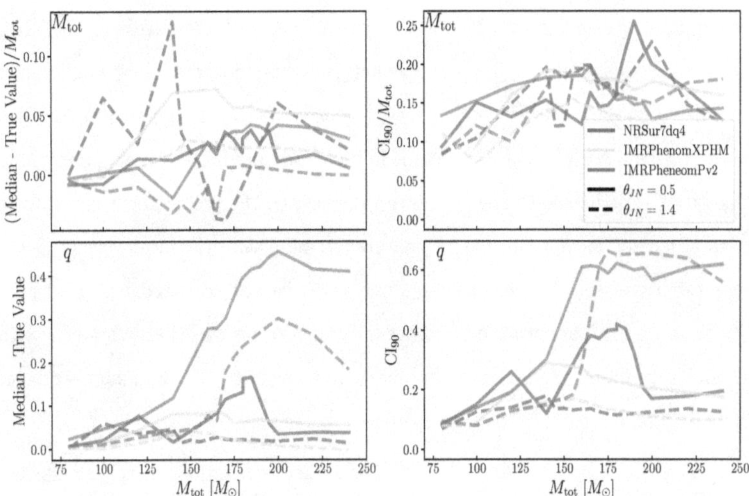

Figure 8-19: Bias and width of the 90% credible interval of the posteriors for the total mass and mass ratio as a function of the true total mass for the original systems with outliers analyzed with three different waveform models: NRSur7dq4 (red), IMRPhenomXPHM (gold), and IMRPhenomPv2 (purple). The systems viewed nearly edge-on are indicated with dashed lines and those observed nearly face-on with solid lines.

for the dramatic increase in the width of the 90% credible interval observed at outlier masses. We find that these three waveform models produce different results.

IMRPhenomPv2 yields the largest bias and 90% credible region for the primary spin parameters for the nearly face-on systems (purple solid lines in Fig. 8-18). It systematically recovers values of χ_p that are smaller than the true value for systems with this inclination angle, and the χ_p posterior is basically unconstrained for total masses of $M_{\mathrm{tot}} \gtrsim 140\ M_\odot$. The IMRPhenonPv2 results for the nearly edge-on systems (purple dashed lines) are closer to those obtained with the waveforms including higher-order modes, although χ_{eff} is systematically biased towards lower values, and the width of the 90% credible interval for χ_{eff} is bigger by ~ 0.2 compared to IMRPhenomXPHM and NRSur7dq4. The trends in the mass ratio posteriors are particularly striking for IMRPhenomPv2, where both the bias and the precision get much worse for total masses $M_{\mathrm{tot}} \gtrsim 140\ M_\odot$. These results emphasize the improvement in the measurability of spin for precessing systems when viewed nearly edge-on compared to face-on and the importance of higher-order modes for such heavy systems with unequal mass ratios, the presence of which helps to place significantly more stringent constraints on the mass ratio of the system.

While IMRPhenomXPHM and NRSur7dq4 demonstrate qualitatively similar behavior for the bias and posterior width at low masses, $M_{\mathrm{tot}} \lesssim 160\ M_\odot$, particularly for χ_A and χ_p, the two disagree at the highest masses in the range of the outliers originally observed in the NRSur7dq4 results, which are not present with IMRPhenomXPHM. The posterior width has a smaller peak at around $M_{\mathrm{tot}} \approx 140\ M_\odot$ for χ_A, χ_p, q, corresponding to a similar feature in the NRSur7dq4 results, but then continues to decrease gradually all the way up to the highest masses. A similar small peak is observed in the bias for χ_{eff}, but no other parameters exhibit significant features in the bias. The results are qualitatively similar for systems observed at both inclinations. The significant differences in the results between these two waveform models that include similar physical effects indicate that the more sophisticated spin treatment of the NRSur7dq4 model could have an important impact on the measurability of spin for high-mass, precessing systems. Furthermore, IMRPhenomXPHM is not tuned to precessing NR simulations. Since NR modifications are more important around the merger, which contributes a larger fraction of the SNR at higher masses, IMRPhenomXPHM is not as reliable as NRSur7dq4 in this regime. In this case, the improvement in the spin constraints seen for IMRPhenomXPHM is artificial since the waveform is agnostic to the physics it fails to capture for the highest-mass systems, and we are using the same model for simulation and recovery.

Chapter 9

A new spin on LIGO-Virgo binary black holes

The content of this chapter was previously published in Physical Review Letters as Ref. [153] in April 2021. ASB performed the analysis, wrote the manuscript, and contributed to project development and conception.

Abstract

Gravitational waves from binary black holes have the potential to yield information on both of the intrinsic parameters that characterize the compact objects: their masses and spins. While the component masses are usually resolvable, the component spins have proven difficult to measure. This limitation stems in great part from our choice to inquire about the spins of the most and least massive objects in each binary, a question that becomes ill-defined when the masses are equal. In this chapter we show that one can ask a different question of the data: what are the spins of the objects with the highest and lowest dimensionless spins in the binary? We show that this can significantly improve estimates of the individual spins, especially for binary systems with comparable masses. When applying this parameterization to the first 13 gravitational-wave events detected by the LIGO-Virgo collaboration (LVC), we find that the highest-spinning object is constrained to have nonzero spin for most sources and to have significant support at the Kerr limit for GW151226 and GW170729. A joint analysis of all the confident binary black hole detections by the LVC finds that, unlike with the traditional parametrization, the distribution of spin magnitude for the highest-spinning object has negligible support at zero spin. Regardless of the parameterization used, the configuration where all of the spins in the population are aligned with the orbital angular momentum is excluded from the 90% credible interval for the first ten events and from the 99% credible interval for all current confident detections.

9.1 Introduction

Gravitational waves from compact binary coalescences (CBCs) carry imprints of the spin angular momenta $\vec{S}$ of the black holes (BHs) or neutron stars (NSs) that originated them. The Advanced LIGO [3] and Virgo [53] detectors can extract this information to obtain key insights about the astrophysics of compact binaries; because the magnitude and orientation of the spins reflect the system's history, such a measurement could reveal the binary's formation mechanism [404, 402]. For instance, we expect the spins of compact binaries formed in isolation to be preferentially aligned with the orbital angular momentum $\vec{L}$ [889, 492, 428, 709, 134, 582, 596, 758, 669, 837], while the same is not true of binaries formed dynamically [812, 625, 587, 707, 140, 758]. Identifying the formation channel of compact binaries is one of the most pressing open problems in astrophysics, making the measurement of component spins a high-value target.

Unfortunately, the ability to measure individual spins with the LIGO and Virgo detectors has been limited, since little information about these quantities is imprinted in the inspiral waveform at leading order [271, 729, 64, 655]. At the population level, current inferences on the black hole spin distribution indicate that most sources have low spin magnitudes when considering the distributions of both the individual component spins [25, 515] and of the spin components aligned with [336, 878, 769, 626] and perpendicular to [326] the orbital angular momentum. In this chapter, we show that we can draw clearer conclusions about the spins of individual objects by using a more suitable basis. Rather than attempting to identify the spin of the heaviest and lightest of the two objects, as is usually done, we infer the properties of the objects with the highest and lowest dimensionless spin. This straightforward reparametrization of the problem can cast a new light on the component spin measurements for near-equal-mass binaries, which appear to be the majority [769, 25, 355]. In the following, we present our proposed reparametrization and demonstrate its impact both on simulated signals and on actual LIGO-Virgo detections.

9.2 Approach

Within general relativity, a CBC signal is fully determined by a set of parameters encoding the intrinsic properties of the binary as well as extrinsic parameters specifying its distance and orientation. The intrinsic parameters correspond to the mass m_i and dimensionless spin $\vec{\chi}_i = \vec{S}_i c/(Gm^2)$ of each component object $i \in \{1, 2\}$, plus additional quantities incorporating matter effects and eccentricity. Virtually all of the literature, including LIGO-Virgo collaboration papers [48, 26, 32, 36], labels the compact objects with respect to their mass, with the index 1 corresponding to the heaviest of the two objects and 2 to the lightest, $m_1 \geq m_2$. However, this choice is suboptimal for systems with similar masses, as it becomes degenerate for $m_1 = m_2$. In that limit, the standard mass-based sorting induces undesired structure in the posteriors for the spin parameters.[1]

To avoid these degeneracies, we instead propose to identify objects by their dimensionless spin magnitude $\chi = |\vec{\chi}|$, and define an equivalent set of quantities m_j and $\vec{\chi}_j$ for $j \in \{A, B\}$, with A referring to the object with the highest spin and B to the lowest, $\chi_A \geq \chi_B$. (In the equal-mass limit, sorting by dimensionless spin is equivalent to sorting by the component angular momenta, $\vec{S}_i$.) This amounts to a coordinate transformation effecting $\chi_A = \max(\chi_1, \chi_2)$ and $\chi_B = \min(\chi_1, \chi_2)$. The mass of the highest-χ component is m_A, just as χ_1 is the spin magnitude of the highest-m component. In the following, we will refer to the usual $\{1, 2\}$ parametrization as *mass sorting*, and to the new $\{A, B\}$ parameterization as *spin sorting*.

---

[1] Assuming a universal equation of state, tidal parameters for binary NSs should be unaffected, since the least massive object should be the most deformable.

Table 9.1: Comparison of the maximum posterior value with uncertainty quoted at the 90% level and the credible level at which the true value is recovered ($\mathrm{CL}_{\mathrm{inj}}$) for the component mass and spin parameters using both the mass and spin sorting for the simulated signal. The credible level is calculated using the highest posterior density method.

Parameter	Inj.	Mass sorting		Spin sorting	
		maxP	$\mathrm{CL}_{\mathrm{inj}}$	maxP	$\mathrm{CL}_{\mathrm{inj}}$
$m_{1/A}$	40 $M_\odot$	$40.90^{+3.02}_{-1.43}$	58.5%	$39.27^{+3.77}_{-2.88}$	45.7%
$m_{2/B}$	40 $M_\odot$	$38.70^{+1.74}_{-2.38}$	67.2%	$40.19^{+3.05}_{-3.05}$	0%
$\chi_{1/A}$	0.8	$0.01^{+0.85}_{-0.01}$	87.1%	$0.77^{+0.21}_{-0.17}$	11.5%
$\chi_{2/B}$	0	$0.80^{+0.19}_{-0.80}$	61.4%	$0.01^{+0.41}_{-0.01}$	0%
$\theta_{1/A}$	1.57 rad	$1.54^{+0.87}_{-0.80}$	0%	$1.59^{+0.29}_{-0.34}$	0%
$\theta_{2/B}$	–	$1.62^{+0.73}_{-0.58}$	–	$1.54^{+1.20}_{-0.81}$	–

9.3 Simulated signal

Before we apply the spin sorting to real detections, we perform Bayesian parameter estimation on a simulated equal-mass binary BH (BBH) system with $\chi_A = 0.8$ and $\chi_B = 0$ to demonstrate the resolving power of the new parameterization. The system has a redshifted total mass of 80 $M_\odot$, and is oriented nearly edge-on with an inclination angle $\theta_{JN} = 80.21°$. The luminosity distance, $d_L = 831.47$ Mpc, is chosen so that the signal is recovered with a network signal-to-noise ratio (SNR) of 30 by the two advanced LIGO instruments plus advanced Virgo, all operating at design sensitivity [3, 53]. The tilt of the spinning object is $\theta_A = 90°$ with respect to the orbital angular momentum, meaning that the spin vector lies entirely in the orbital plane.

We assume standard priors for LIGO-Virgo analyses [906, 26]; these imply a disjoint uniform prior on m_A and m_B, and a uniform two-dimensional prior on χ_A, χ_B. As is the case for m_1 and m_2, the definition $\chi_A > \chi_B$ results in a "triangular" marginal prior for χ_A and χ_B, i.e. a probability density linearly increasing and decreasing, respectively, with the quantity (black histograms in Fig. 9-1). In order to isolate the effect of the chosen parameterization on the recovered posteriors, we do not add noise to the simulated data [895].

In Fig. 9-1, we compare the resulting measurements of the spin magnitudes and tilts using both the mass and spin sortings. The mass sorting induces a bimodal

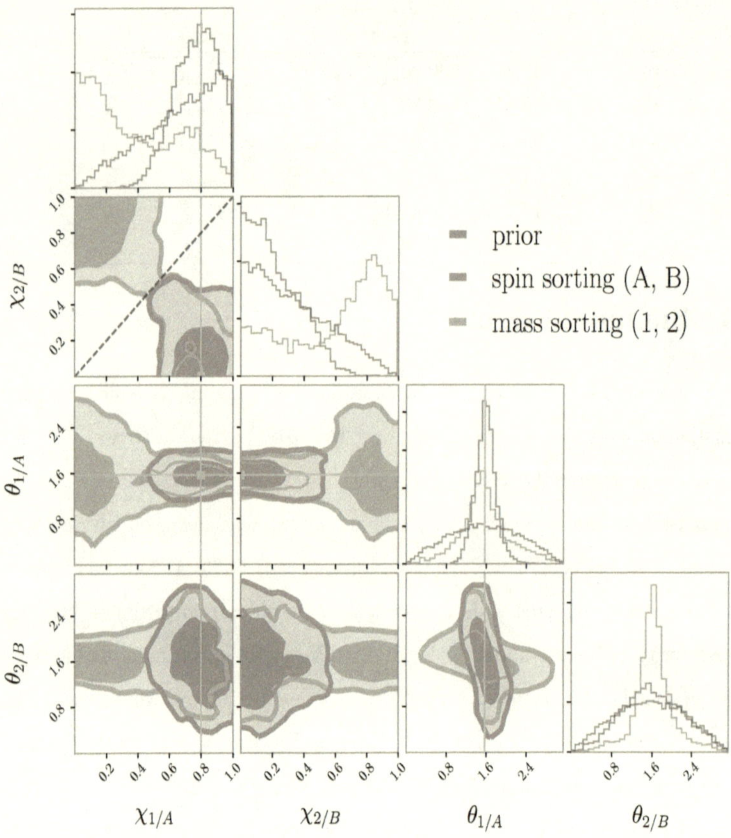

Figure 9-1: Comparison corner plot showing the spin magnitudes and tilts recovered for our simulated equal-mass signal using both the mass sorting in green and the spin sorting in blue. The marginalized one-dimensional priors for the spin sorting are shown in grey. Orange lines mark the true value, and the equal-spin diagonal is shown as a dashed line for reference.

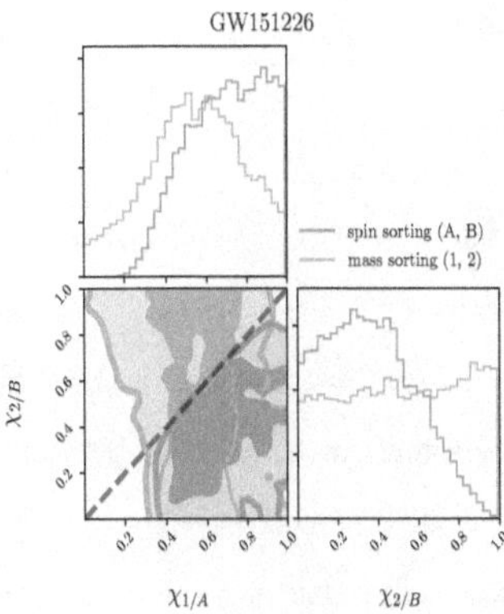

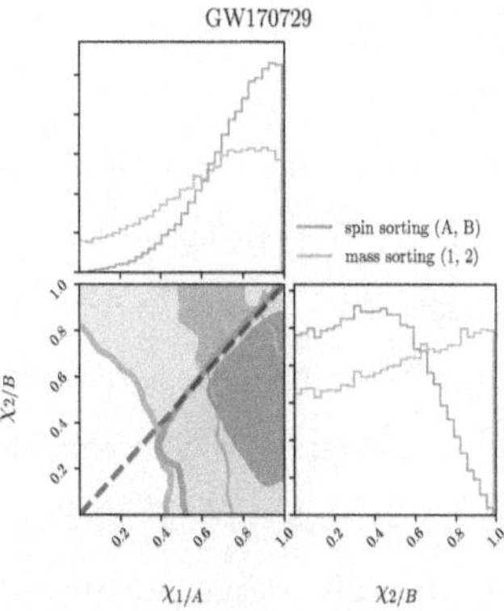

Figure 9-2: Comparison corner plot for the spin magnitudes for the posteriors obtained using the IMRPhenomPv2 waveform for GW151226 and GW170729. The equal-spin diagonal is shown as a dashed line for reference.

posterior in the χ_1, χ_2 plane (symmetric around $\chi_1 = \chi_2$), showing that a high spin could be assigned to either component, while the marginal posteriors on χ_1 and χ_2 are largely unconstrained. The spin sorting breaks this degeneracy, restricting the posterior so that χ_A peaks at the true value and χ_B rails against the lower edge of the prior. A similar degeneracy can be seen in the two-dimensional posterior for θ_1 and θ_2 in the mass sorting, which shows that information is retrieved for the tilt of one of the two objects without identifying which. Switching to the spin sorting, the θ_A posterior is well-constrained, while θ_B returns the prior. Thus, this parametrization makes it clear that we can measure the tilt of the highest-spin object but cannot say anything about the lowest-spin one, as expected given that $\chi_B = 0$, making θ_B irrelevant. In Table 9.1, we show the maximum posterior probability values and associated 90% credible interval for the component masses, spins, and tilt angles obtained using both parameterizations. We also show the credible level at which the true value is recovered. The mass ratio posterior is peaked narrowly around the equal-mass limit, $q = 0.996^{+0.004}_{-0.138}$. The statistical uncertainty for the component masses m_A and m_B is greater than for m_1 and m_2 because there is no imposed ordering on m_A and m_B.

In order to verify that the improved resolution of the spin sorting is robust against changes in the true spin magnitude and tilt and that it extends to systems without exactly equal component masses, we repeat our simulation for a variety of different binary parameters. We find a similar improvement in the resolution of the component spins for systems with lower spins, $\chi_A = 0.2$, with an aligned primary spin, $\theta_A = 0$, and with slightly unequal mass ratios, $q = m_2/m_1 = 0.9$, when each of these parameters is varied independently in simulated signals with the same SNR as the original. When the network SNR is decreased to 12, the component spins cannot be well-measured using either parameterization, meaning that only minor deviations from the priors are observed.

The spin sorting ceases to be useful for systems where the mass ratio is measurably different from unity. Looking at a system with $q = 0.7$ and SNR $= 30$, the spin sorting introduces the same type of degeneracies in the spin parameters as are present in Fig. 9-1 under the mass sorting. This is because the spin of the most massive object is well-defined for systems where the most massive object can be distinguished. Systems with equal spin magnitudes—both nonspinning and highly spinning with $\chi_1 = \chi_2 = 0.8$—similarly do not benefit from the spin sorting, as the spin magnitude posteriors are largely unchanged in this case. (The χ_A posterior for the nonspinning injection peaks at $\chi_A = 0$, even though this region is disfavored by the prior.) However, when analyzing a system with $\chi_1 = \chi_2 = 0.8$ and unequal tilt angles, $\theta_1 = \pi/2$ and $\theta_2 = 0$, the bimodality in the tilt posterior can be resolved by instead sorting by the tilt angles without affecting the spin magnitude posteriors.

9.4   LIGO-Virgo detections

We apply the same reparameterization to the publicly released posterior samples for the first 13 LIGO-Virgo detections [896, 48, 26, 32, 36]. The ten BBH mergers announced in the first LIGO-Virgo catalog (GWTC-1) are all consistent with $q = 1$, although the posteriors all support considerably lower values than the simulated signal in Fig. 9-1. For these systems, we find that the differences between the spin and mass sorting are generally not as significant as for the simulation. This is consistent with the results of the low-SNR simulation discussed above.

For the two events whose posteriors in the mass sorting already indicated a preference for non-zero spins, GW151226 [11] and GW170729 [219], $\chi_A = 0$ is ruled out with 3σ credibility. For GW170729, $\chi_A = \chi_B = 1$ is included within the 90% credible region, while for GW151226, $\chi_A = 1$ is included in the 50% credible region as long as $0.5 < \chi_B < 0.7$. We show spin magnitude posteriors for these events using both parameterizations in Fig. 9-2. For all the BBH systems analyzed, the lower bound of the 90% credible interval for the χ_A posterior is $\chi_A \geq 0.14$, although this is dominated by the triangular prior (grey line in Fig. 9-1). On the other hand, the χ_1 posterior is only constrained to $\chi_1 > 0.14$ for GW151226 and GW170729 under the mass sorting. For the unequal-mass binary GW190412 [36], the spin sorting introduces degeneracies in the spin parameters that were not present in the mass-based sorting, which we expected from our $q < 1$ simulations. In the Supplementary Material, we explore the features of the posteriors for GW190412 and the two binary NS systems and show that waveform systematics proved to be important for some events (including GW150914).

9.5   Population analyses

In order to determine the effects of the spin sorting on the inferred population properties of BH spins, we use the infrastructure of hierarchical Bayesian inference to characterize the underlying distributions of χ_A and χ_B. If the mass-sorted spin magnitudes for individual events, $\chi_{1/2}$, are modeled as being drawn from the same Beta distribution following [946, 25], we can compute the corresponding χ_A and χ_B distributions using order statistics, by assuming they correspond to the maximum and minimum of two draws from the $\chi_{1/2}$ distribution. We use the publicly released posterior samples for the $\chi_{1/2}$ hyperparameters from LVC analyses first including only GWTC-1 events [25, 31], then including all 44 confident BBH detections reported in GWTC-2 [41, 49, 30]. We present our results using the posterior population distribution (PPD), which is the expected distribution for the individual-event parameters of new BBH events inferred from the accumulated set of detections (e.g., [25], see Supplementary Material).

In Figure 9-3 we show the PPDs for $p(\chi_A)$ and $p(\chi_B)$ as well as the 50% and 90% credible bands. The inferred distribution for χ_A (top) peaks at around $\chi_A \sim 0.3$ and has negligible support for $p(\chi_A) = 0$ for both the GWTC-1 and GWTC-2 analyses, indicating that most of the highest-spinning BHs in LIGO-Virgo binaries have nonzero spins. This is very different from the distribution inferred for the mass sorted spins, $p(\chi_{1/2})$, which has considerable support at $\chi_{1/2} = 0$ (see Fig. 8 of [25] and Fig. 10 of [49]). The lower bound of the 50% credible interval for $p(\chi_A)$ is $\chi_A = 0.19$, while the 50% credible interval for $\chi_{1/2}$ extends down to $\chi_{1/2} = 0.06$ for the GWTC-2 analysis. The distribution of spin magnitudes for the lowest-spinning BHs (bottom) is consistent with peaking at $\chi_B = 0$ for both analyses and has more posterior support at $\chi_B = 0$ when including the full GWTC-2 sample. All these distributions vary significantly from those obtained using samples from the spin magnitude prior instead of posterior samples for the 44 confident BBH detections from GWTC-2 (dashed lines in Fig. 9-3).

For the spin tilt angles, we follow [850, 25] and model the distribution as the sum of two populations motivated by the most popular BBH formation channels (eg. [492, 796, 709, 758]): an isotropic component and a preferentially aligned component, where the hyperparameters $\sigma_{1/A}$ and $\sigma_{2/B}$ control the spread in the possible tilt angles around $\theta_{1/A} = \theta_{2/B} = 0$. A nonzero value for σ indicates that not all tilts are aligned. We conduct hierarchical Bayesian inference using this model for the tilt angles under both the mass and spin sorting. In Fig. 9-4, we show the posteriors on the hyperparameters describing the spin tilt population model for the 44 confident BBH detections in GWTC-2. The blue distribution corresponds to inference starting from the mass-sorted single-event posteriors, while the green distribution shows the same for the spin-sorted posteriors. The posterior for σ_A is constrained slightly further away from 0 than that of σ_1, while the opposite is true for σ_B and σ_2. This indicates

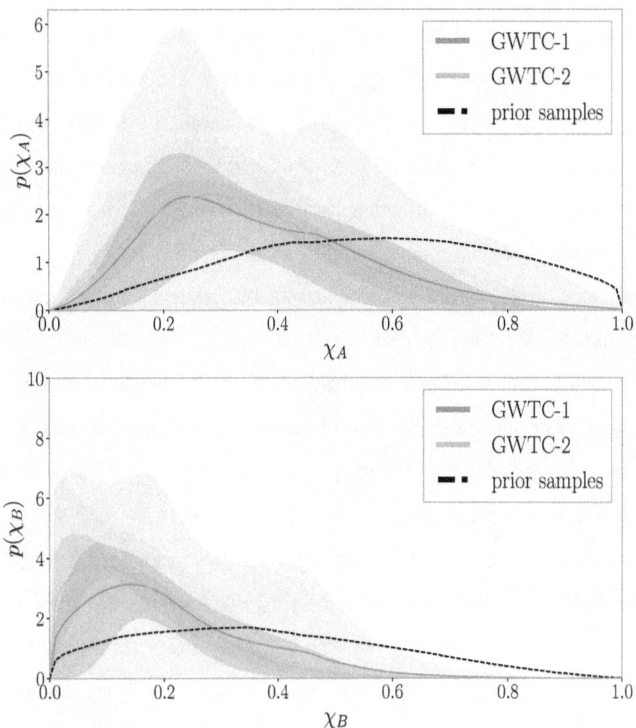

Figure 9-3: PPDs for $p(\chi_A)$ (top) and $p(\chi_B)$ (bottom). The blue curves include only events from GWTC-1, while the orange curves include all 44 confident BBH events in GWTC-2. The shaded regions correspond to the 50% and 90% credible intervals, and the PPDs obtained using prior samples for the individual GWTC-2 events are shown in the dashed lines.

that the posterior information gain relative to the prior is less balanced between the two tilt angles in the spin sorting than the mass sorting.

We highlight that $\sigma_{1/A} = \sigma_{2/B} = 0$ is excluded with $> 99\%$ credibility for both parameterizations when marginalized over spin magnitude. The same is true at 90% credibility when considering only the GWTC-1 events. This means that the preferentially aligned component is more likely to have nonzero width, and hence a fraction of binaries is likely to have in-plane spin components. This result agrees with previous analyses, which find that a fully-aligned population is disfavored by the LVC detections reported in GWTC-1 [878, 25, 946], and that the current detections are consistent with a population of high spins that are significantly misaligned with respect to the orbital angular momentum. [336, 626]. The GWTC-2 results agree with those presented in [49] using two different spin tilt parameterizations from the one we use, which find evidence for general-relativistic spin precession at the population level. We confirm that this feature originates in the data—and is not just an artifact of the Monte Carlo integration performed during hierarchical inference step—by replacing the mass-sorted spin tilt posteriors from individual events with draws from the prior. This results in an uninformative distribution for σ_1, σ_2 consistent with having uniform support across the prior range, including the region around $\sigma_1 = \sigma_2 = 0$ (shown in grey in Fig. 9-4). The region $\sigma_{1/A}, \sigma_{2/B} \geq 1.6$ is also excluded at $> 90\%$ credibility, indicating that a fully isotropic spin distribution corresponding to high values of σ is statistically disfavored. Additional analysis details are provided in the Supplementary Material.

9.6   Conclusion

We have demonstrated the advantages of introducing an alternative labeling for the component objects of a compact binary based on the spins instead of the masses; we denote the object with the largest (smallest) spin magnitude by A (B), such that $\chi_A > \chi_B$. Through analysis of simulated signals, we find that this sorting improves the resolution of the component spins of binaries consistent with having

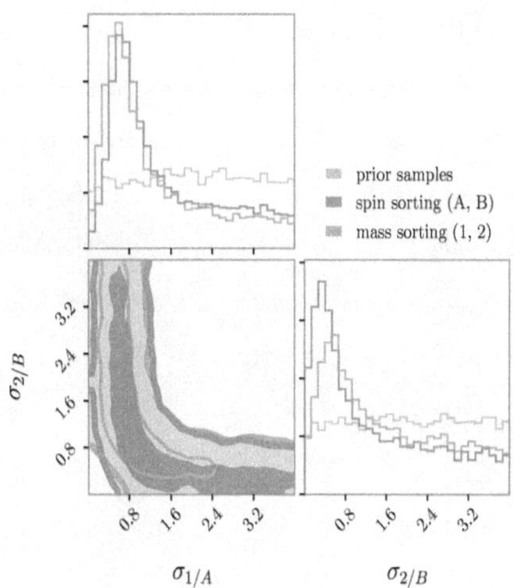

Figure 9-4: Corner plot comparing the inference on the spin tilt hyperparameters using the mass sorted tilts θ_1 and θ_2 to the posteriors obtained using the same population model but for the spin sorted tilts θ_A and θ_B for the GWTC-2 events. The posteriors obtained using prior samples for the individual events are shown in grey.

equal mass, regardless of the magnitude and tilt of the primary spin. When applied to the posteriors for the GWTC-1 events, we find that the most-spinning object is consistent with having extremal spin for the events that were already known to prefer nonzero spins, GW151226 and GW170729. The spin sorting ceases to be useful for systems with measurably unequal masses, as was the case for GW190412.

We characterize the distributions of $\chi_A, \chi_B, \theta_A, \theta_B$ of the 44 confident GWTC-2 BBH events by means of hierarchical Bayesian inference. Modeling the mass-sorted spins as drawn from a Beta distribution, we compute the implied probability densities for χ_A and χ_B and find that $p(\chi_A)$ peaks at $\chi_A \sim 0.3$ and has negligible support at $\chi_A = 0$, while $p(\chi_B)$ and $p(\chi_{1/2})$ have considerable posterior support at $\chi = 0$. We thus conclude, at the population level, that most of the BBHs detected by LIGO-Virgo have at least one component with nonzero spin. When modeling the distributions of spin tilt angles, we find that the configuration where all of the spins in the population are aligned with the orbital angular momentum is excluded from the 99% credible interval for both the mass and the spin sorting.

We end by stressing that the spin sorting does not introduce new information into the analysis. We are not applying different priors, but rather defining a new set of parameters that can be inferred using the same individual-event posterior samples used in the original mass-sorting. This implies that the Bayesian evidence of the data is unchanged. The reparameterization can be done entirely in post-processing and does not affect the posteriors for parameters that do not distinguish between the binary components like the effective aligned and precessing spins, χ_{eff} and χ_{p}. The spin sorting also does not change existing population-level inferences obtained from the mass-sorted component parameters, with the possible exception of analyses relying on marginalized one-dimensional posteriors on the component spin quantities, which cannot encode the parameter degeneracies that we noted for signals consistent with $q \approx 1$. While in some cases BBH formation channels make it more natural to label the component objects based on their mass, thinking about the objects primarily in terms of spin could lead to rich new ways to test astrophysical models moving forward.

9.7 Supplementary material

9.7.1 Parameter estimation methods

To analyze the simulated signals, we perform Bayesian parameter estimation using the standard Gaussian likelihood for gravitational-wave data [761, 906]:

$$\mathcal{L}(d_i|\theta) = \prod_j \frac{2}{\pi T S_n(f_j)} \exp\left[-\frac{2|d_i(f_j) - h(f_j;\theta)|^2}{T S_n(f_j)}\right], \tag{9.1}$$

where T is the duration of the data segment being analyzed, $S_n(f_j)$ is the noise power spectral density of the detector, $d_i(f_j)$ is the strain data for event i, and $h(f_j;\theta)$ is the waveform model for the compact binary source. We simulate the measurement using the `LALInference` algorithm [906] and the numerical relativity surrogate waveform model NRSur7dq4 [903]. In order to obtain posterior samples for the parameters θ using the likelihood in Eq. 9.1, we impose priors that are uniform in the component masses m_1, m_2 between 10 $M_\odot$ and 240 $M_\odot$, with constraints on the total mass between 70 $M_\odot$ and 240 $M_\odot$ and on mass ratio q between 0.2 and 1. The luminosity distance prior is $\propto d_L^2$ over the range 1–7000 Mpc.

For the spin tilt population inference, we simultaneously fit the mass and spin magnitude distributions. We follow [946, 25] and assume that both of the black hole (BH) spin magnitudes under the mass sorting are drawn from the same Beta distribution with hyperparameters α and β,

$$p(\chi_{1/2}|\alpha,\beta) = \frac{\chi_{1/2}^{\alpha-1}(1 - \chi_{1/2})^{\beta-1}}{B(\alpha,\beta)}, \tag{9.2}$$

where $B(\alpha,\beta)$ is the Beta function. We restrict the priors on the Beta function parameters to exclude values of $\alpha, \beta \leq 1$ corresponding to singular Beta distributions. This means that $p(\chi)$ *must* peak within $0 < \chi < 1$, as nonsingular Beta distributions vanish at those values. The Beta distribution described in Eq. 9.2 can also be parameterized in terms of its mean and variance:

$$\mu(\chi) = \frac{\alpha}{\alpha + \beta}, \tag{9.3}$$

$$\sigma^2(\chi) = \frac{\alpha\beta}{(\alpha + \beta)^2(\alpha + \beta + 1)}. \tag{9.4}$$

We choose to sample in $\mu(\chi)$ and $\sigma^2(\chi)$, imposing constraints such that $\alpha, \beta > 1$ to restrict the parameter space to only nonsingular Beta distributions.

The distributions for χ_A and χ_B obtained by applying order statistics to the Beta distribution for $\chi_{1/2}$ are given by:

$$p(\chi_A) = 2\,p(\chi_{1/2}|\alpha, \beta)\,\mathrm{CDF}(\chi_{1/2}|\alpha, \beta), \tag{9.5}$$

$$p(\chi_B) = 2\,p(\chi_{1/2}|\alpha, \beta)\left[1 - \mathrm{CDF}(\chi_{1/2}|\alpha, \beta)\right], \tag{9.6}$$

assuming χ_A is the maximum and χ_B is the minimum of two draws from $p(\chi_{1/2}|\alpha, \beta)$. $\mathrm{CDF}(\chi_{1/2})$ is the cumulative distribution function for $\chi_{1/2}$ given by the regularized incomplete Beta function with parameters (α, β).

We further assume that the primary mass distribution is described by the sum of a truncated power-law with low-mass smoothing and a Gaussian component [851]. The hyperparameters describing this model are the slope α_m, upper and lower cutoffs $m_{\max}$ and $m_{\min}$, the low-mass smoothing parameter δ_m, the peak and width of the Gaussian component μ_m and σ_m, and the mixing fraction between the two components λ_{peak}. The mass ratio distribution is modeled as a power law with slope β_q. This corresponds to Model C from [25], dubbed"power-law + peak". We use this same mass model when computing the spin magnitude distribution using prior samples for the GWTC-2 events, shown in the dashed black line in Fig. 3 in the main text.

Following [850], the distribution for spin tilt angles is given by the sum of an isotropic component and a preferentially aligned component, which is composed of the product of two truncated Gaussians peaked at $\cos t_i = 1$ for each tilt angle:

$$p(\cos t_{1/A}, \cos t_{2/B} | \sigma_{1/A}, \sigma_{2/B}, \xi) = \frac{1-\xi}{4} + \tag{9.7}$$
$$\frac{2\xi}{\pi} \prod_{i \in \{1/A, 2/B\}} \frac{\exp(-(1-\cos t_i)^2/(2\sigma_i^2))}{\sigma_i \mathrm{erf}(\sqrt{2}/\sigma_i)},$$

where the hyperparameter ξ gives the mixture fraction between the two components. The complete population model, $\pi(\theta|\Lambda)$, is given by the product of Eqs. 9.2, 9.7, and the "power-law + peak" mass distribution.

Using the population model $\pi(\theta|\Lambda)$ to describe the distribution of individual-event parameters θ, the likelihood of observing the hierarchical parameters Λ for a data set $\{d\}$ consisting of N_{det} detected events is given by:

$$\mathcal{L}(\{d\}|\Lambda) \propto \prod_{i=1}^{N_{\mathrm{det}}} \frac{\int \mathcal{L}(d_i|\theta)\pi(\theta|\Lambda)}{\alpha(\Lambda)} \tag{9.8}$$

where $\alpha(\Lambda)$ represents the detectable fraction of events assuming the individual-event parameters are drawn from distributions specified by hyperparameters Λ [566, 357, 874, 946, 585, 919]. We use the sensitive spacetime volume estimates released by the LVC in [34], which for the GWTC-2 analysis were determined through injection campaign [41] and for the GWTC-1 events were obtained using simulated data. We calculate $\alpha(\Lambda)$ using the formalism described in [334]. We do not account for the selection biases due to the spin parameters, since those have a much smaller effect than the mass parameters [49].

The likelihood in Eq. 10.7 is evaluated using a Monte Carlo integral over the individual-event parameter posteriors released by the LVC for the binary BH (BBH) events included in GWTC-1 [26, 48] and GWTC-2 [41, 33]. For the GWTC-1 events, we use the samples obtained with the IMRPhenomPv2 waveform model, while for GWTC-2 we use the "Publication" posterior samples presented in [41], which in most cases use a combination of waveform models including the effects of spin precession and higher-order multipoles. For GWTC-2, we only analyze the 44 confident BBH detections with a false alarm rate < 1 yr^{-1} and exclude the events where at least one

Parameter	Prior	GWTC-1	GWTC-2
α_m	U(-4, 12)	$6.41^{+4.87}_{-4.44}$	$2.96^{+0.77}_{-0.63}$
β_m	U(-4, 12)	$5.60^{+5.65}_{-5.73}$	$1.01^{+2.26}_{-1.41}$
m_{max}	$U(30\ M_\odot, 100\ M_\odot)$	$59.80^{+35.92}_{-26.66}\ M_\odot$	$86.30^{+12.10}_{-12.66}\ M_\odot$
m_{min}	$U(2\ M_\odot, 10\ M_\odot)$	$7.40^{+1.35}_{-3.31}\ M_\odot$	$4.71^{+1.38}_{-1.84}\ M_\odot$
δ_m	$U(0\ M_\odot, 10\ M_\odot)$	$2.58^{+5.28}_{-2.38}\ M_\odot$	$4.59^{+4.24.35}_{-3.98}\ M_\odot$
μ_m	$U(20\ M_\odot, 50\ M_\odot)$	$28.60^{+5.71}_{-6.98}\ M_\odot$	$32.42^{+3.52}_{-5.91}\ M_\odot$
σ_m	$U(0.4\ M_\odot, 10\ M_\odot)$	$6.13^{+3.37}_{-4.18}\ M_\odot$	$5.17^{+4.21}_{-3.85}\ M_\odot$
λ_{peak}	$U(0, 1)$	$0.19^{+0.40}_{-0.16}$	$0.07^{+0.13}_{-0.05}$
$\mu(\chi)$	$U(0, 1)$	$0.30^{+0.20}_{-0.15}$	$0.31^{+0.11}_{-0.09}$
$\sigma^2(\chi)$	$U(0, 0.25)$	$0.02^{+0.03}_{-0.02}$	$0.03^{+0.02}_{-0.02}$
ξ	$U(0, 1)$	$0.54^{+0.41}_{-0.47}$	$0.79^{+0.19}_{-0.43}$

Table 9.1: Priors, posterior medians, and 90% credible intervals for the hyperparameters used in our population analysis obtained for spin-sorted component tilt angles including both GWTC-1 events alone and the 44 confident BBH detections in GWTC-2. All the priors are uniform across the specified range and match those used in [49].

of the compact objects has considerable posterior support below 3 $M_\odot$. We use the `dynesty` [828] sampler, as implemented in the `GWPopulation` [849] package, to obtain hyperparameter posterior samples.

The posterior population distribution calculated using the hyperparameter posteriors obtained using the likelihood in Eq. 10.7 is given by:

$$\mathrm{PPD}(\theta|\{d\}) = \int \pi(\theta|\Lambda)p(\Lambda|\{d\})d\Lambda. \tag{9.9}$$

The priors and posterior results for all hyperparameters except $\sigma_{1/A}$ and $\sigma_{2/B}$ for the spin-sorted tilt inference performed using the GWTC-2 BBH events are shown in Table 9.1. The priors are uniform for all parameters and identical for both the mass and spin-sorted tilt analysis. We use uniform priors over the range $(0, 4)$ for the tilt σ parameters. The posterior for the spin tilt mixing fraction ξ peaks at the upper edge of its prior, indicating that a purely isotropic distribution of tilts is statistically disfavored by the data. We obtain posteriors for the mass, spin magnitude, and ξ hyperparameters comparable to those quoted in [49]. The results for these parameters change negligibly under the mass and spin sortings.

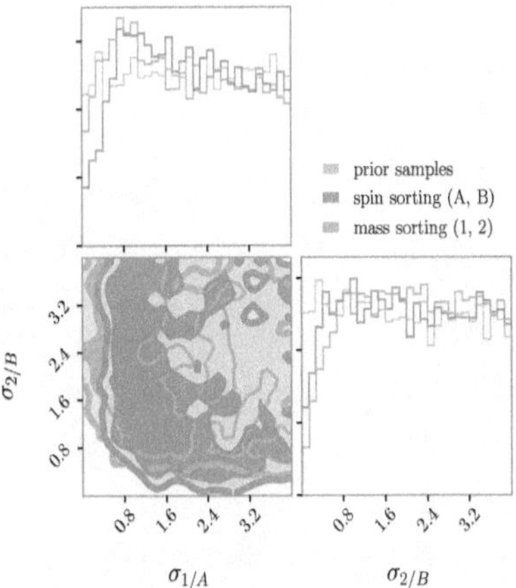

Figure 9-1: Corner plot comparing the inference on the spin tilt hyperparameters using the mass sorted tilts θ_1 and θ_2 and spin sorted tilts θ_A and θ_B for the GWTC-1 events. The posteriors obtained using prior samples for the individual events are shown in grey.

The corner plot of the tilt σ hyperparameters for the GWTC-1 analysis is shown in Fig. 9-1. Even with only the GWTC-1 events, the corner of parameter space at $\sigma_{1/A} = \sigma_{2/B} = 0$ representing a fully-aligned population is excluded at $> 90\%$ credibility for both the mass and spin sortings. The posterior for σ_B is less constrained that that for σ_2—indicating that for this analysis, the information on the tilt angles at the population level is predominantly obtained from the measurement of the highest spinning object, rather than the most massive. A similar trend is present in the σ posteriors for the full GWTC-2 analysis shown in Fig. 4 in the main text.

9.7.2 Additional results for individual sources

For certain sources, we note significant differences in the posteriors obtained with the two different waveform models applied to BBHs in GWTC-1: IMRPhenomPv2 [482, 445, 509], which uses an effective precessing spin model, and SEOBNRv3, which uses a fully precessing spin model [104, 858, 682]. For GW150914, the one-dimensional

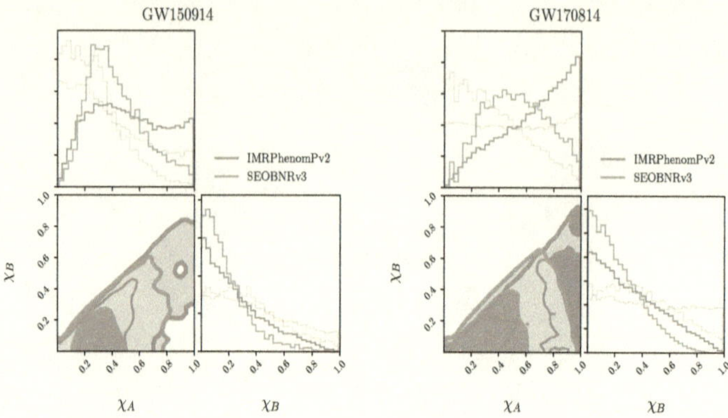

Figure 9-2: Comparison corner plot for the spin magnitude posteriors for GW150914 (left) and GW170814 (right) obtained using both the IMRPhenomPv2 (green) and SEOBNRv3 (purple) waveforms for GW150914 using the spin sorting. The marginalized posteriors obtained using the mass sorting are shown in lighter green and purple.

posterior for χ_A is much more tightly constrained for SEOBNRv3, with $\chi_A = 0.39^{+0.44}_{-0.24}$ compared to $\chi_A = 0.49^{+0.45}_{-0.38}$ for IMRPhenomPv2, a feature which is not as easily recognizable in the χ_1 posterior. A comparison of the spin magnitudes obtained using both waveform models is shown in Fig. 9-2. Similarly for GW170814, the posterior for χ_A turns over at around $\chi_A \sim 0.5$ for SEOBNRv3, but not for IMRPhenomPv2. The two-dimensional χ_A, χ_B posterior recovered with SEOBNRv3 is much more tightly clustered around low spins for GW170814 and also for GW170818, although these features are distinguishable in the χ_1, χ_2 posteriors as well.

For the unequal-mass binary GW190412 [36], the spin sorting introduces degeneracies in the spin parameters that were not present in the mass-based sorting. The one-dimensional χ_A posterior is much less constrained than χ_1, and it features a tail extending to higher spin magnitudes. Unlike for the other BBH signals, which are consistent with $q = 1$, the tilt posteriors for this event change considerably between the mass and spin sortings. There is a clear degeneracy observed between θ_A and θ_B where either one or the other is constrained to lie in the orbital plane, similar to the pattern observed for θ_1 and θ_2 in our simulated signal. The χ_B posterior is more constrained than the χ_2 posterior, peaking at $\chi_B \sim 0.5$ and ruling out $\chi_B \gtrsim 0.7$

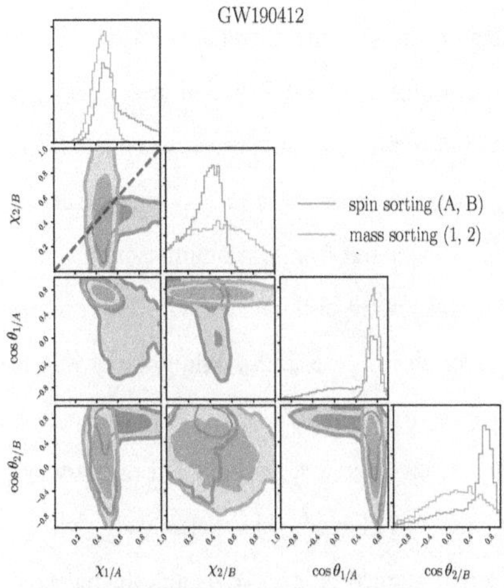

Figure 9-3: Comparison corner plot for the spin magnitudes and tilts for the posteriors obtained using the SEOBNRv4PHM waveform for GW190412.

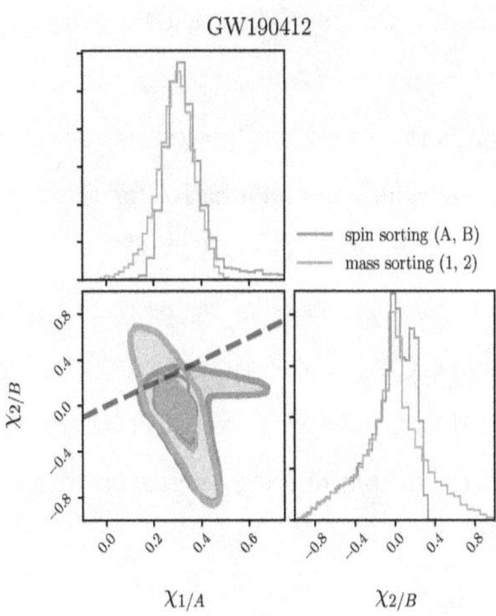

Figure 9-4: Comparison corner plot for the spin magnitudes for the posteriors obtained using the aligned-spin waveform SEOBNRv4HM_ROM for GW190412 using both the spin (green) and mass (blue) sorting. Negative values of χ are included due to anti-alignment.

with 3σ credibility across the various waveforms allowing spin precession. Fig. 9-3 compares the posteriors on the spin magnitudes and tilts for both the spin and mass sorting obtained using a waveform model that allows for spin precession. Based on the application of the spin sorting to this event, we conclude that it is preferable to use the mass sorting when the mass ratio of the binary can be clearly constrained away from equal mass, as expected from our simulations.

When analyzing the posteriors obtained for GW190412 with aligned-spin waveforms, the χ_1, χ_2 posterior shown in Fig. 9-4 exhibits a strong correlation depending on orientation with respect to $\vec{L}$: high, aligned primary spins are allowed when the secondary spin is high and anti-aligned; low, aligned primary spins are allowed when the secondary spin is high and aligned. This correlation, which is due to the strong constraint on the effective aligned spin, χ_{eff}, is simply reflected over the $\chi_A = \chi_B$ boundary when the posteriors are projected into the spin sorting.

For the binary NSs, GW170817 [17] and GW190425 [32], two priors were used to include or exclude high spins (maximum χ_i of 0.99 vs 0.05, respectively), where the high-spin prior allows for the possibility that the binary components are BHs. For the high-spin prior, the posteriors for both $\chi_{1/A}$ and $\chi_{2/B}$ favor low spin values for both events and for both aligned and precessing-spin waveforms. The posteriors for the spin magnitudes are less informative for the low-spin prior, since the prior volume is considerably reduced. For GW190425, we find that the constraints on $\cos\theta_A$ are tighter than those on $\cos\theta_1$ for the precessing-spin waveform IMRPhenomPv2_NRTidal: $\cos\theta_A = 0.24^{+0.60}_{-0.44}$ compared to $\cos\theta_1 = 0.24^{+0.62}_{-0.65}$ for the high-spin prior, with a similar trend for the low-spin prior. Conversely, the posterior for $\cos\theta_B$ broadens slightly compared to that of $\cos\theta_2$, consistent with the behavior observed for the simulated signal.

9.7.3 Spin disk plots

In Figs. 9-5–9-8, we show the spin disk plots for the spin-sorted posterior samples for the first 13 LVC detections (see e.g. Fig. 5 of [13]) calculated using PESummary [474]. The angular direction indicates the misalignment with the orbital angular momentum,

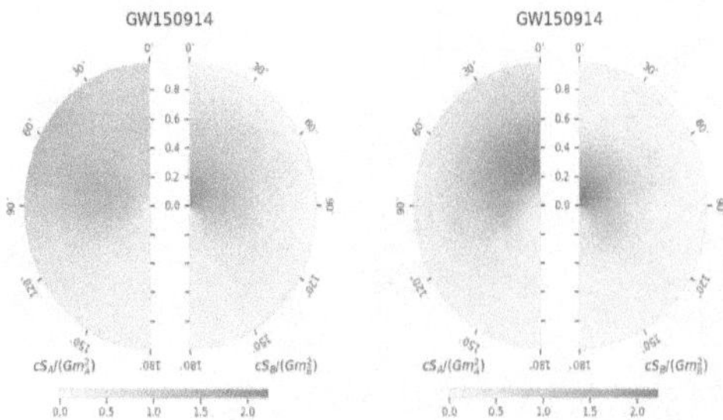

Figure 9-5: Spin disk plots for the spin-sorted posterior samples for GW150914 using the IMRPhenomPv2 waveform on the left and the SEOBNRv3 waveform on the right.

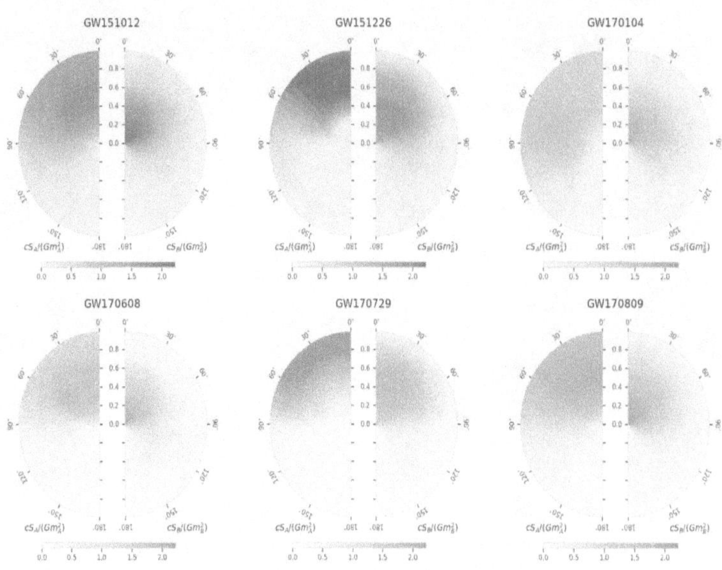

Figure 9-6: Spin disk plots for the spin-sorted posterior samples for current LVC detections.

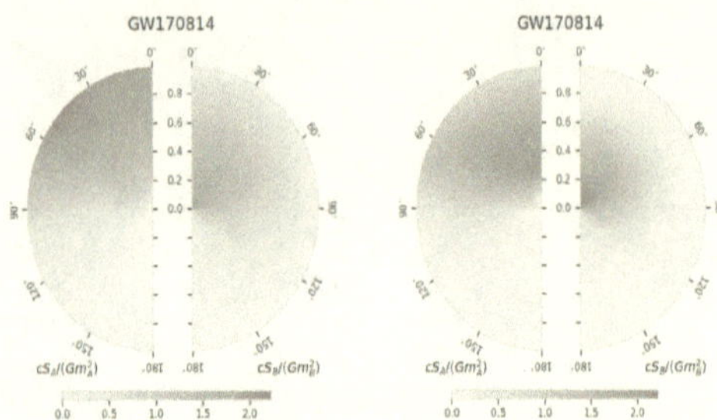

Figure 9-7: Spin disk plots for the spin-sorted posterior samples for GW170814 using the IMRPhenomPv2 waveform on the left and the SEOBNRv3 waveform on the right.

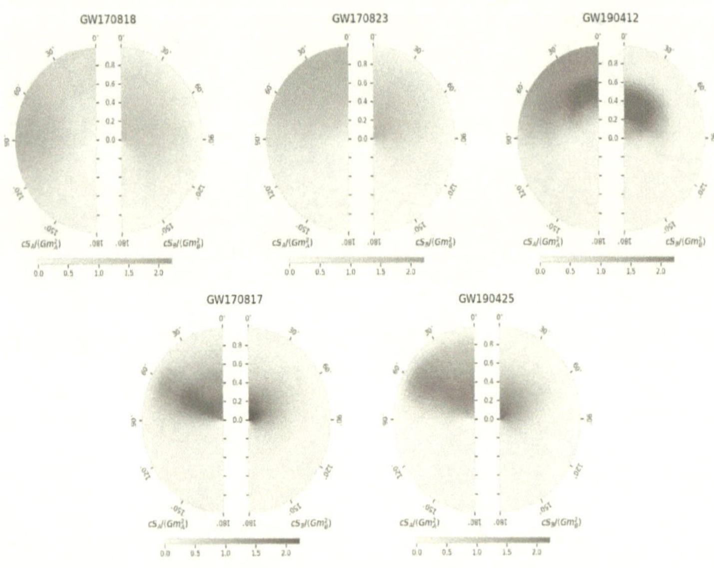

Figure 9-8: Spin disk plots for the spin-sorted posterior samples for current LVC detections.

while the radial direction shows the spin magnitude for the highest-spinning compact object on the left and the least-spinning object on the right. We show the posterior samples obtained using the IMRPhenomPv2 waveform for BBH events reported in GWTC-1, IMRPhenomPv2_NRTidal with the high-spin prior for the two BNS detections, and IMRPhenomPv3 for GW190412. We also show the disk plots for the posteriors obtained with the SEOBNRv3 waveform for GW150914 and GW170814 in Figs. 9-5 and 9-7 to supplement the comparison in Fig. 9-2.

Chapter 10

The binary black hole spin distribution likely broadens with redshift

The content of this chapter was previously published in the Astrophysical Journal Letters as Ref. [149] in June 2022. ASB performed the analysis, wrote the manuscript, and contributed to project development.

Abstract

The population-level distributions of the masses, spins, and redshifts of binary black holes (BBHs) observed using gravitational waves can shed light on how these systems form and evolve. Because of the complex astrophysical processes shaping the inferred BBH population, models allowing for correlations among these parameters will be necessary to fully characterize these sources. We hierarchically analyze the BBH population detected by LIGO and Virgo with a model allowing for correlations between the effective aligned spin and the primary mass and redshift. We find that the width of the effective spin distribution grows with redshift at 98.6% credibility. We determine this trend to be robust under the application of several alternative models and additionally verify that such a correlation is unlikely to be spuriously introduced using a simulated population. We discuss the possibility that this correlation could be due to a change in the natal black hole spin distribution with redshift.

10.1 Introduction

The growing catalog of compact binary mergers detected with gravitational waves [26, 42, 44] has allowed for increasingly precise characterization of the population properties of binary black holes (BBHs) and neutron stars [52], which can shed light on how these systems form and evolve [938, 963, 173]. The latest data from the Advanced LIGO [3] and Virgo [53] observatories has revealed that the BBH mass distribution has substructure [877, 314, 554, 911] beyond a smooth power-law and single peak or break at ~ 40 M$_\odot$ [353, 851, 25]. BBH spins are found to be small but non-zero [946, 769, 626, 396, 153], with some of their tilts misaligned to the orbital angular momentum [850, 25, 52], although these conclusions have been challenged when applying a different population model [768, 394]. The merger rate is found to evolve with redshift at a rate consistent with the star formation rate [572, 357, 52].

In addition to fitting the mass, spin, and redshift distributions independently, previous works have looked for correlations among these parameters [780, 199, 52, 876, 376]. Such correlations can be imprinted via evolutionary processes within a single formation channel or can be caused by the presence of multiple populations arising from distinct formation channels. For example, systems formed dynamically in dense environments via repeated mergers are expected to be more massive and have higher spins [708, 140, 625, 604, 89, 756, 401, 356, 757, 515, 403]. [199] and [52] found statistically significant evidence for a correlation between the distribution of the mass-weighted spin aligned with the orbital angular momentum, $\chi_{\rm eff}$, and the binary mass ratio, q, where the mean of the $\chi_{\rm eff}$ distribution increases for more extreme mass ratios. This sort of correlation is unexpected for most individual BBH formation models, with the possible exception of formation in the disks of active galactic nuclei [604, 838, 603, 605, 846, 199] and super-Eddington accretion during stable mass transfer for systems formed via isolated binary evolution [125, 962], so may hint at the superposition of multiple populations formed via independent channels with unique q vs. $\chi_{\rm eff}$ signatures.

Ref. [780] examined whether the effective spin distribution correlates with various mass parameters of the binary. They found a possible negative correlation between the mean effective spin and each of the chirp mass ($\mathcal{M} = (m_1 m_2)^{3/5}/(m_1 + m_2)^{1/5}$), total mass, and primary mass parameters and a possible positive correlation between the dispersion of the effective spin and mass; neither finding reached high statistical significance ($\sim 80\%$ credibility depending on the mass and correlation parameters).

[52] and [876] also found that the spread in the component spins aligned with the orbital angular momentum may increase with the binary chirp mass. This is explained as an effect of the paucity of events at large chirp masses, which manifests as a degradation of the constraint on the spin distribution, rather than a firm measurement of an increase in the width of the distribution at larger chirp masses. [376] also find initial evidence of an evolution of the χ_{eff} distribution with increasing total mass, which they interpret as evidence for a sub-population of dynamically-formed or primordial black hole binaries [286, 284, 285, 375], as both models predict correlations between mass and spin.

Previous works have also explored potential correlations between the BBH mass and redshift distributions [351, 52]. Heavier black holes are predicted to form from lower-metallicity stellar progenitors at higher redshifts, which could lead to such a correlation [133, 593, 651]. For systems formed via isolated binary evolution, those that undergo a common envelope event tend to be both less massive and have shorter delay times, merging at higher redshifts [901]. Motivated by these theoretical predictions, [351] find that redshift dependence of the maximum black hole mass is required if the primary mass distribution has a sharp cutoff, but a gradual tapering of the primary mass distribution is consistent with no evolution with redshift, a finding which was confirmed with the latest catalog of events by [52].

In this work, we search for correlations between the χ_{eff} distribution and the primary mass and redshift distributions. We find robust evidence of a correlation between χ_{eff} and redshift, where the width of the χ_{eff} distribution increases with redshift. We also find a weaker correlation between the width of the χ_{eff} distribution and primary mass. When allowing the χ_{eff} distribution to correlate with both redshift

Parameter	Description	Prior
α	m_1 power-law index	U(-4, 12)
β	q power-law index	U(-4, 7)
$m_{\max}$	maximum BH mass	$U(30\ M_\odot, 100\ M_\odot)$
$m_{\min}$	minimum BH mass	$U(2\ M_\odot, 10\ M_\odot)$
δ_m	low-mass smoothing parameter	$U(0\ M_\odot, 10\ M_\odot)$
μ_m	PISN peak location	$U(20\ M_\odot, 50\ M_\odot)$
σ_m	PISN peak width	$U(1\ M_\odot, 10\ M_\odot)$
λ	fraction of systems in PISN peak	$U(0, 1)$
λ_z	z power-law index	$U(-2, 10)$

Table 10.1: Mass and redshift hyper-parameter priors for the standard POWER-LAW + PEAK and POWER-LAW REDSHIFT distributions

and primary mass, we find a preference in the data for a correlation with one or the other, although we cannot distinguish which. In Section 10.2, we describe the Bayesian methods and models employed in our analysis. The results on data from the latest GWTC-3 catalog [44] of compact binaries observed by LIGO-Virgo are presented in Section 10.3. Section 10.4 includes a validation of our results using simulated populations and alternative models applied to the real data. We conclude and comment on the potential astrophysical implications of our finding in Section 10.5.

10.2 Methods

We employ the framework of hierarchical Bayesian inference to constrain the hyper-parameters governing the population-level distributions of the masses, spins, and redshifts of the binary black hole systems detected by LIGO and Virgo. We assume the distribution of primary masses, m_1, is described by the sum of a truncated power law with low-mass smoothing and a Gaussian component [851], dubbed the POWER LAW + PEAK model in [49, 52], and that the mass ratio distribution is also a power law, bounded such that the minimum and maximum masses of the secondary component are the same as those of the primary [355]. The merger rate is allowed to evolve with redshift as a power-law, following the model in [357] and [52]. The hyper-parameters governing these mass and redshift distributions are described in Table 10.1.

Our goal is to determine if there is a correlation between the BBH spin distribution and the distributions of masses or redshifts. To this end, we modify the correlated spin model introduced in [199] so that the effective aligned spin [271, 64, 62, 786],

$$\chi_{\text{eff}} = \frac{\chi_1 \cos \theta_1 + q \chi_2 \cos \theta_2}{1 + q},$$

(10.1)

is modeled as a truncated Gaussian on $[-1, 1]$ whose mean and variance can each evolve linearly with primary mass and redshift,

$$\pi_{\text{pop}}(\chi_{\text{eff}} | \Lambda_{\chi_{\text{eff}}}, m_1, z) = \mathcal{N}(\chi_{\text{eff}}; \mu_\chi, \sigma_\chi),$$

(10.2)

$$\mu_\chi(z, m_1) = \mu_0 + \delta\mu_z(z - 0.5) + \delta\mu_{m_1} \left(\frac{m_1}{10 \text{ M}_\odot} - 1 \right)$$

(10.3)

$$\log \sigma_\chi(z, m_1) = \log \sigma_0 + \delta\log \sigma_z(z - 0.5)$$

(10.4)

$$+ \delta\log \sigma_{m_1} \left(\frac{m_1}{10 \text{ M}_\odot} - 1 \right),$$

where we use log to indicate log base 10 everywhere and ln to indicate the natural logarithm. The pivot points of $z = 0.5$ and $m_1 = 10 \text{ M}_\odot$ are chosen to be near the peak of the population distributions of those parameters, but we have verified that the exact values do not affect our results. To avoid unphysically narrow distributions using the model above, we impose a cut such that

$$\pi_{\text{pop}}(\chi_{\text{eff}} | \Lambda_{\chi_{\text{eff}}}) = \begin{cases} 0, & \text{if } \log_{10} \sigma_\chi(z, m_1) < -3 \\ \mathcal{N}(\chi_{\text{eff}}; \mu_\chi, \sigma_\chi), & \text{elsewhere.} \end{cases}$$

(10.5)

The full set of $\Lambda_{\chi_{\text{eff}}}$ parameters are described in Table 10.2. We note that unlike the model of [199], our model in Eq. 10.2 does not allow for correlations between χ_{eff} and mass ratio. However, we amend the model to incorporate mass ratio correlations later in Section 10.3.

Given a set of posterior samples for the binary parameters θ of N individual BBH events, they can be combined to obtain posteriors on the hyper-parameters governing

Parameter	Description	Prior
μ_0	Independent χ_{eff} mean	$U(-1,1)$
$\log \sigma_0$	Log of independent χ_{eff} width	$U(-1.5, 0.5)$
$\delta\mu_q$	q-dependent χ_{eff} mean	$U(-2.5, 1)$
$\delta\log \sigma_q$	Log of q-dependent χ_{eff} width	$U(-2, 1.5)$
$\delta\mu_z$	z-dependent χ_{eff} mean	$U(-2.5, 1)$
$\delta\log \sigma_z$	Log of z-dependent χ_{eff} width	$U(-0.5, 1.5)$
$\delta\mu_{m_1}$	m_1-dependent χ_{eff} mean	$U(-2.5, 1)$
$\delta\log \sigma_{m_1}$	Log of m_1-dependent χ_{eff} width	$U(-2, 1.5)$

Table 10.2: Spin hyper-parameter priors and descriptions for the model in Eq. 10.2

the population-level distributions:

$$p(\Lambda|\{d\}) \propto \mathcal{L}(\{d\}|\Lambda)\pi(\Lambda), \tag{10.6}$$

$$\mathcal{L}(\{d\}|\Lambda) = \frac{1}{\alpha(\Lambda)^N} \prod_i^N \sum_j \frac{\pi_{\mathrm{pop}}(\theta_{i,j}|\Lambda)}{\pi_{\mathrm{PE}}(\theta_{i,j})}. \tag{10.7}$$

The factor of $\pi(\Lambda)$ in Eq. 10.6 represents the prior probability for the hyper-parameters, given in Tables 10.1-10.2. The likelihood in Eq. 10.7 consists of a Monte Carlo integral over the posterior samples j for each individual event i, where $\pi_{\mathrm{pop}}(\theta_{i,j}|\Lambda)$ is the product of the population distributions for the masses, redshift, and χ_{eff} in Eq. 10.2. $\pi_{\mathrm{PE}}(\theta_{i,j})$ is the original prior that was applied during the parameter estimation (PE) of the binary parameters for individual events. Finally, the factor of $\alpha(\Lambda)^N$ accounts for the fact that the observed BBH sources are a biased sample of the underlying astrophysical distribution [566, 874, 585, 919]. This selection effect arises from the dependence of the sensitivity of the detector network on the intrinsic parameters of the source.

We use the GWPOPULATION package [849] and the DYNESTY nested sampler [828] to obtain samples from the hyper-parameter posterior in Eq. 10.6. We include the 69 BBH events reported in GWTC-3 with a false alarm rate (FAR) less than 1 per year [44], as was done for the BBH analyses in [52]. We use the individual-event posterior samples publicly released by LIGO and Virgo [21, 33, 43, 46] obtained with the IMRPhenomPv2 waveform model for events first published in GWTC-1 [26], and using a combination of different waveform models including the effects of spin pre-

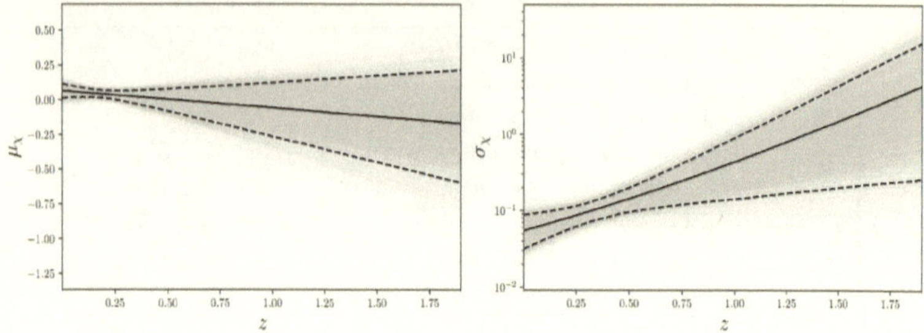

Figure 10-1: Posteriors for the mean, μ_χ, and standard deviation, σ_χ, of the binary black hole effective spin distribution as a function of redshift obtained for GWTC-3 events. The solid black line shows the mean, while the dashed black lines show the 90% credible region.

cession and higher-order modes for events reported in later catalogs[1] [41, 42, 44]. We calculate $\alpha(\Lambda)$ following the method described in [334], using the sensitivity estimates for BBH systems released at the end of the most recent LIGO-Virgo observing run (O3) obtained via a simulated injection campaign [45].

10.3 Results for GWTC-3

We first allow the $\chi_{\rm eff}$ distribution to be correlated with redshift and primary mass individually. We recover mass and redshift hyper-parameter posteriors consistent with those reported in [52]. The evolution of μ_χ and σ_χ as a function of redshift for individual hyper-parameter posterior samples obtained with the model only allowing for redshift correlations is shown in Fig. 10-1. In this case the $\delta\mu_{m_1}$ and $\delta\log\sigma_{m_1}$ parameters are fixed to zero. While μ_χ does not exhibit significant evolution with redshift, σ_χ increases with redshift. The corresponding posteriors on $\Lambda_{\chi_{\rm eff}}$ are shown in Fig. 10-2. The $\delta\mu_z$ posterior is consistent with 0, indicating no evolution of the mean of the $\chi_{\rm eff}$ distribution as a function of redshift, but we find that $\delta\log\sigma_z = 0$ is disfavored at 98.6% credibility. We recover $\delta\log\sigma_z = 0.93^{+0.54}_{-0.54}$, (maximum posterior value and 90% credible interval calculated with the maximum posterior density

---

[1]These correspond to the `PublicationSamples`, `PrecessingSpinIMRHM`, and `C01:Mixed` datasets for GWTC-2, GWTC-2.1, and GWTC-3 events, respectively.

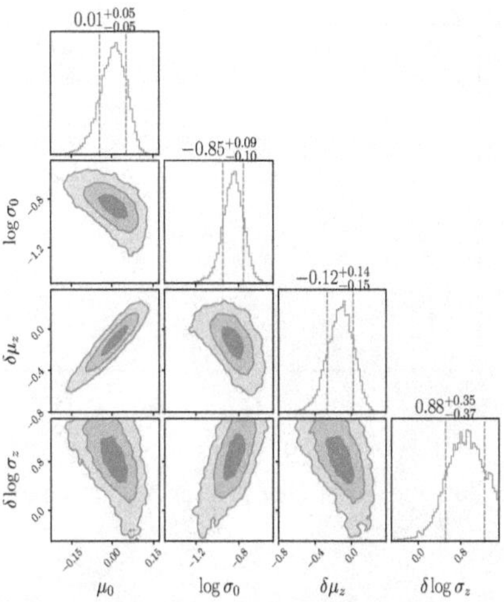

Figure 10-2: Spin hyper-parameter posteriors obtained for GWTC-3 events when allowing for redshift correlations only. The colors indicate the 1, 2, and 3σ 2D credible regions, the dashed lines on the individual histograms show the 1D 1σ credible interval, and the median and 1σ credible interval are printed above each histogram.

method) indicating that the spin distribution broadens with increasing redshift. This correlation can also be visualized by examining slices of the χ_{eff} distribution at fixed redshift, as shown in Fig. 10-3. The thickness of the 90% credible region for each slice is comparable, but the distribution becomes noticeably broader with increasing redshift. This indicates that the apparent correlation between the width of the spin distribution and the redshift is not dominated by increased uncertainty in the constraint on the distribution at higher redshifts. The location of the peak of the χ_{eff} distribution does not vary significantly between the different redshift slices, consistent with the lack of observed evolution in μ_χ.

While these results present compelling evidence for a correlation between χ_{eff} and redshift, it is possible that the BBH mass and redshift distributions are themselves correlated, so that our recovered redshift correlation is actually a manifestation of a mass vs. χ_{eff} correlation viewed through the wrong model. To test this, we now allow the χ_{eff} distribution to be correlated only with primary mass, fixing $\delta\mu_z = 0$, $\delta\log\sigma_z = 0$. The posteriors for the spin hyper-parameters for the primary mass correlation model

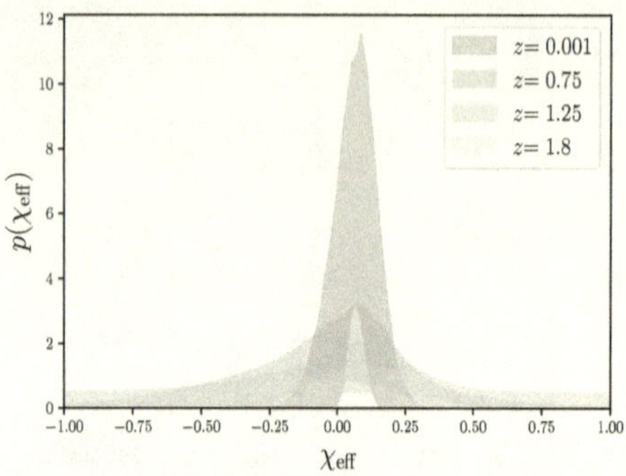

Figure 10-3: 90% credible region for slices of the $\chi_{\rm eff}$ distribution at four different values of redshift.

are shown in Fig. 10-4, and the distributions for $\chi_{\rm eff}$ along different slices in primary mass are shown in Fig. 10-5. We find a similar but less significant trend between the width of the $\chi_{\rm eff}$ distribution and the primary mass, where $\delta \log \sigma_{m_1} = 0.09^{+0.10}_{-0.08}$, and $\delta \log \sigma_{m_1} = 0$ is excluded at 93.5% credibility. While the $\chi_{\rm eff}$ distributions in Fig. 10-5 appear visually to increase both in mean and width with increasing primary mass, the posterior on $\delta \mu_{m_1}$ is still consistent with 0 at 31.1% credibility. In Section 10.4, we comment on the possibility of a correlation between primary mass and $\chi_{\rm eff}$ being spuriously introduced due to mismodeling a true correlation between redshift and $\chi_{\rm eff}$.

In order to determine if there is a preference in whether the redshift or the primary mass drives the evolution of the width of the $\chi_{\rm eff}$ distribution, we now analyze the data with a model that allows for correlations with both parameters. The posteriors on the spin hyper-parameters are shown in Fig. 10-6. We obtain much weaker constraints on $\delta \log \sigma_z$ and $\delta \log \sigma_{m_1}$ individually, but both posteriors have more support for positive than negative values, and the point $\delta \log \sigma_z, \delta \log \sigma_{m_1} = (0,0)$ is disfavored, lying at the 96.2% credibility contour.

Based on these posteriors alone, we cannot confidently identify whether the correlation is dominated by the redshift or the primary mass, but comparing the Bayes factors between the various models that we have considered can provide an indication

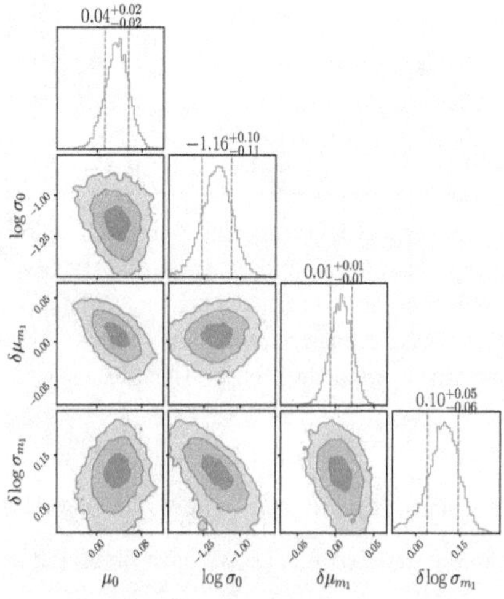

Figure 10-4: Spin hyper-parameter posteriors obtained for GWTC-3 events when allowing for a correlation between χ_{eff} and primary mass only. The colors indicate the 1, 2, and 3σ 2D credible regions, the dashed lines on the individual histograms show the 1D 1σ credible interval, and the median and 1σ credible interval are printed above each histogram.

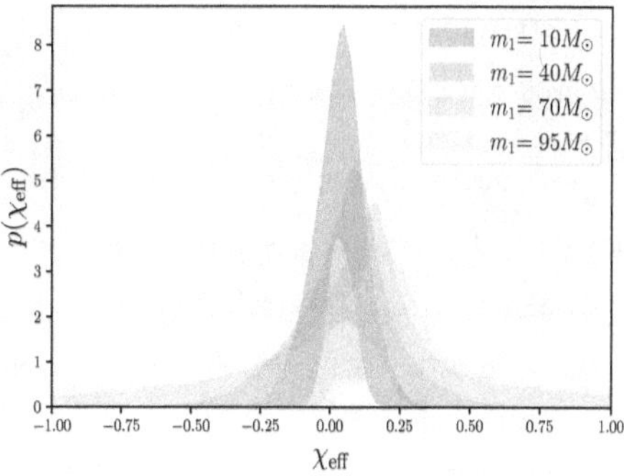

Figure 10-5: 90% credible region for slices of the χ_{eff} distribution at four different values of primary mass.

Correlation	ln(BF)
Redshift, $\delta\mu_z = 0$	1.43
Redshift	0.24
Primary mass	-2.09
Mass ratio	3.75
Redshift & primary mass	-3.63
Redshift & mass ratio	3.20

Table 10.3: Natural log Bayes factors comparing the models allowing for correlations between redshift, primary mass, mass ratio, and χ_{eff} and the base model without any correlations. The redshift-only models include both $\delta\mu_z$ and $\delta\log\sigma_z$ as free parameters and only $\delta\log\sigma_z$ as a free parameter with $\delta\mu_z$ fixed to zero. The models including a correlation with primary mass are disfavored by the data.

of which correlation is statistically preferred. Table 10.3 shows the natural log Bayes factors between the three correlated models we have presented so far and an uncorrelated model, where only μ_0 and $\log\sigma_0$ in Eq. 10.2 are left as free parameters. We repeat the analysis allowing only for redshift correlations with fixed $\delta\mu_z = 0$ in order to gauge the effect of the Occam penalty on the Bayes factor, since this sub-hypothesis is consistent with our initial redshift correlation result. While none of the values are particularly statistically significant, we find that the models including a correlation with primary mass are disfavored relative to the redshift-only models. We emphasize that while we have ensured consistent priors between all the models which are sub-hypotheses of each other, the choice of priors for individual hyper-parameters is necessarily somewhat arbitrary, complicating the interpretation of the Bayes factors[2].

Finally, we want to ensure that the apparent correlation between redshift and χ_{eff} is not falsely introduced by the previously-reported correlation between the mean of the χ_{eff} distribution and mass ratio [199, 52]. We modify the model in Eq. 10.2 to allow for correlations with both redshift and mass ratio rather than redshift and

---

[2]We can use a simple back-of-the-envelope calculation to determine the effect of the narrower width of the prior on $\delta\log\sigma_z$ on the Bayes factor for the redshift correlation models. Since the prior volume for $\delta\log\sigma_z$ is $2/3.5$ times the prior volume for $\delta\log\sigma_q$ and $\delta\log\sigma_{m_1}$, the Bayes factor would be increased for the redshift correlation model by $\sim -\ln 2/3.5 \sim 0.56$. This is comparable to the uncertainty in the Bayes factor due to the Monte Carlo integration in the likelihood in Eq. 10.7.

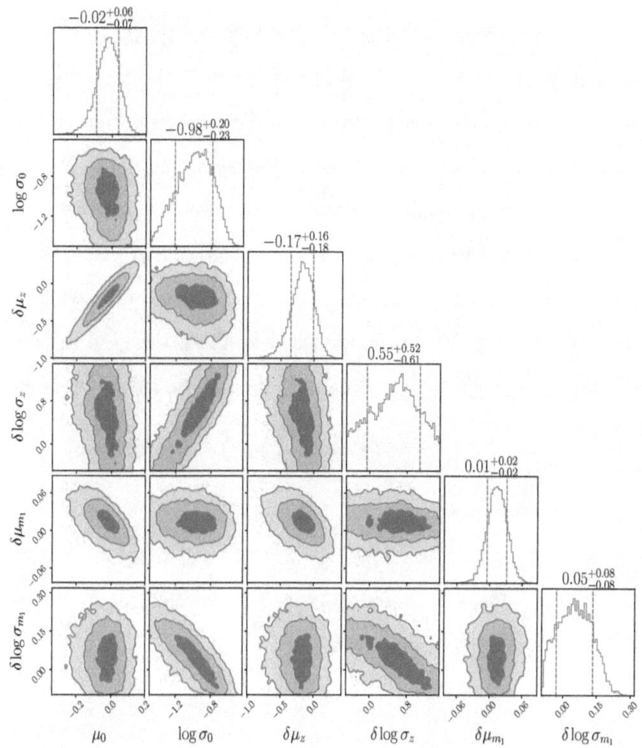

Figure 10-6: Spin hyper-parameter posteriors obtained for GWTC-3 events when allowing for a correlation between $\chi_{\rm eff}$ and both primary mass and redshift. The colors indicate the 1, 2, and 3σ 2D credible regions, the dashed lines on the individual histograms show the 1D 1σ credible interval, and the median and 1σ credible interval are printed above each histogram.

primary mass:

$$\mu_\chi(z, q) = \mu_0 + \delta\mu_z(z - 0.5) + \delta\mu_q(q - 0.7) \tag{10.8}$$

$$\log_{10}\sigma_\chi(z, q) = \log\sigma_0 + \delta\log\sigma_z(z - 0.5) + \delta\log\sigma_q(q - 0.7). \tag{10.9}$$

The priors on $\delta\mu_q$ and $\delta\log\sigma_q$ are given in Table 10.2, and the Bayes factor between this model and the uncorrlated model is also shown in Table 10.3. The results we obtain with this model are consistent with both the previously-reported mass ratio correlation and the redshift correlation presented earlier in this work. The $\delta\mu_q$ posterior peaks at negative values, indicating that the mean of the χ_{eff} distribution shifts towards smaller values as the mass ratio becomes more equal. Simultaneously, the $\delta\log\sigma_z$ posterior peaks at positive values, similar to the posteriors shown in Figs. 10-2, 10-6. Thus, we conclude that the increase in the width of the χ_{eff} distribution that we report here is independent of the previously-established correlation between χ_{eff} and mass ratio.

10.4 Validation of results

The results presented so far suggest that for the 69 BBH events we analyze with FAR < 1 yr^{-1} detected up to the end of O3, the χ_{eff} distribution broadens with increasing redshift. As noted above, though, it is difficult to disentangle a correlation between spin and redshift from a correlation between spin and mass and to distinguish the extent to which the broadening of the χ_{eff} distribution is due to increased uncertainty in spin measurements at high redshifts. We test the robustness of the observed redshift correlation first with a series of simulated populations, then by analyzing the GWTC-3 data with alternative models allowing for a correlation between χ_{eff} and redshift.

10.4.1 Simulated populations

One potential concern is that such a spin-redshift correlation could be introduced due to the degradation of the constraint on χ_{eff} for individual events at farther redshifts, as these sources are generally detected with smaller signal-to-noise ratios (SNRs). To verify whether such a spurious correlation could be introduced, we simulate a population of 69 BBH events detected by a Hanford-Livingston detector network operating at O3 sensitivity [35] drawn from an intrinsically uncorrelated population. The true values of the hyper-parameters describing the population are given in Table 10.4. We consider an event to be "detected" if it has a network optimal SNR ≥ 9. We perform individual-event parameter estimation using the reduced order quadrature implementation [820] of the IMRPhenomPv2 waveform [482, 509, 445] and the DYNESTY nested sampler [828] through the BILBY package [96, 762]. We then recover posteriors on the mass, redshift, and spin hyper-parameters using the three population models applied to the real data in the previous section.

The true values of all hyper-parameters describing this simulated population are recovered within at least the 3σ level, although the posteriors for both $\delta \log \sigma_z$ and $\delta \log \sigma_{m_1}$ tend to prefer negative values. This suggests that it is easier to rule out a distribution that broadens with redshift or mass rather than one that becomes narrower. A broadening distribution implies the presence of highly spinning sources at high mass/redshift, which are easier to detect. Therefore, their absence from the observed population indicates that they are also absent from the underlying population. The converse is not true. If the χ_{eff} distribution instead became narrower, there would be an increasing number of sources with nearly zero spin at high mass/redshift. These sources are harder to detect, so their absence from the observed distribution does not necessarily imply their absence from the underlying distribution.

The χ_{eff} distributions along different slices in redshift and primary mass are shown in Fig. 10-7. These distributions are morphologically distinct from those obtained for the real data in Figs. 10-3 and 10-5. The distributions obtained for this uncorrelated population appear to initially get narrower between the lower two values of redshift

Parameter	Value	Recovery
α	4	$4.74^{+1.33}_{-0.98}$
β	1.5	$1.02^{+2.10}_{-1.44}$
$m_{\max}$	50 M$_\odot$	$46.63^{+46.35}_{-7.87}$ M$_\odot$
$m_{\min}$	5 M$_\odot$	$4.99^{+0.87}_{-1.10}$ M$_\odot$
δ_m	5 M$_\odot$	$4.88^{+4.12}_{-2.12}$ M$_\odot$
μ_m	35 M$_\odot$	$31.40^{+1.59}_{-1.59}$ M$_\odot$
σ_m	4 M$_\odot$	$3.09^{+0.99}_{-2.09}$ M$_\odot$
λ	0.04	$0.012^{+0.019}_{-0.012}$
λ_z	3	$5.15^{+1.83}_{-1.83}$
μ_0	0.05	$0.025^{+0.076}_{-0.088}$
$\log\sigma_0$	-0.85	$-0.82^{+0.22}_{-0.20}$
$\delta\mu_z$	0	$-0.17^{+0.21}_{-0.11}$
$\delta\log\sigma_z$	0	$-0.24^{+0.62}_{-0.47}$
$\delta\mu_{m_1}$	0	$0.023^{+0.035}_{-0.035}$
$\delta\log\sigma_{m_1}$	0	$-0.14^{+0.07}_{-0.08}$

Table 10.4: True and recovered values of the hyper-parameters describing the mass, spin, and redshift distributions from which the simulated population with no correlation was drawn. The correlated population is described by the same hyper-parameters with the exception of $\delta\log\sigma_z = 0.85$. The recovered values are represented by the maximum posterior value and 90% credible interval calculated with the maximum posterior density method obtained with the model that allows for both primary mass and redshift correlations.

Correlation	ln(BF)
Redshift	-1.93
Primary mass	-1.59
Both	-3.35

Table 10.5: Natural log Bayes factors comparing the models allowing for correlations between redshift, primary mass, and χ_{eff} and the base model without any correlations for the simulated population with no correlations.

and primary mass but then broaden again. This indicates that there is no strong evidence for either an increase or a decrease in the width of the χ_{eff} distribution with either parameter. There is more uncertainty in the distributions at higher primary mass and redshift, a feature which is not observed for the real data, further increasing our confidence in our measurement of a broadening in the spin distribution with mass and/or redshift.

The natural log Bayes factors between each of the models allowing for correlations and the model without any correlations for this simulated population are given in Table 10.5. In this case, the redshift-correlation model is disfavored at a level comparable to the primary-mass-correlation model, as expected since the true population does not contain any correlations. This simulation suggests that a positive correlation between the width of the χ_{eff} distribution and either the redshift or primary mass is unlikely to be spuriously introduced by the worsening constraint on χ_{eff} for individual events at higher redshift or primary mass.

We next seek to verify if we are able to successfully recover a correlation in a simulated population similar to the one we find in real data, and whether a redshift correlation can manifest as a primary mass correlation when analyzed with the wrong model. We again generate 69 sources detected at O3 sensitivity now drawn from a distribution with $\delta\log\sigma_z = 0.85$. All the other true hyper-parameter values are the same as for the uncorrelated population given in Table 10.4. We initially analyze this simulated population allowing for mass and redshift correlations in turn. As before, all the hyper-parameter values are recovered within 3σ credibility, although the posterior on $\delta\log\sigma_z$ peaks below the true value. The distributions for χ_{eff} along different slices in mass and redshift shown in Fig. 10-8 are similar to those recovered

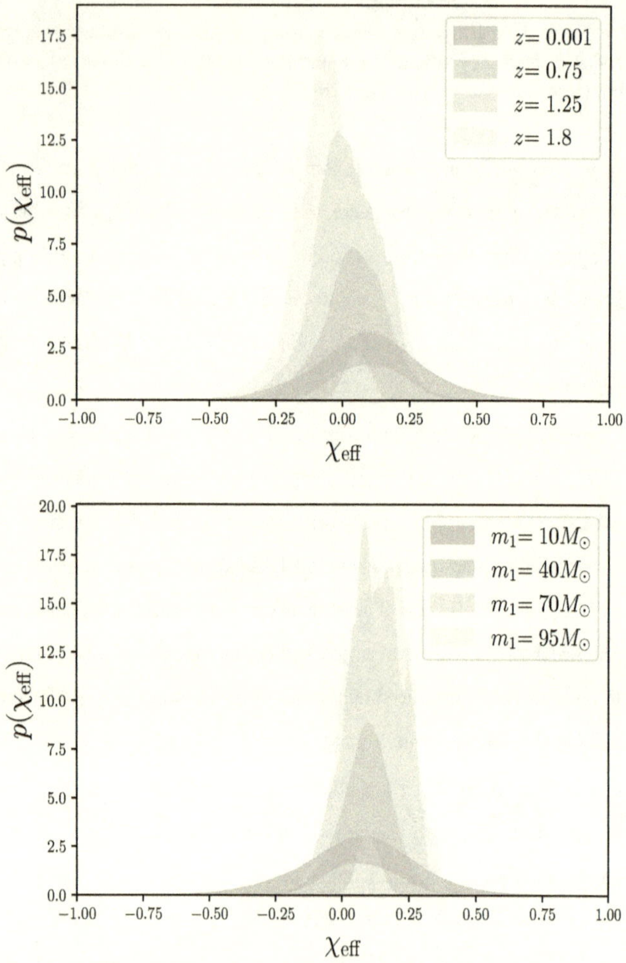

Figure 10-7: 90% credible region for slices of the χ_{eff} distribution at four different values of redshift (top) and primary mass (bottom) for the simulated population with no correlations.

Correlation	ln(BF)
Redshift	-2.45
Primary mass	-4.55
Both	-6.97

Table 10.6: Natural log Bayes factors comparing the models allowing for correlations between redshift, primary mass, and χ_{eff} and the base model without any correlations for the simulated population with $\delta \log \sigma_z = 0.85$.

in the real data. They become consistently wider between increasing values of redshift and primary mass, although they also become more uncertain, which is not observed in the real data. This is because the posteriors for individual events in the simulated data set only extend up to $z = 1.14$, $m_1 = 104\ M_\odot$, while those for real events include values up to $z = 1.90$, $m_1 = 296\ M_\odot$, meaning there is more resolving power for the mass and redshift distributions at large masses and redshifts in the real data.

While the posterior on $\delta \log \sigma_{m_1}$ includes 0 within the 81.4% credible level, this simulation indicates that a true correlation between redshift and χ_{eff} could be perceived as a correlation between primary mass and χ_{eff} if analyzed with the wrong model. It also reinforces the significance of our finding a preference for positive values of $\delta \log \sigma_z$ in the real data, since both simulations tend to recover smaller values of $\delta \log \sigma_z$ compared to the true value. This is further supported by the Bayes factors we recover for the correlated simulation, given in Table 10.6. In this case, the recovered correlation with redshift-only is not significant enough to overcome the Occam penalty for adding parameters relative to the uncorrelated model. However, the relative Bayes factors between the three models considered are similar to those obtained for the real data. This indicates that the redshift-only correlation found in the real data is preferred over the other correlated models we explore at a similar level of statistical significance to the simulated population with a known correlation in $\delta \log \sigma_z$. On the other hand, the relative Bayes factors for the uncorrelated simulation span a much narrower range, revealing that none of the correlated models is strongly preferred over the others in the case of no correlation.

Conversely, it is also possible that a true correlation between primary mass and χ_{eff} could manifest as a correlation between redshift and χ_{eff} if analyzed with the wrong

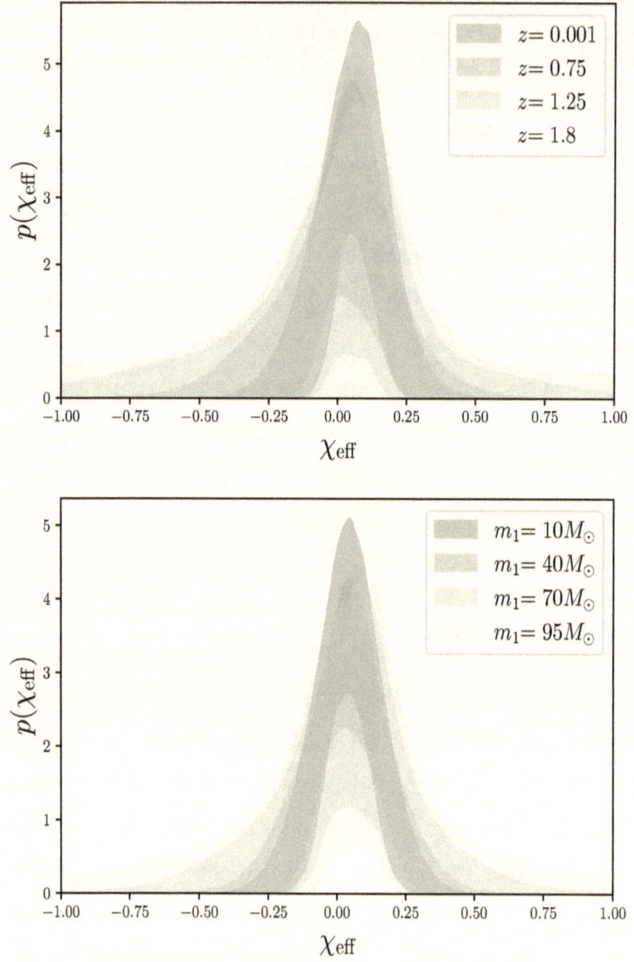

Figure 10-8: 90% credible region for slices of the χ_{eff} distribution at four different values of redshift (top) and primary mass (bottom) for the simulated population with a correlation between the width of the χ_{eff} distribution and redshift.

model. To verify whether this is a potential false source of the redshift-χ_{eff} correlation we observe, we simulate a third population of 69 events detected at O3 sensitivity with hyper-parameters identical to the previous two, except now $\delta \log \sigma_z = 0$ and $\delta \log \sigma_{m_1} = 0.15$. We find that when this population is analyzed with the model allowing for a redshift correlation only, $\delta \log \sigma_z = 0$ is incorrectly ruled out at 99.0% credibility, indicating that a primary mass-χ_{eff} correlation can indeed be confused for a redshift-χ_{eff} correlation. This is similar to the case explored in the previous two paragraphs where $\delta \log \sigma_{m_1} = 0$ was disfavored at the 81.4% credible level for a true correlation between redshift and spin analyzed with the wrong model.

However, we find in general that a correlation between spin and primary mass is easier to confidently identify with 69 events detected at O3 sensitivity. The posterior on $\delta \log \sigma_{m_1}$ disfavors 0 at 99.7% credibility under the model allowing only for a primary mass correlation and at 99.5% credibility under the model allowing for both primary mass and redshift correlations. This is illustrated in the corner plot of the posteriors on $\delta \log \sigma_{m_1}, \delta \log \sigma_z$ in Fig. 10-9 for both the mass-correlated population and the redshift-correlated population previously presented in Fig. 10-8. While it is difficult to experimentally distinguish between mass and redshift evolution of the spin distribution in the case of a redshift-correlated population, the same ambiguity is not present for the mass-correlated population. This further supports the redshift-χ_{eff} correlation we find in the real data, since a correlation between primary mass and χ_{eff} would likely have been unambiguously identified.

10.4.2 Alternative models

While the results of the two simulated populations lend credence to the finding that the width of the χ_{eff} distribution increases with redshift in the real data, we want to verify whether this result is model-driven. To this end, we now analyze the GWTC-3 events with a different model—a mixture of truncated Gaussians with a redshift-

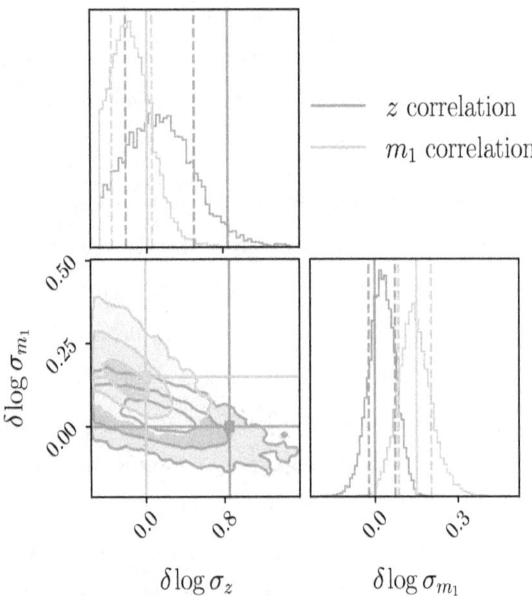

Figure 10-9: Posteriors for $\delta\log\sigma_{m_1}, \delta\log\sigma_z$ obtained under the model allowing for both redshift and primary mass correlations with χ_{eff}, for two different simulated populations. The dark purple shows the results for the population with a simulated redshift correlation, $\delta\log\sigma_z = 0.85$, while the light purple shows the results for the population with a simulated primary mass correlation, $\delta\log\sigma_{m_1} = 0.15$. The shading indicates the 1, 2, and 3σ 2D credible regions and the dashed lines on the individual histograms show the 1D 1σ credible interval. While the posterior for the redshift-correlated population is consistent with primary mass being the sole source of correlation, the opposite is not true.

Parameter	Description	Prior 1	Prior 2	Prior 3
μ_a	Mean of the secondary Gaussian	$U(-1, -0.1) \cup U(0.1, 1)$	$U(-1, 1)$	$U(0.1, 1)$
μ_b	Mean of the bulk Gaussian	0.06	0	0
σ_a	Log-width of secondary Gaussian	$U(-1.5, 0.5)$	$U(-0.7, 1.5)$	$U(-1.5, 0.5)$
σ_b	Log-width of the bulk Gaussian	$U(-1.5, 0.5)$	$U(-2, -0.7)$	$U(-1.5, 0.5)$
$\chi_{a,min}$	Lower bound of the secondary Gaussian	-1	-1	0
f_A	Independent offset in mixing fraction	$\mathcal{N}(0, 1.5)$	$\mathcal{N}(0, 1.5)$	$\mathcal{N}(0, 1.5)$
f_B	z-dependent offset in mixing fraction	$\mathcal{N}(0, 1.5)$	$\mathcal{N}(0, 1.5)$	$\mathcal{N}(0, 1.5)$

Table 10.7: Spin hyper-parameter priors and descriptions for the model in Eq. 10.12.

dependent mixing fraction,

$$\pi(\chi_{\text{eff}}|\Lambda_{\chi_{\text{eff}}}) = \mathcal{N}(\chi_{\text{eff}}; \mu_b, 10^{\sigma_b})(1 - f(z)) \tag{10.10}$$

$$+ \mathcal{N}(\chi_{\text{eff}}; \mu_a, 10^{\sigma_a})f(z), \tag{10.11}$$

where

$$f(z) = \frac{1}{1 + \exp\left[f_A + f_B(z - 0.5)\right]}. \tag{10.12}$$

We choose three different priors on the μ and σ parameters for each Gaussian, outlined in Table 10.7, as proxies for different astrophysical scenarios.

When modeling the effective spin distribution with a single, redshift-independent Gaussian, it is found that the mean effective spin is $\chi_{\text{eff}} = 0.06$ [626, 52]. For our first prior choice, we assume that this result encapsulates the bulk of the population, fixing $\mu_b = 0.06$, and let the second mixture component model the departure from this assumption as we look to higher redshifts. We allow this secondary sub-population to peak at either positive or negative values of χ_{eff}, but the absolute value of the peak must be ≥ 0.1, so as to minimize the degeneracy between the bulk and this second sub-population. The evolution of the mixture fraction with redshift for individual hyper-parameter posterior samples is shown in Fig. 10-10. The posterior on the mixture fraction differs substantially from the prior, which is symmetric about $f = 0.5$ for all redshifts. This result indicates that the contribution of the Gaussian defined by the parameters μ_a and σ_a becomes more significant as redshift increases. The posterior for μ_a is 3.4 times more likely to be positive than negative, and the positive part of the posterior rails against the boundary at $\mu_a = 0.1$, indicating that it may be trying

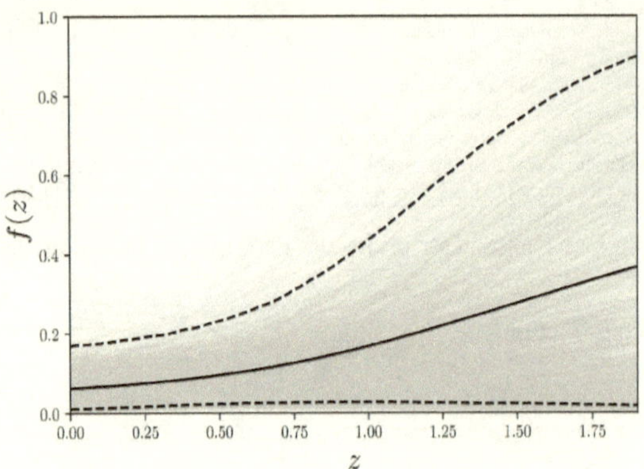

Figure 10-10: Posteriors for the mixture fraction in Eq. 10.12 obtained for GWTC-3 events under the first prior choice in Table 10.7. The solid black line shows the mean, while the dashed black lines show the 90% credible region.

to replicate the bulk distribution rather than to extract an independent component. σ_a is only weakly constrained to be > -1.02 at 90% credibility. Nonetheless, the χ_{eff} distributions at different redshift slices shown in the top of Fig. 10-11 for this model and prior choice paint a similar picture to the linear evolution model presented in Section 10.3; although skewed towards positive values of χ_{eff}, the distributions become wider with increasing redshift.

The second prior choice targets two populations with χ_{eff} distributions characterized by different widths. The peak of the narrower bulk distribution is now fixed to $\mu_b = 0$ to capture the BBH population formed dynamically with small spin and the majority of field binaries predicted to have negligible spin [387]. Meanwhile the posterior on the peak of the broader Gaussian, μ_a, is largely uninformative. The posteriors for the width parameters and the mixture fraction are very similar to those obtained with the first prior, with $\sigma_a > -0.25$ at 90% credibility and σ_b more strongly constrained to peak at $-1.23^{+0.17}_{-0.17}$. These results do not provide significant evidence for two distinct populations, but the χ_{eff} distributions at different redshift slices shown in the middle panel of Fig. 10-11 still exhibit some broadening with increasing redshift.

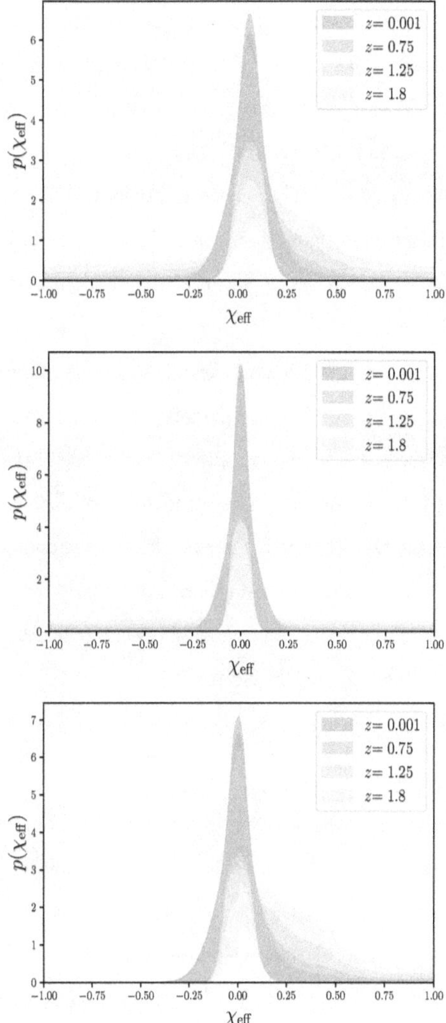

Figure 10-11: 90% credible region for slices of the χ_{eff} distribution at four different values of redshift for the three different prior choices in Table 10.7 (1-3 from top to bottom).

The final prior choice is designed to look for a redshift-dependent excess of preferentially-aligned BBH systems. The mean of the bulk distribution is still fixed to $\mu_b = 0$, as expected for a population of dynamically-formed systems with isotropic spin tilts [708, 89, 756, 401, 757, 515, 403]. The mean of the aligned population is restricted to $0.1 < \mu_a < 1$, in order to represent systems formed via isolated binary evolution in the field [402, 725, 961, 133, 127, 125, 126], and the corresponding Gaussian is truncated on $[0, 1]$ instead of $[-1, 1]$. The χ_{eff} distributions at different redshift slices for this prior choice are shown in the bottom panel of Fig. 10-11. The population model is asymmetric by construction, so the mean of the distribution also increases with redshift along with the width.

The posterior on the mixture fraction is again very similar to the one obtained with the first prior choice in Fig. 10-10. Since the mixture fraction increases with redshift for all three prior choices, the second Gaussian added on top of the bulk distribution always contributes more at high redshifts, consistent with the conclusion that the χ_{eff} distribution broadens with increasing redshift. These results indicate that the apparent correlation between the width of the χ_{eff} distribution and the redshift is not driven by our choice of linear evolution model in Section 10.2, but can be observed under a variety of population models and priors.

10.5   Discussion and Conclusion

In this work, we have found weak but robust evidence for an increase in the width of the χ_{eff} distribution of BBH systems with increasing redshift. We have verified that this correlation is unlikely to be spuriously introduced by the increased uncertainty in the χ_{eff} posteriors for individual high-redshift sources using a simulated population and that this trend remains present using alternative population models. We also observe a less significant correlation between the width of the χ_{eff} distribution and the primary mass, although we find that such a correlation can be falsely recovered if a population with a χ_{eff}-redshift correlation is instead analyzed assuming only a χ_{eff}-primary mass correlation. When allowing for correlations with both redshift and primary mass, we find evidence for an increase in the width of the χ_{eff} distribution with one or the other but cannot distinguish which.

A correlation between χ_{eff} and redshift might be explained by one of two broad scenarios. First, the correlation could be indicative of multiple sub-populations arising from distinct formation channels, each of which occupy a different region in the (χ_{eff}, z) plane. The third model defined in Section 10.4.2, for example, serves as a proxy for this scenario. In this case, the symmetric χ_{eff} component centered at zero might serve to capture binaries assembled dynamically in dense stellar environments [708, 89, 756, 401, 757, 515, 403], while the preferentially-positive χ_{eff} component corresponds to a sub-population of systems formed via isolated binary evolution in the field [402, 725, 961, 133, 127, 125, 126]. However, such a mixture between two sub-populations generally leads to a shift in the *mean* of the χ_{eff} distribution with redshift. Under the more generic model explored in Section 10.3, we much more strongly prefer evolution of the *width* with redshift, although evolution of μ_χ is not strictly ruled out.

The second possibility, broadly, is that a spin-redshift correlation arises due to evolutionary processes operating within a single population. Within isolated binary evolution, for example, tidal interactions may be responsible for correlating black hole spins and redshifts. Some authors predict that isolated black holes are naturally born slowly rotating due to efficient angular momentum transport from stellar cores [832, 388, 387]. Spin can nevertheless be introduced via tidal torques exerted by the first-born black hole on its companion, prior to the companion's core collapse [961, 725, 125, 127, 386]. In this scenario, the shortest-period binaries both acquire the largest spins and merge most promptly, correlating the spins of binary black holes and the redshifts at which they merge [725, 387]. This effect may be enhanced by the lower metallicities present at high redshifts, which diminish the strength of stellar winds and hence prevent binary orbits from widening and avoiding tidal spin-up. [126] corroborate this prediction using population synthesis simulations, finding that an increasing fraction of systems undergo tidal spin-up in close orbits at high redshift. Because there are systems that do not meet the criteria for efficient tidal spin-up at all redshifts, they find that the spin distribution both broadens and increases in mean.

Another possible explanation comes from the effect of metallicity on the efficiency of angular momentum transport in the envelope of the stellar progenitors of the BBH system. Because stellar winds are weaker at low metallicity (higher redshift), this leads to a less efficient removal of the angular momentum stored in the stellar envelope and hence a possibly higher spin of the resulting black hole [725, 387]. However, this naive picture is complicated by the interplay between the strength of stellar winds, the extent of the hydrogen-burning region inside the star, and the efficiency of elemental mixing within the star. As such, [133] find a non-monotonic relationship between the metallicity and the black hole spin for a given carbon-oxygen-core mass of the progenitor. This relationship further depends on the stellar evolution model employed in the binary evolution simulation. Because of the uncertainty in the processes that impart natal spin to the black hole, we cannot place meaningful constraints on these models.

While both of the possibilities described above lead to systems with larger spin magnitudes at increasing redshift, we must also explain the increased number of systems with negative values of χ_{eff} due to the symmetric broadening of the distribution. One potential explanation within the framework of isolated binary evolution is that natal kicks are stronger at higher redshift, leading to more significant misalignment of the BH spin relative to the orbital angular momentum upon birth. It is not clear how such an effect would arise, however. Furthermore, even with moderately strong ($\sim 100\,\mathrm{km\,s^{-1}}$) natal kicks, the median of the χ_{eff} distribution for systems formed in the field is still much higher than what we recover [402]. Alternatively, the mixing fraction between systems formed via isolated evolution and dynamical assembly may remain constant with redshift, but the spin *magnitudes* of systems formed via both channels can increase due to one or both of the scenarios previously discussed. In this case, the orientations of the spins in systems formed dynamically would remain isotropic, but their magnitudes would be larger at higher redshift, leading to a broadening of the distribution but not a shift in the mean.

One caveat with our analysis is that we restrict any correlations to be with χ_{eff}, rather than also allowing for mass-redshift correlations. It is likely that the complex processes leading to the formation of BBH systems would produce correlations between all of these parameters, rather than just with spin [725, 125, 386, 962]. However, it is difficult to encapsulate all possible correlations in a simple phenomenological model without dramatically increasing the dimensionality of the parameter space. We leave the exploration of simultaneous correlations between all parameters to future work.

Chapter 11

Measuring the primordial gravitational-wave background in the presence of astrophysical foregrounds

The content of this chapter was previously published in Physical Review Letters as Ref. [155] in Dec. 2020. ASB performed the analysis, wrote the manuscript, and contributed to project development.

Abstract

Primordial gravitational waves are expected to create a stochastic background encoding information about the early Universe that may not be accessible by other means. However, the primordial background is obscured by an astrophysical foreground consisting of gravitational waves from compact binaries. We demonstrate a Bayesian method for estimating the primordial background in the presence of an astrophysical foreground. Since the background and foreground signal parameters are estimated simultaneously, there is no subtraction step, and therefore we avoid astrophysical contamination of the primordial measurement, sometimes referred to as "residuals." Additionally, since we include the non-Gaussianity of the astrophysical foreground in our model, this method represents the statistically optimal approach to the simultaneous detection of a multi-component stochastic background.

11.1 Introduction

Detection of a cosmological gravitational-wave background from the early Universe is one of the most ambitious goals of gravitational-wave astronomy. There are several scenarios which may give rise to primordial backgrounds, including inflationary scenarios and phase-transition scenarios [576]. Inflationary models in general produce a gravitational-wave background through the amplification of vacuum fluctuations [439, 887, 309, 440]. In the slow-roll inflation model, the dimensionless energy density of the background,

$$\Omega_{\text{gw}}(f) \equiv \frac{1}{\rho_c} \frac{d\rho_{\text{gw}}}{d\ln f}, \tag{11.1}$$

is expected to be $\Omega_{\text{gw}}(f) \approx 10^{-15}$ across many orders of magnitude in frequency f [576]. Here, $d\rho_{\text{gw}}$ is the gravitational-wave energy density between f and $f + df$ while ρ_c is the critical energy density for a flat universe.

Such a low value of Ω_{gw} is unlikely to be directly detected by all but the most ambitious space-based gravitational-wave detectors [259, 506]. However, in models with either non-standard inflation or non-standard cosmology, it is possible to generate inflationary backgrounds accessible by current detectors [589]. Alternatively, it may be possible for the inflaton to decay non-perturbatively through parametric resonance. This process, known as preheating, may produce a potentially detectable gravitational-wave background through explosive particle production, peaking as high as $\Omega_{\text{gw}} \approx 10^{-11}$ [511, 715]. In reality, the physics of inflation is highly uncertain. Indirect detection, via the observation of B-modes in the cosmic microwave background, provides an alternative means of observing inflationary gravitational waves [58].

Phase transitions in the early Universe may produce gravitational waves if they are strongly first-order [494, 490, 207, 211]. The peak frequency of the gravitational-wave energy density spectrum f_0 is related to the energy scale of the transition T_* [576,

490, 69]:

$$f_0 \approx 170\,\mathrm{Hz} \left(\frac{T_*}{10^9\,\mathrm{GeV}} \right). \tag{11.2}$$

Thus, the detection of a primordial background from a phase transition by either an audio-band or millihertz gravitational-wave detector, such as LIGO [3] or LISA [78], respectively, would probe physics at energy scales inaccessible by colliders, corresponding to a time when the Universe was only $\gtrsim 10^{-14}\,$s old. The energy density created from phase transitions depends on model-dependent details, but numerical simulations and scaling arguments suggest that $\Omega_{\mathrm{gw}}(f_0) \approx 10^{-12\pm2}$ for a strongly first-order transition [414]. This is just below the projected sensitivity of advanced detectors operating at design sensitivity [3, 53], but well within the range of space-based detectors and proposed third-generation terrestrial detectors [873, 744, 577].

Astrophysical foregrounds are interesting in their own right since they contain valuable information about the population properties of compact binaries at high redshifts [590, 918, 198, 779]. However, recent observations of merging compact binaries [12, 17, 26, 32, 36, 39] by the Advanced LIGO [3] and Virgo [53] detectors imply that primordial backgrounds are masked by much brighter astrophysical foregrounds [10, 20]. Binary black holes (BBHs) and binary neutron stars (BNSs) each produce astrophysical foregrounds of $\Omega_{\mathrm{gw}}(f = 25\,\mathrm{Hz}) \approx 10^{-9}$ with $\alpha = 2/3$ [973, 942, 764, 975, 20]. Some fraction of these astrophysical foregrounds is *resolvable* with current detectors, meaning that some of the events contributing to the background are unambiguously detectable. The most ambitious proposed detectors will resolve essentially every compact binary in the visible Universe [259, 506, 742]. These astrophysical foregrounds are non-Gaussian because the signals do not combine to create a random signal, characterized only by its statistical properties. Rather, BBHs merge every $2 - 10\,$minutes while BNSs merge every $4 - 62\,$s [20]. While there are likely to be many BNS signals in the LIGO/Virgo band at any given time, they are nonetheless distinguishable based on their different coalescence times [742, 924].

Previous proposals to disentangle the primordial background from the astrophysical foreground utilize the concept of subtraction. The idea, pioneered in [265], is to measure the parameters of each resolved compact binary in order to subtract the gravitational waveform from the data. Inevitably, this results in "residuals," or systematic error from imperfect subtraction. However, the residuals can be "projected out" using a Fisher matrix formalism, which reduces the level of contamination [446]. While [265] considered the subtraction problem in the context of the ambitious Big Bang Observer [259], more recent work has explored the possibility of carrying out subtraction using the third-generation detectors Einstein Telescope [718] and Cosmic Explorer [744] that are planned to come online in the next decade [742, 804].

One limitation of the subtraction paradigm is that weak, unresolved signals are not subtracted and therefore contaminate the measurement of the primordial background, introducing a systematic error. While BBH mergers will be more easily resolvable, sub-threshold BNS mergers will impact the sensitivity of third-generation ground-based gravitational-wave detectors to the primordial background [777].

Other analyses have proposed methods for the simultaneous measurement of stochastic gravitational-wave backgrounds with different spectral shapes [890, 690]. However, none of these methods account for the non-Gaussianity of the astrophysical foreground, resulting in a decrease in the sensitivity of the search.

Here, we present a Bayesian formulation in which the primordial background and the astrophysical foreground are measured simultaneously. Our method estimates the astrophysical foreground from both resolved and unresolved binaries, which ensures that our measurement of the primordial background is free from bias. The method can therefore also include the contributions from high signal-to-noise ratio compact binaries. Because our likelihood models the non-Gaussianity of the astrophysical foreground as in [822], this method serves as the scaffolding for a unified, statistically optimal approach (yielding the minimum unbiased credible interval posterior) to the simultaneous detection of compact binaries and the primordial background.

11.2 Formalism

We seek to measure a cosmological stochastic background described by two power-law parameters:

$$\Omega_{\mathrm{gw}}(f) = \Omega_\alpha \left(\frac{f}{25\,\mathrm{Hz}} \right)^\alpha . \tag{11.3}$$

Here, α is a power-law index while Ω_α is the amplitude. The power-law model is chosen for consistency with cross-correlation searches for the stochastic background (e.g. [73, 10, 20, 761]), but the subsequent formalism can be applied to any spectral shape. The background is obscured by a foreground of merging compact binaries, each described by a vector of fifteen parameters θ including properties such as the component masses and the sky location. We only consider BBH mergers for this analysis and assume that the population distribution for the binary parameters $\pi(\theta)$ is known to curtail the computational cost and additional complications for longer-duration BNS signals, but later discuss how the method can be generalized to relax these assumptions. Since we want our formalism to include sub-threshold events, the number of compact binary signals in the data is, by assumption, unknown.

Following [761] and [822], the likelihood of observing frequency-domain strain data, $s_{i,k}$, with a Gaussian stochastic background characterized by the parameters (Ω_α, α) and a compact binary coalescence with signal $h_k(\theta)$ is derived by marginalizing over the random Gaussian strain perturbation of the background:

$$\mathcal{L}(s_{i,k}|\theta, \Omega_\alpha, \alpha) = \frac{1}{\det(\pi T \mathbf{C}_k(\Omega_\alpha, \alpha)/2)} \exp\left(-\frac{2}{T} (s_{i,k} - h_k(\theta))^\dagger \mathbf{C}_k^{-1}(\Omega_\alpha, \alpha) (s_{i,k} - h_k(\theta)) \right),$$

$$\tag{11.4}$$

Here, we assume that the data is divided into segments of duration T labeled with index i. The frequency dependence is denoted with the index k such that $s_{i,k} = s_i(f_k)$. The strain data in each segment, $s_{i,k}$, and the binary signal model, $h_k(\theta)$, are vectors with one entry for each detector in some network:

$$s_{i,k} = \begin{pmatrix} s_{i,k}^{(1)} \\ s_{i,k}^{(2)} \end{pmatrix} \quad h_k(\theta) = \begin{pmatrix} h_k^{(1)}(\theta) \\ h_k^{(2)}(\theta) \end{pmatrix}. \tag{11.5}$$

At least two detectors are required to search for stochastic backgrounds modeled as excess cross-power, since the auto-power of one detector cannot distinguish between instrumental noise and signal, but the framework presented here can be extended to include multiple detector baselines.

The frequency-dependent covariance matrix, $\mathbf{C}_k$, includes contributions from both the detector noise power spectral density (PSD) $P_I(f_k)$ and the primordial background energy density:

$$\mathbf{C}_k = \begin{pmatrix} P_1(f_k) + \kappa_{11}(f_k)\Omega_{\text{gw}} & \kappa_{12}(f_k)\Omega_{\text{gw}} \\ \kappa_{21}(f_k)\Omega_{\text{gw}} & P_2(f_k) + \kappa_{22}(f_k)\Omega_{\text{gw}} \end{pmatrix}, \tag{11.6}$$

where

$$\kappa_{IJ}(f_k) \equiv \gamma_{IJ}(f_k) \frac{3H_0^2}{10\pi^2 f^3} \tag{11.7}$$

converts the primordial background energy density Ω_{gw} into a (signal) strain power spectral density [73, 627]. The variable $\gamma_{IJ}(f_k)$ is the overlap reduction function for detector pair IJ, encoding the geometry of the detector network [232, 358, 761]. It is normalized to $\gamma_{II} = 1$ for coincident and coaligned detectors with perpendicular arms. Additionally, H_0 is the Hubble constant. Combining data from many frequency bins, the likelihood is the product of the individual-frequency likelihoods:

$$\mathcal{L}(s_i|\theta, \Omega_\alpha, \alpha) = \prod_{k}^{m} \mathcal{L}(s_{i,k}|\theta, \Omega_\alpha, \alpha). \tag{11.8}$$

For an astrophysical non-Gaussian foreground, we are interested in determining the fraction of segments containing a signal, ξ, (which we call the "duty cycle" following [822]) rather than the binary parameters, θ, for a particular segment. We say that a segment "contains" a binary signal if the time of coalescence falls inside the segment. In this case, the likelihood in Eq. 11.4 can be marginalized over the binary parameters θ to obtain

$$\mathcal{L}\left(s_i | \Omega_\alpha, \alpha, \xi\right) = \xi \mathcal{L}_S(s_i | \Omega_\alpha, \alpha) + (1 - \xi)\mathcal{L}_N(s_i | \Omega_\alpha, \alpha), \tag{11.9}$$

where we have defined the marginalized signal and "noise" likelihoods as:

$$\mathcal{L}_S(s_i | \Omega_\alpha, \alpha) = \int d\theta \, \mathcal{L}(s_i | \theta, \Omega_\alpha, \alpha) \, \pi(\theta) \tag{11.10}$$

$$\mathcal{L}_N(s_i | \Omega_\alpha, \alpha) = \mathcal{L}(s_i | \theta = 0, \Omega_\alpha, \alpha). \tag{11.11}$$

The $\theta = 0$ appearing in the expression for $\mathcal{L}_N$ indicates that the noise likelihood is functionally identical to the signal likelihood if we set the compact binary signal, $h_k(\theta)$, equal to zero. Readers should understand the phrase "noise likelihood" to refer to noise $+$ a low-level Gaussian stochastic background but no binary signal. We assume that the probability of observing one BBH merger event in a single segment is much less than one, so that the probability of observing two events is negligibly small, which is a reasonable assumption for BBH mergers [20, 822, 742]. We discuss how this assumption can be relaxed later.

For an ensemble of N data segments, $\{s\}$, the total likelihood is given by multiplying the likelihoods for individual segments:

$$\mathcal{L}\left(\{s\} | \Omega_\alpha, \alpha, \xi\right) = \prod_i^N \mathcal{L}(s_i | \Omega_\alpha, \alpha, \xi). \tag{11.12}$$

This joint likelihood function for $(\Omega_\alpha, \alpha, \xi)$ defined in Eq. 11.12 is the product of N single-segment likelihood functions, each of which contains a signal sub-hypothesis (with probability ξ) and a noise sub-hypothesis (with probability $1 - \xi$).

To obtain joint posteriors on $(\Omega_\alpha, \alpha, \xi)$, we apply Bayes theorem:

$$p(\Omega_\alpha, \alpha, \xi | \{s\}) = \frac{\pi(\Omega_\alpha, \alpha, \xi)}{\mathcal{Z}} \mathcal{L}(\{s\} | \Omega_\alpha, \alpha, \xi), \tag{11.13}$$

where $\mathcal{Z}$ is the Bayesian evidence given by marginalizing the total likelihood over the stochastic parameters,

$$\mathcal{Z} = \int d\Omega_\alpha \, d\alpha \, d\xi \, \mathcal{L}(\{s\}|\Omega_\alpha, \alpha, \xi) \, \pi(\Omega_\alpha, \alpha, \xi), \tag{11.14}$$

and $\pi(\Omega_\alpha, \alpha, \xi)$ is the prior.

11.3 Demonstration

We demonstrate this formalism with mock data. Assuming a two-detector network of the LIGO Hanford and Livingston observatories operating at design sensitivity [3], we simulate data for 101 non-overlapping segments each with a duration of 4 s. Each segment contains uncorrelated Gaussian noise[1] colored by the noise PSD $P(f_k)$ of the interferometers as well as correlated Gaussian noise colored by the signal power spectral density of the primordial background. The correlated noise is simulated such that the cross power spectral density is given by $\kappa_{IJ}(f_k)\Omega_{\mathrm{gw}}(f_k)$ for a cosmological background characterized by $(\log \Omega_\alpha = -6, \alpha = 0)$, where we use $\log \equiv \log_{10}$ throughout. While this amplitude is several orders of magnitude higher than that expected for primordial backgrounds, we have chosen this value so that our simulated cosmological signal corresponds to an unambiguous primordial-background detection with signal-to-noise ratio (SNR) of ~ 5.4 for 404 seconds of data observed with advanced LIGO. The value of $\alpha \approx 0$ is expected for the background due to slow-roll inflation [233].

Next, we randomly assign BBH mergers to 11 of our simulated segments for a corresponding duty cycle of $\xi = 11/101 = 0.11$. This duty cycle is higher than would be expected based on the current estimates of the BBH merger rate [20], but is just chosen for the purposes of our demonstration. The chirp mass,

$$\mathcal{M} = \frac{(m_1 m_2)^{3/5}}{(m_1 + m_2)^{1/5}}, \tag{11.15}$$

---

[1]While terrestrial detectors may suffer correlated noise from Schumann resonances [872], studies suggest that this noise can be subtracted with Wiener filtering [871].

is drawn from a uniform prior over the range $(13, 45)$ $M_\odot$. The prior for the symmetric mass ratio,

$$\eta = \frac{m_1 m_2}{(m_1 + m_2)^2},\qquad(11.16)$$

is uniform over $(0.09876, 0.25)$. The sky locations and component spin orientations are distributed isotropically, with spin magnitudes ranging uniformly from 0 to 0.8, and the luminosity distance prior is $\propto d_L^2$ between 500 and 5000 Mpc. This results in a range of network optimal SNRs between 2.06 and 12.17 with a median of 3.54. Only the signal with the highest SNR corresponds to a confident detection. The rest of the simulated events have network optimal SNR < 7 that would not be individually detected with high confidence.

If we were using real LIGO data instead of simulated Gaussian noise, the noise power spectral density in the covariance matrix in Eq. 11.6 would have to be estimated from the data itself. Estimates of the PSD include both of the terms on the diagonal of the covariance matrix, $P(f_k) + \kappa_{II}(f_k)\Omega_{\mathrm{gw}}$, since auto-power due to detector noise cannot typically be distinguished from the persistent Gaussian background [761]. This results in a decrease in the sensitivity of the search, which we mimic in our demonstration by fixing the diagonal terms to the sum of the known noise PSD and the signal power from the simulated cosmological background. Hence, the diagonal terms do not contribute to the estimation of the (Ω_α, α) parameters, although simultaneously fitting a parameterized PSD model as in [240] could be a possible future extension.

Evaluating the likelihood in Eq. 11.12 poses a computational challenge due to the product over N single-segment likelihoods. To overcome this issue, we use likelihood reweighting [695] to evaluate the marginalized signal likelihood (Eq. 11.10) and the noise likelihood (Eq. 11.11) on a grid in (Ω_α, α). For each segment we use the `cpnest` [909] nested sampler as implemented in the `Bilby` [96, 762] package to obtain posterior samples for the binary parameters using the likelihood in Eq. 11.4 under the assumption that there is no Gaussian background present: $\Omega_\alpha = 0$. The priors for the binary parameters are the same as those used to generate the BBH injections previously described. We use the IMRPhenomPv2 waveform model [445, 482, 509] for the compact binary signal, $h_k(\theta)$, in both the simulation and recovery.

The marginalized signal likelihood for each segment at a particular value of (Ω_α, α) is calculated via a Monte Carlo integral over the n posterior samples obtained in the original sampling step:

$$\mathcal{L}_S(s_i|\Omega_\alpha, \alpha) = \frac{\mathcal{Z}_{0,i}}{n} \sum_j^n \frac{\mathcal{L}(s_i|\theta_j, \Omega_\alpha, \alpha)}{\mathcal{L}(s_i|\theta_j, \Omega_\alpha = 0)}, \tag{11.17}$$

where $\mathcal{Z}_{0,i}$ is the evidence calculated by the sampler using the likelihood where $\Omega_\alpha = 0$:

$$\mathcal{Z}_{0,i} = \int d\theta\, \mathcal{L}(s_i|\theta, \Omega_\alpha = 0)\pi(\theta). \tag{11.18}$$

The noise likelihood in Eq. 11.11 can be directly evaluated on the same grid in (Ω_α, α) as the reweighted signal likelihood. We use a 50×50 grid ranging from $\log \Omega_\alpha \in [-8, -4]$ and $\alpha \in [0, 4]$.

Once we have obtained the marginalized signal and noise likelihoods for each segment using reweighting, we calculate the joint likelihood in Eq. 11.9 on a grid in ξ, with 100 values ranging from [0,1]. The full likelihood in Eq. 11.12 is then calculated by multiplying the individual three-dimensional grids from each data segment. Fig. 11-1 shows the marginalized likelihoods for the cosmological background parameters (Ω_α, α) as well as ξ obtained using all 101 simulated data segments. We recover values for all three parameters that are consistent with the true values used in the simulation: $\log \Omega_\alpha = -5.96^{+0.08}_{-0.16}$, $\alpha = 0.49^{+1.14}_{-0.49}$, and $\xi = 0.08^{+0.09}_{-0.05}$, where the uncertainty is the 90% credible interval calculated using the highest probability density method.

In addition to successfully measuring the parameters characterizing both the astrophysical foreground and the cosmological stochastic background simultaneously, we also calculate a Bayes factor comparing the cosmological signal hypothesis to the no-signal hypothesis. This quantifies to what extent the model where $\Omega_\alpha = 0$

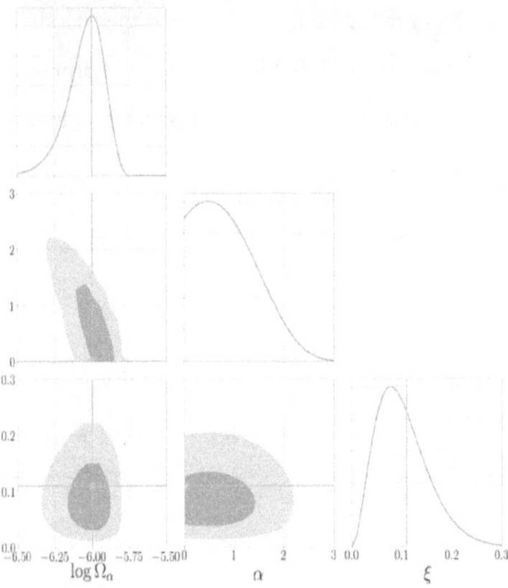

Figure 11-1: Corner plot for the combined posterior for $(\Omega_\alpha, \alpha, \xi)$, with the orange lines showing the true values used in the simulated data. The 90% credible region is shown in light blue and the 50% credible region in dark blue.

is statistically disfavored compared to the model where (Ω_α, α) can take on any of the values on our grid. In the high-SNR limit, the natural log of the Bayes factor is proportional to the square of the SNR familiar from frequentist cross-correlation searches, $\ln \mathrm{BF}_N^S \sim \mathrm{SNR}^2/2$ (see, e.g., [874]). The "signal" evidence for a non-zero cosmological background is given by Eq. 11.14. We set the priors on α and $\log \Omega_\alpha$ to be uniform across the ranges covered by the grid. The "noise" evidence is evaluated by integrating Eq. 11.12 assuming that $\Omega_\alpha = 0$:

$$\mathcal{Z}_N = \int d\xi \prod_i \xi \mathcal{Z}_{0,i} + (1 - \xi)\mathcal{Z}_{N,i}, \tag{11.19}$$

where $\mathcal{Z}_{N,i}$ is the likelihood in Eq. 11.4 evaluated with *both* $\Omega_\alpha = 0$ and $h_k(\theta) = 0$. We obtain $\ln \mathrm{BF}_N^S = \ln \mathcal{Z}_S - \ln \mathcal{Z}_N = 11.16$, which is consistent with the naive scaling based on SNR for a signal with SNR $= 5.41$.

11.4 Discussion

In this chapter we have demonstrated a new method for simultaneously detecting two distinct stochastic gravitational-wave backgrounds—a non-Gaussian astrophysical foreground from sub-threshold merging BBHs and a Gaussian cosmological background. Our method models both contributions simultaneously, so that subtraction of the foreground is not required. Additionally, this is the statistically optimal framework for detecting a stochastic background consisting of both a Gaussian and non-Gaussian component, resulting in significant improvements in the estimated time-to-detection of the astrophysical foreground compared to other methods for multi-component analyses, as described in [822]. However, in the absence of a non-Gaussian foreground, we find that there is no statistical advantage to using the fully Bayesian method compared to the standard cross-correlation method. Based on the comparison of the signal-to-noise Bayes factor and SNR for the presence of the Gaussian background, the two methods yield a similar level of statistical confidence, to the extent that it is possible to compare frequentist and Bayesian detection statistics.

In our demonstration, we assume that the sampling priors chosen for the BBH parameters, $\pi(\theta)$, match the true population distribution. In order to avoid biases that would be introduced due to a mismatch between the population distribution and the sampling prior, our method could be amended to instead measure these population priors simultaneously with the cosmological background parameters and duty cycle, following the formalism described in [821]. This would amount to adding additional hyper-parameters to the marginalized signal likelihood in Eq. 11.17:

$$\mathcal{L}_S(s_i|\Lambda, \Omega_\alpha, \alpha) = \frac{\mathcal{Z}_0}{n} \sum_j^n \frac{\mathcal{L}(s_i|\theta_j, \Omega_\alpha, \alpha)\pi(\theta_j|\Lambda)}{\mathcal{L}(s_i|\theta_j, \Omega_\alpha = 0, \alpha)\pi_0(\theta_j)}. \tag{11.20}$$

The hyper-parameters Λ describe the shape of the distribution $\pi(\theta|\Lambda)$, while the original prior used in the first step of sampling, $\pi_0(\theta)$, must also be divided out. The hyper-parameters do not enter the noise likelihood in Eq. 11.11 because the noise model assumes that each segment contains only the cosmological background with no binary signal.

Evaluating the marginalized signal likelihood in Eq. 11.20 using the same grid-based reweighting technique becomes computationally prohibitive, since the hyper-parameters Λ drastically increase the dimensionality of the grid. One possible solution that has been applied to similar problems in gravitational-wave astronomy could be to build a high-dimensional interpolant [457, 948]. Another promising approach could be to factorize the problem into two separate calculations, first carrying out population studies ignoring the stochastic background then using the inferred posterior predictive distributions for $\pi(\theta|\Lambda)$ as priors for the $\Omega > 0$ run. We leave exploration of these approaches to future work.

Another simplifying assumption we make in our demonstration is that only merging BBHs contribute to the astrophysical foreground, while in reality there will also be a foreground from binary neutron star and neutron star-black hole mergers. While our assumption that there is only one binary signal in a 4 s analysis segment is valid for BBHs, the rate of BNS mergers is higher, meaning that there are typically ~ 15 unresolved BNS signals in the LIGO band at any given time [20].

Because we need to model multiple populations of merging binaries simultaneously to avoid contamination from residual power, one possible solution would be to treat the number of binary mergers in a given segment as a free parameter using a trans-dimensional Markov Chain Monte Carlo algorithm, fitting the binary parameters for multiple mergers along with the cosmological background parameters all at once [432]. Another possible method is to analyze overlapping stretches of data that are offset by a shorter 0.2 s window, constraining the coalescence time prior to this window so that at most one binary system merges during this short "segment", allowing us to keep the same definition of ξ presented above. Preliminary tests suggests that the presence of other binary signals, merging at times outside of the segment, have a negligible effect on inferences about the binary merging during the segment. By marginalizing over the BNS parameters in many short segments, it should be possible to calculate the likelihood of a much longer span of data given the stochastic parameters. We estimate that it would take about $\sim 10^5$ CPUs to perform the BNS analysis in real time [350], followed by ~ 10 GPUs to perform the hierarchical inference including the uncertainty in the population distribution using the likelihood interpolation method for each individual segment [849]. We leave investigation of these methods to future work.

The formalism we describe and demonstrate assumes that the uncorrelated detector noise is Gaussian, while it is known that interferometric gravitational-wave data suffers from non-Gaussian noise transients called glitches [8]. This assumption can be relaxed via the introduction of additional duty cycle parameters to the likelihood in Eq. 11.4, characterizing the fraction of segments that contain a glitch in each detector, as described in [822]. This would increase the computational cost for each data segment analyzed, but the method is embarrassingly parallelizable, so the overall wall time for running the analysis does not increase significantly.

We also note that limitations in the accuracy of the waveform model describing the compact binary signal can leave behind coherent residual power that could bias the inference of the Gaussian background parameters. Based on current estimates of the uncertainty in numerical relativity waveforms [723], this level of contamination would likely not affect cosmological backgrounds probeable with proposed third-generation detectors, but improvements to waveform modeling would be necessary to recover unbiased parameter estimates for the weakest background models. The subtraction-projection methods for background detection would also be affected by waveform systematics, but our method could be modified to account for marginalizing over different waveform models [97] or parameterizing the waveform uncertainty [315].

While we demonstrate our method for the simultaneous detection of a stochastic background with both Gaussian and non-Gaussian components in the context of a cosmological background and an astrophysical foreground of BBH mergers, this formalism can be applied to any analogous problem. For example, this method could be applied to simultaneously measure both individual compact binary mergers or a foreground of these sources in the frequency band of the space-based LISA detector [78] on top of the white dwarf confusion noise background [141, 55]. Our model can also be extended to include multiple Gaussian backgrounds with different spectral shapes through the addition of extra terms in the covariance matrix defined in Eq. 11.6. One such example is the contamination from correlated magnetic noise in a ground-based detector network [232, 872, 871], which has a unique overlap reduction function [461].

Chapter 12

Conclusion

This book has focused on applications of Bayesian inference to learn about astrophysics and cosmology with gravitational waves. Connected by this broad theme, the analyses presented here include the development of new methods for increasingly detailed analysis of gravitational-wave data for both compact object binaries and the stochastic background, the identification of sources of systematic error in population-level analyses, and the characterization of the constraining power of the current detector network for key astrophysical observables.

Chapter 1 introduced the main astrophysical and cosmological sources of gravitational waves and the analysis methods used for their detection and characterization. In Chapter 2, I expanded upon the standard Bayesian framework for parameter estimation of individual compact-object binaries to include marginalizing over the uncertainty in the estimate of the detector noise (the power spectral density). I found that PSD uncertainty is a systematic that causes variations in the individual parameter posteriors on a similar scale to those introduced by waveform systematics. Taking this uncertainty into account will be especially important upon entering the high-precision regime with the next generation of ground-based gravitational-wave detectors.

Following the introduction of the Bayesian methods used throughout this book, Part I centered on multimessenger observations of compact-object binaries including a neutron star. Gravitational-wave and electromagnetic observations of these merger events carry complementary information on the astrophysical sites of heavy-element nucleosynthesis, the neutron star equation of state, and the central engine launching the most energetic electromagnetic explosions in the Universe. I have developed methods to facilitate counterpart detection and to simultaneously analyze both the gravitational-wave data and electromagnetic data to maximize the insight that can be gained into the astrophysical processes at work in these systems.

In Chapter 3, I showed that the estimate of the binary chirp mass obtained by matched-filter gravitational-wave search pipelines in low latency is accurate enough to distinguish between different merger outcomes and could thus be used to inform electromagnetic follow-up strategies. While the LVK does not presently release mass information in low latency, a policy informed by my work would be especially relevant in upcoming observing runs, where the BNS detection rate could be as high as several per week [22].

In Chapter 6, I presented an end-to-end simulation of the expected performance of the Wide-field Infrared Transient Explorer (WINTER), a new seeing-limited time-domain survey instrument dedicated to compact-object merger follow-up during future LVK observing runs. This simulation included the detection and sky localization of a realistic population of binary neutron star mergers with gravitational waves and found that WINTER will discover up to ten new kilonovae during O4. WINTER will be online in time to begin observing at the start of O4, with a strategy optimized for detection based on the simulation results—focusing on following up sources within 300 Mpc for up to seven nights.

I have also leveraged hierarchical Bayesian inference to learn about the *population* properties of compact object mergers including a neutron star and their counterparts. In Chapter 5, I developed a Bayesian method to constrain the geometry and emission profile of gamma-ray burst jets using multimessenger observations of the gravitational-wave and prompt gamma-ray signals that bypasses the traditionally-required observation of the rare "jet break" feature in the afterglow light curve. This method could provide informative constraints on the jet geometry during the fifth observing run expected to begin in 2027 based on predicted multimessenger detection rates.

In Chapter 7, I placed constraints on the neutron star equation of state while simultaneously measuring the neutron star-black hole binary mass and spin distributions using the non-detection of an electromagnetic counterpart for any of the candidate NSBH events detected so far. The EoS constraints derived from NSBH nondetections with this multimessenger method are unlikely to be competitive compared to those obtained with gravitational-wave measurements of BNS tidal deformability and pulsar observations. However, this method can be expanded to include electromagnetic selections effects in order to include counterpart detections. The development of a robust framework for multimessenger population inference will allow for the simultaneous characterization of the neutron star EoS and the counterpart properties, like the composition and geometry of kilonovae.

The neutron star mass distribution is interesting even in the absence of multimessenger information, as it can be used to probe supernova physics and the expansion rate of the universe. In Chapter 4, I found that mismodeling the BNS spin distribution can severely bias the inferred mass distribution and showed that allowing for large spin magnitudes misaligned to the orbital angular momentum on the population level is the safest choice. Based on the results of this work, I will advocate that future LVK population analyses including neutron stars use the individual-event parameter estimates obtained under the high, precessing spin prior to avoid biases.

Part II of this book focused on binary black hole spin. Spin is a promising tracer of BBH formation and evolutionary pathways, since the two main theoretical formation channels predict different observational signatures in the spin distribution. Despite these unique observational signatures, the formation channels of binary black holes remain largely unknown because the component spin magnitudes and tilt angles relative to the orbital angular momentum are difficult to measure for individual systems, leading to large uncertainties in their inferred population-level distributions. The work presented in this book will facilitate improved constraints on the component spin parameters for individual systems and contributes to our understanding of the astrophysical processes shaping the BBH population as a whole.

In Chapter 9, I demonstrated that measuring the spin of the most and least highly spinning black holes (spin sorting) instead of the most and least massive black holes in the binary as is customary (mass sorting) improves the constraints on the spin magnitudes and tilts for near equal-mass systems, which appear to be the majority of the astrophysical population. This method was applied in Ref. [52] to the population of 69 confident BBHs included in GWTC-3, leading to improved constraints on the population-level distributions of χ_A and χ_B. Spin sorting should be particularly useful when comparing gravitational-wave and electromagnetic measurements of black hole spins, as the electromagnetically-bright population of stellar-mass black holes should be the highest-spinning binary components. It can also be useful when looking for signatures of mass ratio reversal, since the more massive black hole should also be the most highly spinning in binaries that undergo this evolutionary process.

The spins of the most massive binary black holes are particularly difficult to measure due to the short duration of the signal in the sensitive frequency band of current detectors. In Chapter 8, I performed a systematic study on the measurability of spin in such heavy systems, finding that most of the spin information actually comes from the post-merger part of the signal rather than the inspiral. This work suggests that there is more to be learned about spin from high-mass BBHs—which are the majority of the detected population—than previously thought.

While the standard approach to inferring population properties has been to assume that the mass, spin, and redshift distributions of binary black holes are independent, in Chapter 10, I found a correlation between the distribution of the mass-weighted spin aligned to the orbital angular momentum and the redshift. The broadening of the spin distribution with redshift may indicate that the efficiency of processes that determine the natal spin of black holes formed from stellar collapse, like angular momentum transport, changes over cosmic time. The much larger BBH population expected after O4 should reveal if the tentative (2.45σ) evidence for this correlation found after O3 grows to become more significant.

Finally, in Chapter 11, I devised the statistically-optimal method for the simultaneous detection of the cosmological stochastic background in the presence of a much louder astrophysical foreground. By exploiting the statistical differences between the two classes of signals, I showed that this method is unbiased with no loss of sensitivity to the Gaussian cosmological background compared to the standard cross-correlation search in the absence of a foreground. This method should be expanded in the future to accommodate a mixed foreground of all three types of merging binaries including overlapping BNS signals, laying the foundation for a fully Bayesian pipeline for the analysis of the source-rich data promised by next-generation gravitational-wave detectors with vastly improved sensitivities. Such a framework will be necessary for the achievement of one of the most ambitious goals of gravitational-wave astronomy—the characterization of the primordial gravitational-wave signal.

The improved detector sensitivity expected in the upcoming fourth observing run will greatly expand the known population of compact-object binaries, with predicted detection rates of one BBH per day and roughly one BNS per week. I look forward to applying the multimessenger inference methods I have developed in this book to a population of joint gravitational-wave and electromagnetic observations of compact-object mergers including a neutron star. I am also excited for the evolving picture of binary black hole spin to become much clearer with several hundred detections, so that we can begin to understand which evolutionary processes contribute to the underlying BBH population and how they differ from those shaping the population of electromagnetically-bright stellar-mass black holes. Finally, I anticipate the detection of the stochastic gravitational-wave background of unresolvable compact-object binaries in the next couple years using accelerated detection strategies including the method developed in this book. This signal would allow us to probe the properties of high-redshift compact-object mergers without waiting for the next-generation of ground-based gravitational-wave detectors expected to come online in the next decade, allowing us to learn how the properties of black hole and neutron star binaries evolve over cosmic time.

www.ingramcontent.com/pod-product-compliance
Lightning Source LLC
La Vergne TN
LVHW042345190726
843493LV00005B/925